AF581794

Dietary Curcumin and Health Effects

Dietary Curcumin and Health Effects

Editors

Roberta Masella
Francesca Cirulli

MDPI • Basel • Beijing • Wuhan • Barcelona • Belgrade • Manchester • Tokyo • Cluj • Tianjin

Editors
Roberta Masella
Istituto Superiore di Sanita
Rome
Italy

Francesca Cirulli
Istituto Superiore di Sanita
Rome
Italy

Editorial Office
MDPI
St. Alban-Anlage 66
4052 Basel, Switzerland

This is a reprint of articles from the Special Issue published online in the open access journal *Nutrients* (ISSN 2072-6643) (available at: https://www.mdpi.com/journal/nutrients/special_issues/Dietary_Curcumin_and_health_effects_).

For citation purposes, cite each article independently as indicated on the article page online and as indicated below:

LastName, A.A.; LastName, B.B.; LastName, C.C. Article Title. *Journal Name* **Year**, *Volume Number*, Page Range.

ISBN 978-3-0365-6087-8 (Hbk)
ISBN 978-3-0365-6088-5 (PDF)

© 2023 by the authors. Articles in this book are Open Access and distributed under the Creative Commons Attribution (CC BY) license, which allows users to download, copy and build upon published articles, as long as the author and publisher are properly credited, which ensures maximum dissemination and a wider impact of our publications.
The book as a whole is distributed by MDPI under the terms and conditions of the Creative Commons license CC BY-NC-ND.

Contents

About the Editors

Roberta Masella

Istituto Superiore di Sanità, Research director at Centre for Gender specific Medicine. Dr. Masella has been working on the role of dietary components in the pathogenesis, prevention, and therapy of non-communicable diseases characterized by nutritional risk factors and associated with inflammation and oxidative stress, such as obesity, type 2 diabetes, atherosclerosis, and cancer. The research projects she has managed studied the relationship between adipose tissue dysfunctions and the onset of pathologies, as well as that between dietary components and cell functions. She collaborated on nutritional intervention studies to compare the effects of diets differing in types of FAs on clinical and metabolic parameters. In the last few years, she studied the crosstalk between adipose tissue and immune cells in obese and colon cancer subjects, demonstrating the effects of dietary fatty acid on the inflammatory process activation in human visceral adipose tissue in obesity and colon cancer. Furthermore, she has long-standing experience on the mechanisms of action of dietary polyphenols and their effects on cell functions.

Francesca Cirulli

Francesca Cirulli is a Senior Researcher and group leader at the Center for Behavioural Sciences and Mental Health at the Istituto Superiore di Sanità in Rome, Italy. Her research investigates the role of lifestyle, dietary, and social factors in determining stress vulnerability and resilience and the molecular and cellular mechanisms underlying it. She is currently investigating how developmental stressors, such as a high-fat diet or psychosocial stress, can have adverse consequences on brain development and behavior. Epigenetic markers, neuroimmune regulations and microbiota signatures in preclinical studies and clinical cohorts are assessed to derive mechanisms and novel targets for prevention and treatment of mental disorders. She is currently leading a clinical trial (EPICURO) to study the role of dietary curcumin on inflammatory mediators and cognitive decline in elderly with metabolic syndrome.

Preface to "Dietary Curcumin and Health Effects"

Curcumin is a pleiotropic compound found in the rhizome of Curcuma longa (turmeric). Curcuma longa has been widely used as a spice for a long time, especially in Asian countries; however, the interest in this compound has been growing, and it is largely consumed as dietary component and supplement all around the world. The great interest in curcumin is due to a number of potential biological activities that this compound has demonstrated over time, and these are well documented by the 8601 papers—of which 229 are clinical trials—published in the last five years (retrieved from PubMed on 12 January 2022 using 'curcumin' as keyword). It is important to underline that since the majority of such studies were carried out in cellular and animal models, conclusive evidence of the real effectiveness of curcumin as preventive and therapeutic compound is still far from being reached, although the number of clinical trials addressing the potential therapeutic role of curcumin in a number of pathological and non-pathological conditions is growing every day. It is safe to say that, despite its reported benefits, one of the major drawbacks of ingesting curcumin is its poor bioavailability; thus, major efforts are needed to overcome this problem. The Special Issue "Dietary Curcumin and Health Effects" aimed to collect the most advanced evidence on the relationship between curcumin and health, with the final objective to improve research and move the field forward.

Roberta Masella and Francesca Cirulli
Editors

 MDPI

Editorial

Curcumin: A Promising Tool to Develop Preventive and Therapeutic Strategies against Non-Communicable Diseases, Still Requiring Verification by Sound Clinical Trials

Roberta Masella [1,*] and Francesca Cirulli [2]

[1] Centre for Gender-Specific Medicine, Istituto Superiore di Sanità, Viale Regina Elena 299, 00161 Rome, Italy
[2] Center for Behavioral Sciences and Mental Health, Istituto Superiore di Sanità, Viale Regina Elena 299, 00161 Rome, Italy; francesca.cirulli@iss.it
* Correspondence: roberta.masella@iss.it

Citation: Masella, R.; Cirulli, F. Curcumin: A Promising Tool to Develop Preventive and Therapeutic Strategies against Non-Communicable Diseases, Still Requiring Verification by Sound Clinical Trials. *Nutrients* **2022**, *14*, 1401. https://doi.org/10.3390/nu14071401

Received: 11 February 2022
Accepted: 17 February 2022
Published: 28 March 2022

Publisher's Note: MDPI stays neutral with regard to jurisdictional claims in published maps and institutional affiliations.

Copyright: © 2022 by the authors. Licensee MDPI, Basel, Switzerland. This article is an open access article distributed under the terms and conditions of the Creative Commons Attribution (CC BY) license (https://creativecommons.org/licenses/by/4.0/).

Curcumin is a pleiotropic compound found in the rhizome of Curcuma longa (turmeric). Curcuma longa has been widely used as a spice for a long time, especially in Asian countries; however, the interest in this compound has been growing and it is largely consumed as dietary component and supplement all around the world. The great interest in curcumin has been due to a number of potential biological activities this compound has shown over time, and it is well documented by the 8601 papers—of which 229 are clinical trials—retrieved in PubMed on 12 January 2022, published in the last five years using 'curcumin' as keyword. It is important to underline that, since the majority of such studies were carried out in cellular and animal models, conclusive evidence of the real effectiveness of curcumin as a preventive and therapeutic compound is still far from being reached, although the number of clinical trials addressing the potential therapeutic role of curcumin in a number of pathological and non-pathological conditions is growing every day. It is safe to say that, despite its reported benefits, one of the major drawbacks of ingesting curcumin is its poor bioavailability, thus major efforts are needed to overcome this problem.

The Special Issue, "Dietary Curcumin and Health Effects", was aimed at collecting the most advanced evidence on the relationship between curcumin and health, with the final objective to improve research and move the field forward.

The Special Issue provides twelve contributions, of which three are original articles, seven are narrative reviews, one is a systematic review and one is a meta-analysis, that all together offer a multifaceted and multidisciplinary overview that allows us to draw a picture as intriguing and fascinating as possible on the benefits of curcumin and also to suggest potential research fields still to be explored. In this vein, the review by Filardi et al. [1] is focused on reviewing the studies on the possible use of curcumin during pregnancy to prevent and/or reduce pregnancy-related complications, such as gestational diabetes mellitus, hypertension, and preeclampsia. Unfortunately, the information currently available on these issues is still limited and fragmentary and mainly coming from in vitro and animal studies. However, results from these studies, together with the knowledge about the hypoglycemic and anti-inflammatory effects already gathered, suggest possible positive effects of curcumin consumption during gestation, when the mother and the fetus undergo significant physiological immune-metabolic alterations that may have consequences on both maternal and fetal tissues. On the other hand, the authors also highlight that curcumin has been demonstrated to negatively affect the blastocyst stage, implantation and post-implantation embryo development in healthy animals; thus, the use of curcumin in pregnancy must be carefully evaluated and it should be avoided as self-medication until further studies are carried out to finally define the real safety and effectiveness during pregnancy.

As regards the protective effect of curcumin against T2D and insulin resistance occurrence, the paper by Thota et al. [2] reports data from a case–control study carried out in subjects at high risk of developing T2D. The group receiving curcumin supplementation was characterized by lower plasma insulin level with respect to the placebo group, and a similar trend also in the HOMA2-IR parameter indicating a better glucose control. It is worth noting that the authors showed an interesting amelioration of two parameters associated with insulin resistance, i.e., glycogenosynthase kinase-3β and islet amyloid peptide. The study, although preliminary and carried out on a small sample size, suggests a potential mechanism of action through which curcumin supplementation might exert an adjuvant activity against T2D risk factors. The therapeutic effects of curcumin on glycemic and lipid profile in T2D were further reviewed by Altobelli et al. [3] This systematic review and meta-analysis shows a significant reduction in glycosylated hemoglobin, HOMA, and LDL together with a general improvement of glucose metabolism in uncomplicated T2D patients treated with curcumin.

Among the biological activities exerted by curcumin, the anti-oxidative and anti-inflammatory properties, reported in many studies, make it a potentially effective tool in preventing and counteracting chronic-degenerative diseases, very often associated with obesity and aging, such as cardiovascular diseases, T2D, metabolic dysfunctions, neurodegenerative diseases, and cancer, all of them characterized by the presence of oxidative and inflammatory processes.

In this regard, data are available on the anti-inflammatory activities of curcumin from preclinical and clinical studies addressed not only to demonstrate the protective effect of curcumin administration in patients suffering from pathologies, such as neurodegenerative diseases, metabolic diseases and cancer, but also to shed light on the molecular mechanisms responsible for these effects. Curcumin has been demonstrated to be capable of modulating signaling pathways, regulating relevant cellular activities and resulting in a modulation of inflammatory and oxidative processes. In particular, a downregulation of pro-inflammatory transcription factors, such as NF-κB, together with the increase in the activity of those regulating the antioxidant defense system, such as Nrf2, have been elucidated in in vitro and animal models. These biomolecular activities are most likely responsible for the modulation of inflammatory biomarkers, e.g., pro/anti-inflammatory cytokines and inflammasome, and enzymatic activities, e.g., COX2 and glutathione-related enzymes, detected in humans after consumption of curcumin by diet or supplements, which may be responsible for the decreased systemic inflammation detected at systemic level in patients suffering from several pathologies. Special attention is given to the beneficial effects curcumin may exert in obese patients. Obesity is recognized as a major risk factor for almost all non-communicable diseases, including cardiovascular diseases, type 2 diabetes, neurodegenerative diseases and some types of cancer, and represents a major health problem as its prevalence is consistently increasing all around the world. Obesity is characterized by the expansion of fat depots and the dysfunction of adipose tissue leading to altered secretory activities that induce a systemic low-grade chronic inflammation, that play a pivotal role in the pathogenesis of obesity and its clinical complications. Varì et al. [4] review the most recent studies carried out in humans, providing evidence for the role of curcumin in targeting specific molecular pathways that appear dysregulated in obesity. In addition to a number of studies demonstrating that the administration of curcumin decreases circulating inflammatory markers in overweight and obese patients, several data are reported that suggest an activity of curcumin in modulating specific factors such as adiponectin and leptin, relevant adipokines secreted by the adipose tissue that regulate the functions of organs and systems such as the brain, liver, and immune system. In overweight or obese people, these adipokines are imbalanced, leading to alterations of metabolic and inflammatory pathways. Curcumin has been shown to be able to restore the normal function of these adipokines, particularly of adiponectin, which plays a central role in the regulation of the metabolism and immune functions. It is worth noticing that a number of studies have been carried out to specifically evaluate the molecular mechanisms targeted by curcumin in adipocytes, unravelling its ability to modulate specific kinase and transcription factors responsible, in turn, for the regulation of cellular activities.

Metabolic dysfunction that is associated with oxidative stress and inflammation may greatly accelerate the onset and worsen the progression of cognitive dysfunctions by promoting brain ageing and reducing lifespan. Berry et al. [5] integrate current knowledge on the mechanisms underlying cognitive decline, highlighting the potential role of curcumin in this highly disabling and prevalent condition in the aging population which greatly affects physical health and quality of life. Natural antioxidant agents such as curcumin have pleiotropic protective effects and appear ideal to prevent or treat conditions such as Alzheimer's disease or cognitive decline, whose origin is multifactorial. Preclinical animal models indicate a main effect of curcumin on cognitive function also thanks to its anti-inflammatory and antioxidant properties. While preclinical studies are mostly all in favor of a positive effect of curcumin in counteracting cognitive decline and age-associated brain dysfunction, clinical studies report discordant effects, highlighting the difficulty in translating basic research to the clinic.

Results of published clinical studies, however, show promise for curcumin's use as a therapeutic for cognitive decline. Kuszewski et al. [6] in this issue, provide another example of positive curcumin effects on brain function. They conducted an exploratory analysis of the effects of curcumin and fish oil supplementation on mental wellbeing in middle-aged and older adults. Previous studies had already indicated that curcumin supplementation may significantly reduce fatigue and enhance mood in non-depressed older adults. In this study, curcumin was found to improve vigor, compared to placebo, and reduced subjective memory complaints. However, combining curcumin with fish oil did not result in additive effects. This exploratory analysis indicates that regular supplementation with either curcumin or fish oil (limited to APOE4 non-carriers) has the potential to improve some aspects of mental wellbeing in association with better quality of life.

Curcumin is not only considered for preventive purposes, but also for therapeutic purposes in cancer therapy, which requires a killing effect on cancer cells. Wong et al. [7] further indicate the potential for curcumin in modulating the core pathways involved in glioblastoma cell proliferation, apoptosis, cell cycle arrest, autophagy, paraptosis, oxidative stress, and tumor cell motility. Glioblastoma is the most common and aggressive form of malignant primary adult brain tumor. In their review, the authors discuss curcumin's anticancer mechanism through modulation of Rb, p53, MAPK, P13K/Akt, JAK/STAT, Shh, and NF-κB pathways, which are commonly involved and dysregulated in preclinical and clinical glioblastoma models providing the rationale for future studies. Curcumin is also well known for its potential role in inhibiting cancer by targeting epigenetic machinery, affecting DNA methylation. Fabianowska-Majewska et al. [8] review evidence for curcumin to modulate epigenetic events that are dysregulated in cancer cells and possess the potential to prevent cancer or enhance the effects of conventional anti-cancer therapy. More in detail, the review discusses the potential epigenetic mechanisms of curcumin in reversing altered patterns of DNA methylation in breast cancer, which is the most commonly diagnosed cancer and the leading cause of cancer death among women worldwide.

One main aspect that needs to be addressed by future research is the low bioavailability of curcumin. This is clearly addressed by Scazzocchio et al. [9], who explore the metabolic bases of the chemical instability and the poor detection of curcumin in plasma and urine. This could be also due to the frequent lack of curcumin derivative detection that may lead, therefore, to an underestimation of its absorption.

However, great efforts have been made in recent years to improve curcumin's bioavailability by addressing these various mechanisms. To overcome the low bioavailability of curcumin due to its insolubility in water, Beltzig et al. [10] administered curcumin in different water-soluble formulations, including liposomes or embedded into nanoscaled micelles. These authors successfully demonstrate that the effective concentration on different cell lines, including primary cells, was far above the curcumin concentration that can be achieved systemically in vivo, leading the authors to conclude that native curcumin and curcumin administered as food supplement in a micellar formulation are not cytotoxic/genotoxic, indicating a wide margin of safety opening new avenues in therapeutic strategies.

The review by Scazzocchio et al. [9] also highlights the influence of intestinal microbiota on the expression of curcumin effects. A reciprocal influence exists, in fact, between microbiota and curcumin, with curcumin being able to shape the composition of microbiota that, in turn, influences curcumin absorption and activity by producing a number of metabolites showing different degrees of biological activity. From this point of view, thus, what takes place in the gastrointestinal tract after curcumin ingestion can help to explain, at least partially, the paradox of the very low bioavailability of curcumin and its wide effect on health.

In the review by Marzena Jabczyk et al. [11], the data on the relationship between curcumin and microbiota are further expanded, considering the association between curcumin-induced changes in gut microbiota and attenuation of a number of diseases. The modulation of gut microbiota induced by curcumin treatment seems to be associated with an improvement of liver diseases, such as non-alcoholic fatty liver, as well as of metabolic health restoring a beneficial microbiota profile that is found altered in the course of metabolic disorders such as obesity and diabetes. Interestingly, the authors also reported data on the association of curcumin consumption, gut microbiota modifications, and amelioration of exercise performance in mice. Finally, studies on the inhibitory effect of curcumin on metabolism and proliferation of *Streptococcus mutans* in oral microbiota appear to be of some relevance as this bacterium is considered a main etio-pathological agent of dental caries.

Because curcumin preferentially accumulates in the gastrointestinal tract after oral administration, it is reasonable to hypothesize that this polyphenol may exert some influence in this district. Curcumin, indeed, not only modifies the composition of the microbiota but might also enhance the function of the intestinal barrier. This would have a significant role in the prevention and/or therapy of a number of intestinal inflammatory pathologies, the Inflammatory Bowel Diseases (IBD), characterized by exacerbated inflammatory processes due to alteration of the intestinal barrier integrity. The review by M. Roque Coelho et al. [12] highlights the effectiveness of curcumin as complementary therapy for ulcerative colitis. Unfortunately, there were only six studies that met the eligibility criteria to enter in the systematic review; however, almost all of them showed positive outcomes after the intervention with curcumin, and only one reported some mild side effects. In conclusion, the reported findings suggest that curcumin may be a safe, effective co-adjuvant in the therapy for ulcerative colitis, also inducing symptom remission.

The main concerns that arise from the evaluation of the preventive/therapeutic effects of curcumin, are, first of all, the still-small number of randomized placebo-control studies available; in addition, generally, small sample sizes have been considered, different protocols and different formulations as well as different routes of administration of curcumin used. Thus, it is quite difficult to compare the results and to provide a standard protocol for the use of this promising natural compound. For these reasons, and taking into account the growing use of curcumin by the general population, further investigations to confirm and expand current findings are mandatory and research on curcumin and its effects on human health have to be fostered and promoted.

Conflicts of Interest: The authors declare no conflict of interest.

References

1. Filardi, T.; Varì, R.; Ferretti, E.; Zicari, A.; Morano, S.; Santangelo, C. Curcumin: Could This Compound Be Useful in Pregnancy and Pregnancy-Related Complications? *Nutrients* **2020**, *12*, 3179. [CrossRef] [PubMed]
2. Thota, R.N.; Rosato, J.I.; Dias, C.B.; Burrows, T.L.; Martins, R.N.; Garg, M.L. Dietary Supplementation with Curcumin Reduce Circulating Levels of Glycogen Synthase Kinase-3β and Islet Amyloid Polypeptide in Adults with High Risk of Type 2 Diabetes and Alzheimer's Disease. *Nutrients* **2020**, *12*, 1032. [CrossRef] [PubMed]
3. Altobelli, E.; Angeletti, P.M.; Marziliano, C.; Mastrodomenico, M.; Giuliani, A.R.; Petrocelli, R. Potential Therapeutic Effects of Curcumin on Glycemic and Lipid Profile in Uncomplicated Type 2 Diabetes—A Meta-Analysis of Randomized Controlled Trial. *Nutrients* **2021**, *13*, 404. [CrossRef] [PubMed]

4. Varì, R.; Scazzocchio, B.; Silenzi, A.; Giovannini, C.; Masella, R. Obesity-Associated Inflammation: Does Curcumin Exert a Beneficial Role? *Nutrients* **2021**, *13*, 1021. [CrossRef] [PubMed]
5. Berry, A.; Collacchi, B.; Masella, R.; Varì, R.; Cirulli, F. Curcuma Longa, the "Golden Spice" to Counteract Neuroinflammaging and Cognitive Decline—What Have We Learned and What Needs to Be Done. *Nutrients* **2021**, *13*, 1519. [CrossRef] [PubMed]
6. Kuszewski, J.C.; Howe, P.R.C.; Wong, R.H.X. An Exploratory Analysis of Changes in Mental Wellbeing Following Curcumin and Fish Oil Supplementation in Middle-Aged and Older Adults. *Nutrients* **2020**, *12*, 2902. [CrossRef] [PubMed]
7. Wong, S.C.; Kamarudin, M.N.A.; Naidu, R. Anticancer Mechanism of Curcumin on Human Glioblastoma. *Nutrients* **2021**, *13*, 950. [CrossRef] [PubMed]
8. Fabianowska-Majewska, K.; Kaufman-Szymczyk, A.; Szymanska-Kolba, A.; Jakubik, J.; Majewski, G.; Lubecka, K. Curcumin from Turmeric Rhizome: A Potential Modulator of DNA Methylation Machinery in Breast Cancer Inhibition. *Nutrients* **2021**, *13*, 332. [CrossRef] [PubMed]
9. Scazzocchio, B.; Minghetti, L.; D'Archivio, M. Interaction between Gut Microbiota and Curcumin: A New Key of Understanding for the Health Effects of Curcumin. *Nutrients* **2020**, *12*, 2499. [CrossRef] [PubMed]
10. Beltzig, L.; Frumkina, A.; Schwarzenbach, C.; Kaina, B. Cytotoxic, Genotoxic and Senolytic Potential of Native and Micellar Curcumin. *Nutrients* **2021**, *13*, 2385. [CrossRef] [PubMed]
11. Jabczyk, M.; Nowak, J.; Hudzik, B.; Zubelewicz-Szkodzińska, B. Curcumin and Its Potential Impact on Microbiota. *Nutrients* **2021**, *13*, 2004. [CrossRef] [PubMed]
12. Coelho, M.R.; Romi, M.D.; Ferreira, D.M.T.P.; Zaltman, C.; Soares-Mota, M. The Use of Curcumin as a Complementary Therapy in Ulcerative Colitis: A Systematic Review of Randomized Controlled Clinical Trials. *Nutrients* **2020**, *12*, 2296. [CrossRef] [PubMed]

MDPI

Article

Cytotoxic, Genotoxic and Senolytic Potential of Native and Micellar Curcumin

Lea Beltzig, Anna Frumkina, Christian Schwarzenbach and Bernd Kaina *

Institute of Toxicology, University Medical Center, Obere Zahlbacher Straße 67, 55131 Mainz, Germany; Lea.beltzig@uni-mainz.de (L.B.); frumkian@uni-mainz.de (A.F.); schwarzenbach@uni-mainz.de (C.S.)
* Correspondence: kaina@uni-mainz.de; Tel.: +49-6131-17-9217

Abstract: Background: Curcumin, a natural polyphenol and the principal bioactive compound in *Curcuma longa*, was reported to have anti-inflammatory, anti-cancer, anti-diabetic and anti-rheumatic activity. Curcumin is not only considered for preventive, but also for therapeutic, purposes in cancer therapy, which requires a killing effect on cancer cells. A drawback, however, is the low bioavailability of curcumin due to its insolubility in water. To circumvent this limitation, curcumin was administered in different water-soluble formulations, including liposomes or embedded into nanoscaled micelles. The high uptake rate of micellar curcumin makes it attractive also for cancer therapeutic strategies. Native curcumin solubilised in organic solvent was previously shown to be cytotoxic and bears a genotoxic potential. Corresponding studies with micellar curcumin are lacking. Methods: We compared the cytotoxic and genotoxic activity of native curcumin solubilised in ethanol (Cur-E) with curcumin embedded in micells (Cur-M). We measured cell death by MTT assays, apoptosis, necrosis by flow cytometry, senolysis by MTT and C12FDG and genotoxicity by FPG-alkaline and neutral singe-cell gel electrophoresis (comet assay). Results: Using a variety of primary and established cell lines, we show that Cur-E and Cur-M reduce the viability in all cell types in the same dose range. Cur-E and Cur-M induced dose-dependently apoptosis, but did not exhibit senolytic activity. In the cytotoxic dose range, Cur-E and Cur-M were positive in the alkaline and the neutral comet assay. Genotoxic effects vanished upon removal of curcumin, indicating efficient and complete repair of DNA damage. For inducing cell death, which was measured 48 h after the onset of treatment, permanent exposure was required while 60 min pulse-treatment was ineffective. In all assays, Cur-E and Cur-M were equally active, and the concentration above which significant cytotoxic and genotoxic effects were observed was 10 μM. Micelles not containing curcumin were completely inactive. Conclusions: The data show that micellar curcumin has the same cytotoxicity and genotoxicity profile as native curcumin. The effective concentration on different cell lines, including primary cells, was far above the curcumin concentration that can be achieved systemically in vivo, which leads us to conclude that native curcumin and curcumin administered as food supplement in a micellar formulation at the ADI level are not cytotoxic/genotoxic, indicating a wide margin of safety.

Keywords: curcumin; bioavailability; micelles; apoptosis; comet assay; genotoxicity; senolytics

Citation: Beltzig, L.; Frumkina, A.; Schwarzenbach, C.; Kaina, B. Cytotoxic, Genotoxic and Senolytic Potential of Native and Micellar Curcumin. *Nutrients* **2021**, *13*, 2385. https://doi.org/10.3390/nu13072385

Academic Editors: Roberta Masella and Francesca Cirulli

Received: 28 May 2021
Accepted: 5 July 2021
Published: 13 July 2021

Publisher's Note: MDPI stays neutral with regard to jurisdictional claims in published maps and institutional affiliations.

Copyright: © 2021 by the authors. Licensee MDPI, Basel, Switzerland. This article is an open access article distributed under the terms and conditions of the Creative Commons Attribution (CC BY) license (https://creativecommons.org/licenses/by/4.0/).

1. Introduction

Considerable interest has been addressed to natural compounds that have a health-beneficial potential. Plants are especially rich in bioactive natural compounds [1,2]. In particular, plant (poly)phenols [3] were reported to have a high therapeutic and preventive potential, combined with negligible adverse effects [4,5]. One of these compounds is curcumin (1,7-bis (4-hydroxy-3-methoxyphenyl)-1,6-heptadiene-3,5-dione), a naturally occurring polyphenol present in the rhizome of *Curcuma longa* L. and other curcuma species [6]. Curcuma, which contains a mixture of curcuminoids of which curcumin is the main component, has a long tradition as a food supplement in Asia. Moreover, it is being used in traditional Chinese medicine for the treatment of a wide variety of diseases, including inflammation-related disorders and neurodegenerative diseases [7]. Whether the curcuma-based traditional Chinese medicine formulae are effective in the treatment of diseases is a matter of dispute [8].

With the identification of curcumin as the active constituent of curcuma, numerous molecular-biological studies have been performed, and several molecular targets have been identified, including NF-kB, MAP-kinase, p53, NRF2, AKT, COX-2 and EGFR [9–11]. Some of these targets are key nodes in inflammation and cancer, supporting the rationale that curcumin may be beneficial in cancer prevention. Clinical studies with curcumin point to its anti-rheumatic, anti-cancer, anti-diabetic and wound-healing potential [12–14].

A drawback in the use of curcumin, and a major point of criticism regarding its health effects, is the low bioavailability of the polyphenol [15,16] due to its low water solubility and rapid metabolism [17,18]. Thus, curcumin administered in solid form cannot be taken up systemically, and solubilization in alkaline media results in rapid decomposition of the compound [19]. Since the low bioavailability clearly hampers the application of curcumin as a bioactive agent, several strategies have been pursued to overcome these limitations, including the use of adjuvants for inhibiting its metabolism, micronization, binding to cyclodextrin and embedding it into liposomes (for a review, see [20]). The incorporation of curcumin into micelles was found to be the hitherto most successful strategy to enhance its oral bioavailability, resulting in reasonable blood plasma levels [21–23], and even significant concentrations were found in brain tumors of patients ingesting micellar curcumin prior to tumor resection [24].

In view of the widespread use of curcumin as a food, the question of a possible genotoxic effect deserves particular attention. Curcumin does not bind to DNA and does not form adducts. It does not bear mutagenic activity, as confirmed in the Ames test [25]. In mammalian cells, the effects of curcumin are complex. Thus, curcumin is an anti-oxidant [26], and therefore it can exhibit protective effects by weakening the activity of ROS-generating genotoxins, which explains the protection against radiation-induced DNA damage [27]. However, in the high (micromolar) dose range, curcumin was reported to act as a radical generator and pro-oxidant: it causes a ROS burst and concomitantly oxidative DNA damage. Thus, it has been shown that ROS production increased progressively at curcumin exposure concentrations of 10–100 μM [28], and curcumin induces 8-oxo-guanine in HepG2 cells at concentrations >2.5 μg/mL (6.8 μM) [29]. In the same concentration range (2.5–10 μg/mL), curcumin induced micronuclei in rat PC12 cells [30]. Based on these data, curcumin administered in high (cytotoxic) doses can be regarded as a substance with a weak genotoxic potential. Curcumin was also shown to induce cellular senescence [31,32].

Here, we used the micellar formulation of curcumin (Cur-M), which was previously shown to enhance significantly the curcumin bioavailability in humans [21] and compared it with curcumin solubilised in ethanol (Cur-E) as to its cytotoxic and genotoxic potential. We measured dose dependently viability, apoptosis, necrosis and genotoxic effects using the FPG-alkaline and neutral comet assays. We also assessed the senolytic activity of Cur-M, i.e., the ability to kill specifically senescent cells. We show that curcumin induces cyto- and genotoxicity in a narrow, high dose range between 10 and 60 μM. There was no difference between Cur-E and Cur-M and between cancer and normal cells. Genotoxic effects vanished upon post-incubation in the absence of curcumin, indicating they are transient and subject to repair. Micelles not containing curcumin were negative in all assays, i.e., they do not bear a cytotoxic or genotoxic potential. Curcumin was not senolytic. Overall, the data show that the increased bioavailability of micellar curcumin is not linked to a higher cytotoxic and genotoxic potential compared to native curcumin administered in an organic solvent.

2. Materials and Methods

2.1. Reagents

Native curcumin (Diferuloylmethan) from *Curcuma longa* (Turmeric) was purchased from Sigma-Aldrich, Germany (CAS 458-37-7) and was dissolved in ethanol at a stock concentration of 10 mM. Micellar curcumin (NovaSol containing highly purified curcumin, ≥95%; Kancor Mane, Kerala, India) and control micells containing water were produced on Tween-80 basis and kindly provided by AQUANOVA AG (Darmstadt, Germany). The micellar stocks were always freshly prepared by weighting and dilution in PBS of a given amount of micells, giving a stock solution of 10 mM. Stocks were stored in the dark at 4 °C for no more than 3 days. Temozolomide (TMZ) was a gift from Dr. G. Margison (Manchester, UK) and was handled as described [33]. For other chemicals and reagents, see the corresponding method section.

2.2. Cell Lines and Culture Conditions

The primary human cell lines HUVEC, HUASMC and hPC-PL were purchased from PromoCell. EA.hy926, a cell line created by fusion of the A149 human lung carcinoma and the HUVEC cell line, and the human glioblastoma lines LN229 and A172, were purchased from American Type Culture Collection (ATCC). The human fibroblast cell line VH10T is a diploid telomerase-immortalized line, which was a kind gift from Prof. L. Mullenders, Leiden. The primary cell lines were cultured in endothelial cell growth medium 2, smooth muscle cell growth medium 2 and pericyte growth medium, purchased from PromoCell (Heidelberg, Germany), respectively. EA.hy926, LN229 and A172 were cultured in DMEM or DMEM GlutaMax (Gibco, Life Technologies Corporation, Paisley, UK) supplemented with 10% fetal calf serum (FCS; Gibco, Life Technologies Corporation, Paisley, UK), and for EA.hy926 with 1% HAT. All cells were maintained at 37 °C in a humidified atmosphere containing 5% CO_2. To ensure exponential growth during the whole experiment period, cells were seeded 48 h prior to treatment, and cell densities were chosen accordingly.

2.3. MTT Assay

Cells were seeded in 96-well plates and 2 days later treated with curcumin dissolved in ethanol or packed in micelles for 48 h. For the thiazolyl blue tetrazolium bromide (MTT) assay, cells were washed with PBS, and incubated for 2 h with 100 µL DMEM without phenol red containing 0.5 mg/mL MTT reagent (BIOMOL, Hamburg, Germany). The staining solution was carefully removed, 100 µM 0.04 N HCL were added to each well, and plates were incubated on a shaker (200 rpm) at room temperature for 10 min. Plates were measured in triplets using the Berthold microplate reader at OD_{570} and expressed as absorbance relative to the non-treated control.

2.4. Apoptosis and Necrosis

The amount of apoptotic and necrotic cells was measured by flow cytometry using annexin V (AV) and propidium iodine (PI) staining as described [34]. In brief, cells, including the supernatants, were collected, centrifuged and stored on ice. They were incubated for 15 min at RT in 50 µL annexin binding buffer (10 mM Hepes, 140 mM NaCl, 25 µM $CaCl_2$) containing 2.5 µL AV/FITC (Miltenyi Biotec GmbH, Bergisch Gladbach, Germany). For PI staining, 10 µL PI from a 50 µg/mL stock solution (Sigma-Aldrich, Steinheim, Germany) were added to each sample. Cells were kept in the dark until measurement. Before data acquisition using the FACS Canto II flow cytometer (Becton Dickinson GmbH, Heidelberg, Germany), cells were diluted in an adjusted amount of annexin binding buffer. Data were analysed using the Flowing Software 2 (Perttu Terho, Turku Center for Biotechnology, University of Turku, Finland). Apoptotic cells were defined as AV+/PI− cells, late apoptosis AV+/PI+ and necrosis as AV−/PI+. A very low amount of AV−/PI+ cells were found in all our assays. To make sure that necrosis was not completely neglected, we classified the AV+/PI+ population as late apoptosis/necrosis (see Supplementary Materials, Figure S1).

2.5. Senescence

Temozolomide-induced senescence was measured via flow cytometry using C12FDG staining. Immediately before harvest, cells were treated with 300 μM chloroquine for 30 min to reduce endogenous SA-β-galactosidase activity. Thereafter, C12FDG was added (33 μM), and cells were incubated for an additional 90 min. Cells were washed and resuspended in PBS for measurement, while cells were kept on ice in the dark.

2.6. Determination of Senolytic Activity

Senescent cells were obtained by treatment of glioblastoma cancer cells (LN229 and A172) with temozolomide. To this end, 1.5×10^5 exponentially growing cells were trypsinized, seeded per 10 cm dish and treated 2 days later with 50 μM temozolomide. After 8–10 days, the population contained about 80% senescent cells, determined by SA-ßGAL cytochemistry and C12FDG flow cytometry essentially as reported previously [34]. Measurement occurred in a FACS Canto II flow cytometer (Becton Dickinson GmbH, Heidelberg, Germany). Data were analysed using the Flowing Software 2 (Turku Center for Biotechnology, University of Turku, Finland). Senescent cells were harvested by trypsinization and seeded in microwells (5×10^3/well/100 μL) for the MTT assay and 6-well plates (2×10^5/well/2 mL) for the AV/PI and senescence measurements. The same occurred with cells from an exponentially growing population. Two days later, proliferating and senescent cells were treated with Cur-E, Cur-M or control micelles and, as a control, with the senolytic drug ABT-737 (Sigma-Aldrich, Germany). Cells were further incubated at 37 °C, and the amount of viable cells attached onto the plates was determined by the MTT assay, C12FDG and AV/PI staining as described above

2.7. Comet Assays (Single Cell Gel Electrophoresis, SCGE)

To assess the amount of SSBs and DSB, as well as the oxidative DNA damage, alkaline comet assays with and without FPG and neutral comet assays were performed as described [35]. In brief, cells were collected by trypsinization, resuspended in ice-cold PBS (Bio&Sell, Leipzig, Germany) and embedded in ultra-pure low-melting agarose (Invitrogen, Carlsbad, CA, USA. The cell suspension was then evenly spread onto slides that were pre-coated with agarose, dried and submersed in precooled alkaline or neutral lysis buffer (2.5 M NaCl, 100 mM EDTA, 10 μM Tris, 1% sodium lauroyl sarcosinate, 1% Triton X-100, 10% DMSO, pH 10 and 7.5, respectively) for 1 h at 4 °C. For the assay with FPG, slides were pre-incubated in FPG-buffer (40 mM HEPES, 0.5 mM EDTA, 100 mM KCL, 0.2 mg/mL BSA, pH 8) for 5 min after which cells were incubated with 1 μg/mL fapy-DNA glycosylase (FPG) (kind gift from Prof. Epe, Mainz) for 40 min at 37 °C. The slides were then incubated for 20 min in alkaline (300 mM NaOH, 1 mM EDTA, pH >13) or neutral running buffer (90 mM Tris, 90 mM boric acid, 2 mM EDTA, pH 7.5) for denaturation and equilibration. Cells were then electrophoresed with 300 mA for 15 min at 4 °C. Slides from the alkaline comet assay were then washed in neutralization buffer (400 mM Tris, pH 7.5). All slides were rinsed with water, fixed in 100% ethanol for 5 min and dried. For analysis, cells were stained with propidium iodide (Sigma-Aldrich) and evaluated using a fluorescence microscope (Microphot-FXA, Nikon, Tokyo, Japan) and the Comet IV software (Perceptive Imaging, Liverpool, UK). As a positive control, cells were incubated for 30 min with 300 μM tBOOH.

2.8. ROS

The induction of ROS was measured by an increase in DCF-fluorescence with the flow cytometer using the FACS Canto II flow cytometer (Becton Dickinson GmbH, Heidelberg, Germany). Cells were incubated for 30 min at 37 °C with 10 μM H_2DCFDA (Invitrogen, Carlsbad, CA, USA) in serum-free medium (Gibco, Life Technologies Corporation, Paisley, UK). Cells were collected by trypsinization and re-suspended in cold PBS (Bio&Sell, Leipzig, Germany). Fluorescence was measured immediately. As a positive control, cells were incubated for 30 min with 300 μM tert-butyl hydroperoxide (tBOOH).

2.9. Statistics

All data are given as the mean with the standard error of the mean (SEM), and differences between group means were statistically evaluated using the Two-Way ANOVA, unpaired Student's *t*-test or Mann–Whitney test, as appropriate, and considered significant at $p < 0.05$.

3. Results

3.1. Effect of Cur-E and Cur-M on the Viability of Cells

First, we studied the effect of Cur-E and Cur-M on the cell's viability, which was comparatively analysed on different cell systems, including human telomerase-immortalized fibroblasts VH10T, cancer cells (the glioblastoma line LN229), human endothelial cells (the line EA.hy926), human primary vascular endothelial cells (HUVEC), human primary smooth muscle cells (HUASM) and human primary pericytes (hPC-PL). As shown in Figure 1, in all cell systems, Cur-E and Cur-M were equally cytotoxic, reducing the viability in a concentration range between 5 and 50 μM. Below 5 μM, neither Cur-E nor Cur-M exerted a significant reduction in viability in all cell types, except HUVEC (for D_{50} doses see Supplementary Materials, Figure S2). Freshly isolated human monocytes, macrophages and T cells responded with curcumin concentrations >15 μM (Supplementary Materials, Figure S3).

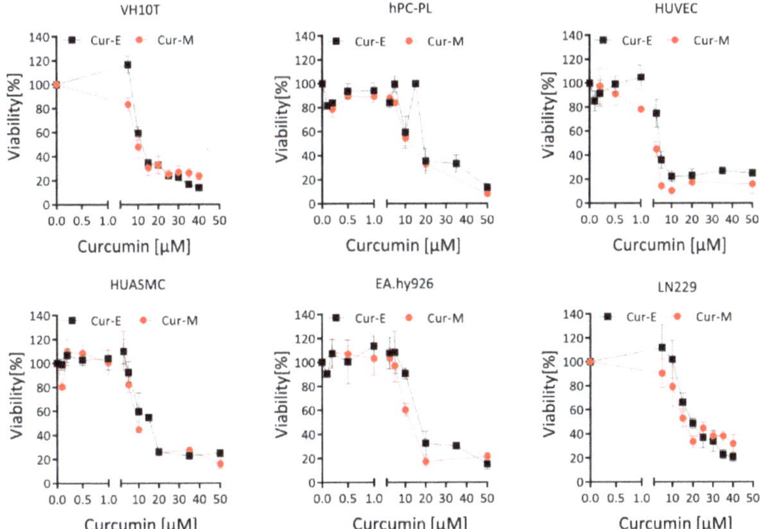

Figure 1. Effect of Cur-E and Cur-M on the viability of different primary cells and the tumor cell line LN229. Dose-dependent toxicity of Cur-E and Cur-M was determined in the MTT assay upon 72 h treatment of LN229, VH10tert, EA.hy926, HUVEC, HUASMC and hPC-PL. Data are the mean +/− SEM of 3 independent experiments.

3.2. Cur-E and Cur-M Induce Cell Death by Apoptosis

The reduction in viability observed in the MTT assay can be due to different mechanisms, including metabolic impairment, arrest of proliferation and genuine cell death. To compare in more detail the cytotoxic potential of the curcumin formulations, we measured apoptosis and necrosis by AV/PI flow cytometry. The data shown in Figure 2 demonstrate that Cur-E and Cur-M are effective in inducing apoptosis. The dose range was similar in human fibroblasts VH10T and the glioblastoma cell lines LN229 and A172. The threshold concentration at which a significant increase was observed was 20 μM, while at 10 μM the levels were still insignificant above the background. In VH10T cells, Cur-E was slightly more effective in inducing cell death than Cur-M. In the glioma cell lines, this difference was not observed. Control micelles (Mic) administered in the same amount did not show any cytotoxic effects (Figure 2). Treatment for 1 h and post-incubation for 48 h was ineffective in inducing apoptosis (Supplementary Materials Figure S4).

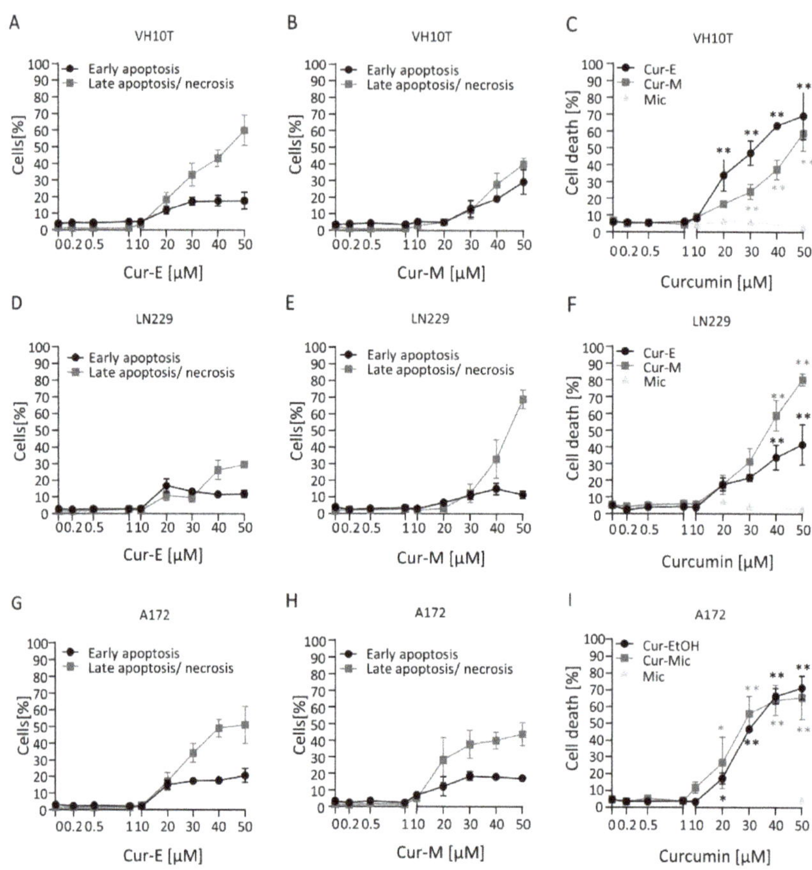

Figure 2. Induction of cell death following curcumin treatment. Proliferating VH10T (**A**–**C**), LN229 (**D**–**F**) and A172 (**G**–**I**) cells were treated with the indicated concentrations of Cur-E or Cur-M for 48 h. Apoptosis, late apoptosis/necrosis and total cell death were measured by AV/PI flow cytometry. Data are the mean of 3 independent experiments ± SEM. Statistical analysis was performed using the Two-way ANOVA. Significant elevation above the control was observed with 20 μM curcumin. Water micelles were ineffective. * $p < 0.05$; ** $p < 0.01$.

3.3. Cur-E and Cur-M Have No Senolytic Activity

Many genotoxic agents induce not only apoptosis, but also cellular senescence (CSEN). A potential for curcumin to induce CSEN was shown previously (for a review, see [36]). Here, we pursued to determine the senolytic potential of curcumin, i.e., the selective ability of a compound to kill senescent cells. The methylating anticancer drug temozolomide is a potent inducer of CSEN in p53 functionally active glioblastoma cells [37]. We induced CSEN in LN229 and A172 cells as previously described [38] and assessed whether curcumin is able to kill preferentially senescent cells. For comparison, we used the well-known senolytic drug ABT-737 [39]. We compared the effect of Cur-M on proliferating versus senescent cells. The data revealed that Cur-M was clearly more effective in reducing the viability of proliferating than senescent LN229 (Figure 3A) and A172 (Figure 3B) cells. This is in sharp contrast to ABT-737, which showed a clear senolytic effect on glioblastoma cells, being more effective on senescent cells (Figure 3A,B). The data were substantiated comparing Cur-E and Cur-M, which were equally effective in reducing the viability of proliferating versus senescent cells (Figure 3C,D).

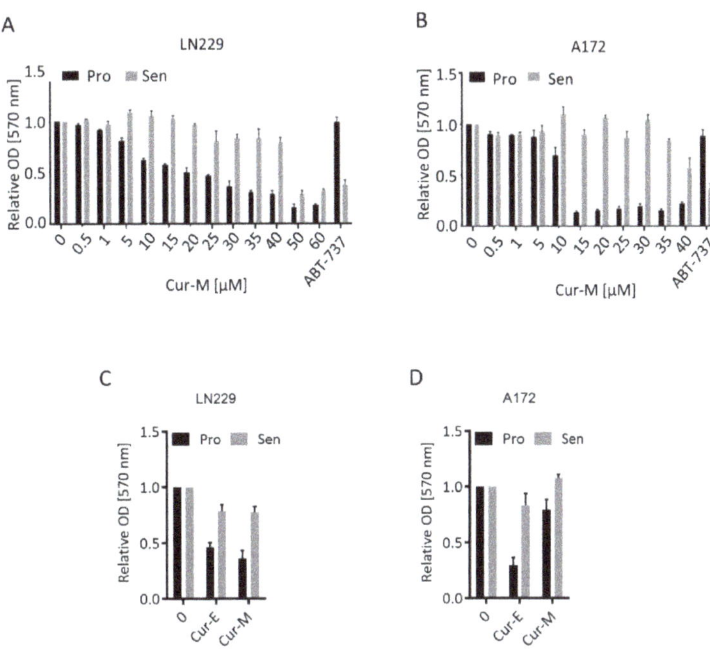

Figure 3. Effect of curcumin on the viability of proliferating and senescent cells. Proliferating (Pro) and senescent (Sen) LN229 (**A**,**C**) and A172 (**B**,**D**) cells were treated with the indicated dosages of Cur-E or Cur-M. (**A**,**B**) Dose response of proliferating and senescent LN229 (**A**) and A172 (**B**) cells upon curcumin treatment. (**C**,**D**) Comparison of Cur-E and Cur-M in reduction in viability in LN229 (**C**) and A172 (**D**) cells. Viability (MTT OD_{570}) of non-treated control was set to 1. The senolytic drug ABT-737 served as a positive control.

To substantiate the data further, we measured the relative amount of senescent and apoptotic cells in the population eight days after temozolomide treatment and, following addition of Cur-E or Cur-M to the medium, two days later. The data revealed that curcumin (10 µM) had no impact on the yield of temozolomide-induced senescent cells and the total cell death (early and late apoptosis/necrosis) level in the population (Figure 4A,B). Overall, the data show that curcumin in a subtoxic concentration has no senolytic activity and does not impact the senescence level, irrespective of whether it is administered as Cur-E or Cur-M.

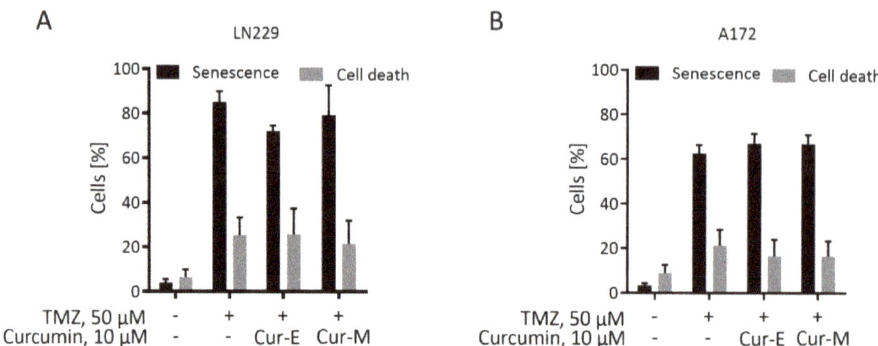

Figure 4. Curcumin does neither reduce the senescence level nor induces apoptosis in a senescent population. Temozolomide-induced senescent LN229 (**A**) and A172 (**B**) cells were treated with Cur-E and Cur-M in a concentration that was not apoptosis-inducing on proliferating cells. Senescence and cell death were measured by C12FDG and AV/PI staining, respectively, in a flow cytometer. Mean ± SEM, statistical analysis through Two-way ANOVA test. Differences between control, Cur-E and Cur-M were not significant.

3.4. Analysis of Genotoxic Effects of Curcumin

The observation that curcumin is able to induce ROS in human cells ([29,40] and Supplementary Materials Figure S5) prompted us to study the genotoxic potential of the compound, focusing on a comparison of Cur-E and Cur-M. First, we measured the amount of DNA damage in human VH10T fibroblasts as a function of the concentration of curcumin after 24 h of exposure of exponentially growing cells. We used the alkaline comet assay, which measures mostly single-strand breaks, and the FPG-modified alkaline assay, which detects FPG-cleavage sites such as 8-oxo-G [41]. As shown in Figure 5, curcumin was positive in the alkaline comet assay at concentrations >20 µM. In the more sensitive FPG-comet assay, already 20 µM curcumin provoked a significant effect in all cell lines. There was no difference between Cur-E and Cur-M (Figure 5A,B), and control micelles (Mic) did not display any genotoxic effect (Figure 5C and Supplementary Materials Figure S6). Similar data were obtained with LN229 cells upon treatment with Cur-E, Cur-M and control micelles (Figure 5D–F). In a low concentration range between 0.2 and 5 µM, curcumin did not induce strand-breaks, as determined in the alkaline comet assay (Supplementary Materials Figure S7).

Curcumin was positive in the alkaline comet assay already after short-term exposure. Thus, after 1 h of treatment, effects were observed, irrespective of whether treatment occurred with Cur-E or Cur-M (Figure 6; repair time zero). This was confirmed in a repair experiment, where VH10T cells were treated with curcumin (40 µM) for 1 h and post-incubated for 1, 2, 3 and 4 h before harvest. The data show that repair of damage clearly occurred in the 4 h post-treatment period. Again, there was no clear difference between Cur-E and Cur-M (Figure 6), and micelles without curcumin were without any effect (Figure 6). In sum, the data revealed that Cur-E and Cur-M bear genotoxic potential at a high dose level (≥20 µM). The lesions appear to be short lived and are subject to repair.

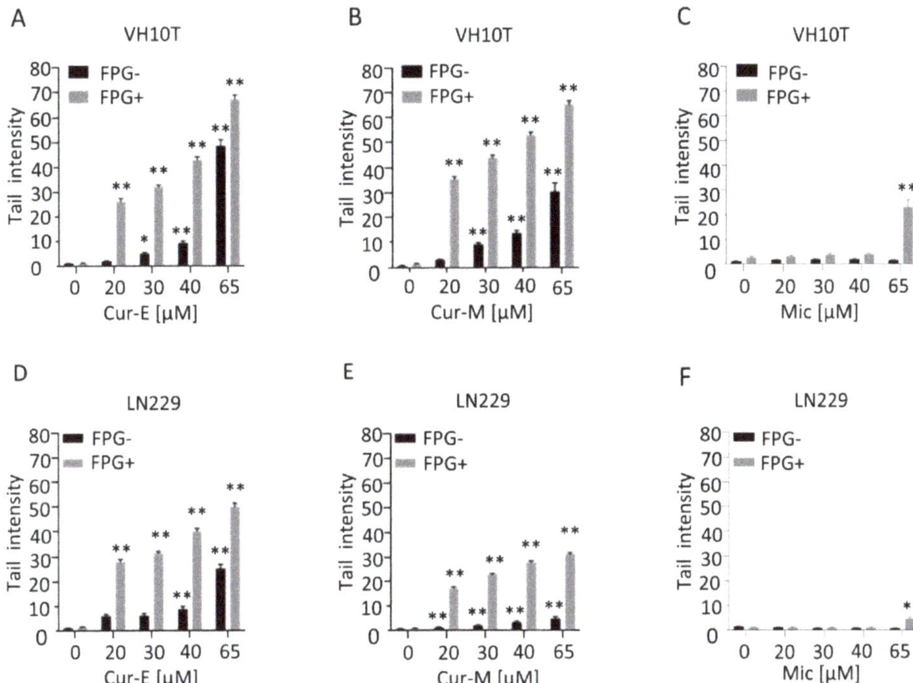

Figure 5. Curcumin induces DNA damage in a dose-dependent manner. Proliferating VH10T (**A**–**C**) and LN229 (**D**–**F**) cells were treated with the respective concentrations of curcumin. (**A**,**D**), curcumin was solubilised in ethanol (Cur-E) or (**B**,**E**), administered as micelles (Cur-M). As control, cells were also treated with water-micelles (Mic, **C**,**F**). Alkaline comet assay with and without FPG was performed 24 h post-treatment. Asterisks indicate significant difference to control * $p < 0.05$; ** $p < 0.01$.

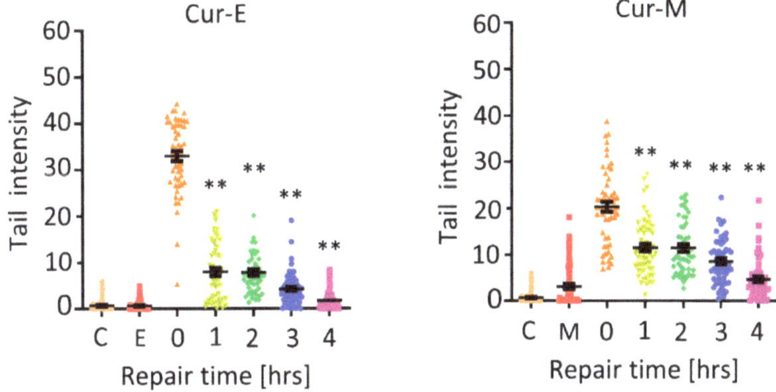

Figure 6. Induction and repair of DNA damage following treatment of VH10T cells with Cur-E and Cur-M. C, untreated control; E, ethanol control; M, micelles control; treatment with 40 μM occurred for 1 h (repair time 0 h) and post-incubation for 1, 2, 3 and 4 h. Statistical analysis was performed using the Two-way ANOVA. ** $p < 0.01$.

The finding of short-lived lesions indicates that curcumin has to be present in the medium in order to elicit a genotoxic effect in the long-term setting. This was proven by an experiment in which cells were treated for 1, 2, 4 and 24 h followed by a recovery time of 23, 22, 20 and 0 h, respectively. Cells were harvested after a total incubation time of 24 h and analysed by the alkaline comet assay. As shown in Figure 7A (for representative images) and Figure 7B,C (for quantification), treatment for 1, 2 and 4 h did not give rise to DNA damage compared to treatment over the whole incubation period of 24 h. This supports the notion that the lesions are short-lived and subject to repair. In order to provoke long-term effects, it is obviously necessary that curcumin is permanently present in the medium. This is especially the case for the observed toxic effects; short-term treatment (1 h) even at a high dose of curcumin did not result in apoptosis induction while long-term exposure (48 h) did (Figure 7D).

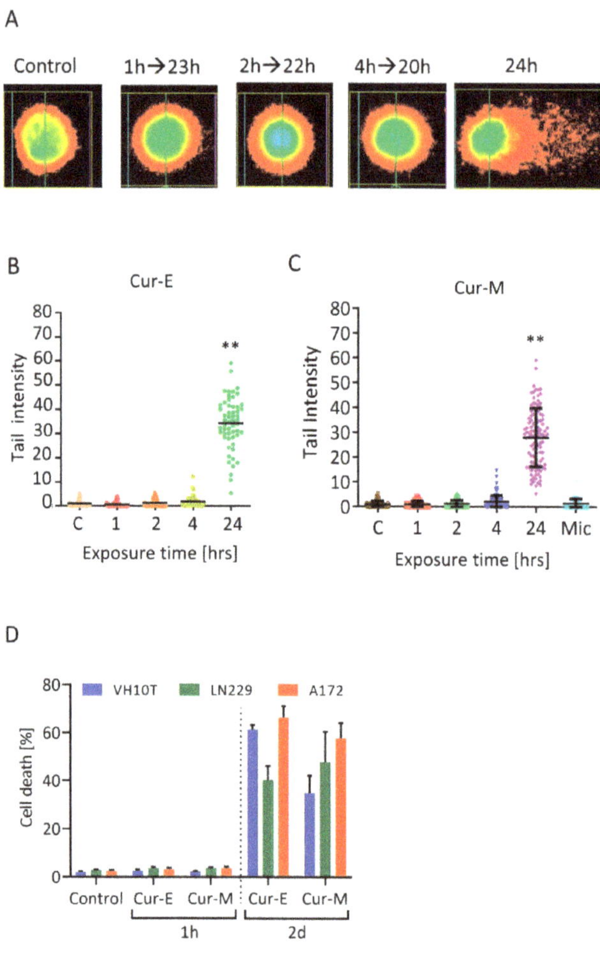

Figure 7. Effect of Cur-E and Cur-M in short- and long-term exposure experiments. (**A**) Representative images of stained single cells. (**B**,**C**) Effect of Cur-E and Cur-M in VH10T cells following exposure for the indicated times and post-exposure. Harvest occurred for all treatments 24 h after the onset of treatment with 40 μM curcumin. FPG comet assay. Statistical analysis was performed using the Two-way ANOVA. (**D**) Death (apoptosis, necrosis) of VH10T, LN229 and A172 cells following treatment of exponentially growing populations with Cur-E or Cur-M (40 μM) for 1 h and post-incubated 47 h in curcumin free medium or treated for the whole period of 48 h, until cell harvest. Cells were measured by AV/PI flow cytometry. Data are the mean of 3 experiments ± SEM. ** $p < 0.01$.

To elucidate whether curcumin bears a potential to induce DSBs, we made use of the neutral comet assay (SCGE, representative pictures Figure 8A). Treatment for 1 h (Figure 8B) or 24 h (Figure 8C) resulted in a dose-dependent increase in tail intensity, and the lowest effective dose was 10 µM curcumin. Again, Cur-E and Cur-M were equally effective in inducing effects in the neutral SCGE, and micelles without curcumin (Mic) were ineffective (Figure 8B,C). Treatment with a concentration of 40 µM curcumin for 1 h and a recovery of 23 h did not result in significant DSBs, while treatment over a 24 h period resulted in a significant yield of DSBs (Figure 8D). These data indicate recovery of cells through repair of DSBs that were induced with a high curcumin concentration during a 1 h exposure period.

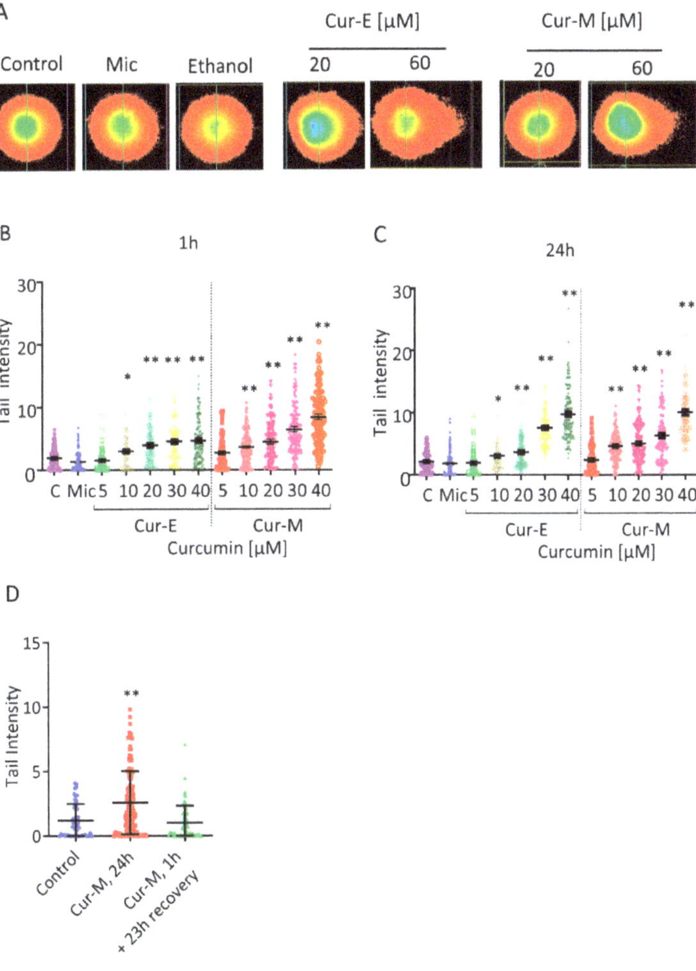

Figure 8. Effect of Cur-E and Cur-M on VH10T cells measured in the neutral comet assay. (**A**) Examples of stained cells following single cell gel electrophoresis (24 h treatment). (**B**) Quantification upon 1 h treatment. (**C**) Quantification upon 24 h treatment. (**D**) Cells were not treated (control) or treated for 1 h and post-incubated for 23 h (Cur-M, 1 h) or for 24 h (Cur-M, 24 h) with Cur-M. Then, 100 cells were evaluated per treatment level. C, non-treated control; Mic, micelles only corresponding to 40 µM. Statistical analysis was performed using the Two-way ANOVA test. * $p < 0.05$; ** $p < 0.01$.

4. Discussion

Curcumin has been used as a food ingredient in Asia for thousands of years and is also extensively applied in traditional Chinese medicine (phytomedicine), especially for the treatment of inflammation-related and neurodegenerative diseases [7,8]. Today, it is widely used worldwide as a spice and food additive (E100). Native curcumin has a low bioavailability [15,16] due to its poor water solubility and rapid metabolism in the liver [17,18]. It is soluble in an alkaline environment (although unstable under these conditions) and in ethanol and dimethyl sulfoxide (DMSO), but these are solvents that are suitable for experimental purposes only [19]. When preparing dishes containing curcumin (e.g., "curry"), turmeric is usually suspended in oil under heat. From this suspension, in which micellar structures form, curcumin can better be absorbed in the gastrointestinal tract. To overcome uptake limitations, a formulation of curcumin embedded in micelles on PEG80 basis was developed and shown to have superb uptake properties [21–23,42–45]. Micellar curcumin was recently also tested in a feeding study and demonstrated to be safe and effective as an anticancer agent in a colorectal mouse model [46].

Although curcumin solubilised in DMSO was extensively studied (for a review, see [14]), comparative analyses of curcumin administered as micelles and solubilised in an organic solvent are lacking. Thus, it is unclear whether micellar curcumin and the micelles itself have properties that are different from curcumin solubilised in an organic solvent. To provide an answer to this question, we rigorously compared Cur-E and Cur-M as to their cytotoxic, senolytic and genotoxic effects on normal and cancer cells in vitro.

First, we show that Cur-E and Cur-M reduce the viability of human cells irrespective of its origin, including human diploid fibroblasts, human primary pericytes, endothelian cells, smooth muscle cells, established endothel cells and human glioblastoma cells. In these cell models, the concentration range in which a decline in viability was observed was 10–60 µM, and Cur-E and Cur-M caused a similar response. The cytotoxic effect was confirmed showing that Cur-E and Cur-M induce cell death by apoptosis in human fibroblasts and cancer (glioblastoma) cells. Apoptosis increased dose-dependently, and the lowest concentration inducing a significant effect was 20 µM (48 h treatment). Again, Cur-E and Cur-M were similarly effective.

Curcumin-induced apoptosis was reported in various cell systems, including NIH3T3 and L929 mouse fibroblasts, human colon carcinoma cells (HT-29), human breast carcinoma cells (MCF-7) and rat glioma cells (C-6)] [28,47–50]. Induction of apoptosis was preceded by a ROS burst, which followed the activation of the mitochondrial apoptosis pathway through a release of AIF and cytochrome C from the mitochondria, activation of caspases and PARP-1 cleavage (as an indicator of apoptosis). The p53-p21-CDC2 signaling pathway was also shown to be activated, leading to cell cycle arrest via Rb dephosphorylation and downregulation of cyclin D1 and cyclin D3 [50]. It is interesting that, in all studies, the induction of apoptosis occurred in a narrow dose range of 10–80 µM curcumin (dissolved in DMSO). Apoptotic effects occurred in MCF-7 cells as early as 24 h after treatment with curcumin and reached a maximum 48 h after administration [50]. Cytotoxicity is therefore an immediate, acute effect of curcumin. Our data are in line with this. Irrespective of the cell type, cell death by apoptosis occurred with an exposure concentration of >10 µM. This was the case in a variety of human primary cells and cancer cell lines. According to our data, 10 µM can be considered as a cytotoxic threshold dose for human cells, which applies for both Cur-E and Cur-M.

Curcumin was shown to be able to induce cellular senescence in different cell systems [31,32,51]. It is unclear, however, whether it bears senolytic activity, which is defined as killing senescent, but not proliferating, cells. To our knowledge, there is no data to show that curcumin is able to kill especially cancer cells in which senescence was induced following chemotherapy. We tested this assumption and found that Cur-E and Cur-M killed proliferating cells more efficiently than senescent cells. This is in sharp contrast to ABT-737, which is a well-known senolytic agent [39] and used as a positive control in our experiments. Our data suggest that curcumin, irrespective of the formulation, is not a senolytic agent. We are aware of the limitations of this study as we assessed the effect in only two cancer cell lines and under limited treatment conditions. More detailed studies are clearly required in order to come to a more generalized conclusion (see also [36]).

The finding that curcumin is able to induce ROS (our data and [40]) prompted us to investigate in detail the question of DNA damage induction. Treatment with Cur-E and Cur-M induced in human fibroblasts dose-dependently DNA breaks which was most obvious in the FPG-modified comet assay, in which ROS-induced DNA damage, such as 8-oxo-G, is recognized by the FAPY-DNA glycosylase and converted into single-strand breaks that can be detected in the assay. The alkaline comet assay was also positive without FPG; however, higher curcumin concentrations were required to elicit the same effect. The observed effects are not a byproduct of cytotoxicity, as positive effects in the FPG comet assay can be observed already after 1 h of exposure. This indicates that oxidative DNA damage is induced immediately after curcumin treatment, which is in line with a previous report [40]. It is important to note that removal of Cur-E or Cur-M from the medium returned the DNA tail moment to the basal level, indicating efficient repair of DNA lesions. Thus, short-term treatment did not result in permanent DNA damage, indicating the effects are transitory and not stable. The finding indicates that, in assessing the genotoxic potential, it is important to take into consideration not only the exposure concentration, but also the period of exposure. This applies also to the endpoint apoptosis, which was measured 2 days after addition of Cur-E or Cur-M to the medium. The continuous presence of curcumin in the medium was necessary for eliciting a significant toxic and genotoxic effect on cells.

To determine whether curcumin has the potency to induce DSBs, we made use of the neutral comet assay. In this assay, curcumin was weakly positive with a concentration $\geq$10 µM. Similar effects were observed when cells were treated for 1 or 24 h, and a recovery period after a 1 h treatment resulted in their complete disappearance. The effects measured in the neutral comet assay might be due to overlapping ROS-induced single-strand breaks, overlapping single-strand breaks and base-excision repair patches, or they represent directly induced DSBs. The effect was also transient; it vanished after post-exposure recovery of cells in the absence of curcumin. It should be noted that curcumin was reported to be negative in the γH2AX foci assay [52].

Curcumin is negative in bacterial test systems [25] and not mutagenic. It is considered an anti-oxidant in the low dose range, causing protection against some genotoxins, e.g., radiation-induced DNA damage [27]. However, in the high dose range (>5 µM), curcumin acts as a radical generator and pro-oxidant, causing a ROS burst and the associated oxidative DNA damage [40]. Thus, our data are in line with a report showing that curcumin induces DNA strand breaks and 8-oxo-guanine in HepG2 cells at concentrations >2.5 µg/mL (6.8 µM) [29]. In the concentration range of 2.5–10 µg/mL, curcumin induced micronuclei in rat PC12 cells [30], likely resulting from chromosome breaks. A study on human peripheral lymphocytes showed that curcumin does not induce sister-chromatid exchanges (SCEs), which are a highly sensitive indicator of DNA damage. However, chromosomal acentric fragments were detected in the high concentration range (50 µg/mL = 135 µM). Interestingly, no chromatid-type aberrations (chromatid gaps, breaks and translocations) that are typical for chemical genotoxins were observed [53]. Since curcumin induces neither direct genotoxic DNA damage (adducts) nor SCEs, it is reasonable to conclude that the observed clastogenicity is based on S-phase independent effects. Curcumin administered at high and cytotoxic concentrations (>10 µM) can therefore be regarded as a substance with a weak radiomimetic activity, at least in cell culture models. Overall, the genotoxic effects observed can all be explained by the intracellular ROS induction after a high dose of curcumin.

It is important to point out that the concentration range in which cytotoxic and genotoxic effects occur is nearly identical in all cell systems used. The overlapping dose range of cytotoxicity and genotoxicity (between 10 and 80 µM continuous treatment) implicates that cells harboring chromosome damage will be eliminated by apoptosis or necrosis.

Under these circumstances, it is unlikely that curcumin-induced genotoxic changes will cause long-term effects. Actually, long-term feeding experiments with native curcumin and micellar curcumin did not provide evidence for adverse effects on rodents [54]. The frequently reported tumor-suppressing effect of curcumin [55] supports the view that the cytotoxic and genotoxic effects observed in the micromolar concentration range in vitro are irrelevant under nutritional conditions. On the other hand, they could become important if curcumin is administered locally at a very high, cytotoxic dose level (>20 μM target concentration) in a therapeutic setting, since a toxic effect on tumor cells was sought. An investigation of various cell lines showed that apoptosis induction by curcumin occurred preferentially in transformed, but not primary, cell lines, from which it was concluded that curcumin has a selectively toxic effect on tumor cells. Our data do not confirm this notion, as curcumin was nearly equally toxic in human diploid fibroblasts, primary endothel and smooth muscle cells, monocytes, macrophages and glioblastoma cancer cells.

In summary, micellar curcumin is taken up efficiently into cells, which makes it ideally suitable for experimental purposes since organic solvents can be avoided. Curcumin administered in a micellar formulation had a cytotoxic and genotoxic profile similar to curcumin dissolved in ethanol (similar data were obtained with curcumin dissolved in DMSO). Thus, the data demonstrate that micellar curcumin does not bear toxic and genotoxic properties that are different from native curcumin. From the available data, we define a cytotoxic and genotoxic threshold for curcumin at a concentration of 10 μM, which seems to be independent of the formulation and cell system (our data and data in the literature). Micelles without curcumin were completely devoid of cytotoxic and genotoxic activity. Therefore, micelles, which have a nanosized structure, can be considered as safe transport vehicles. Following uptake of Cur-M in the acceptable daily intake (ADI) range (max 3 mg/kg BW), the concentration of curcumin in the blood serum of humans was estimated to be lower than 100 nM (ref. [56] and J. Frank, personal communication). This is far below the cytotoxic and genotoxic level observed in vitro. It would be interesting to see whether beneficial effects of curcumin, e.g., anti-inflammatory responses, are evoked at curcumin concentrations that are below the toxic threshold.

It is interesting to note that weak genotoxic effects associated with ROS production have been described also for other phytochemicals. Thus, many natural products induce mild oxidative stress in cells [57,58], which may evoke a hormetic cellular stress response, i.e., cellular adaptation leading to protection against challenging genotoxic stress [59]. A main product of ROS in the DNA is 8-OxoG, which appears to be induced at high dose levels (>10 μM) by curcumin [29]. Although 8-oxo-G is a mispairing lesion, recent studies showed that 8-OxoG also acts as a gene regulator. It activates signaling pathways, which may contribute to the beneficial effects described above [60].

5. Conclusions

Overall, curcumin reduces cell viability and induces apoptosis as well as genotoxicity in a narrow, overlapping concentration range between 10 and 60 μM. For inducing cytotoxicity through apoptosis, long-term treatment is required; short-term exposure (60 min) was insufficient to elicit toxic effects. Genotoxic effects, measured in the alkaline and comet assay, vanished after post-exposure of cells in curcumin-free medium, indicating that lesions were transiently induced and repaired. Cur-E and Cur-M did not show senolytic activity on cancer cells, in which senescence was induced by the chemotherapeutic temozolomide. Cur-M was not more effective than Cur-E in inducing DNA damage, as measured in comet assays, and micelles were completely devoid of these effects, indicating that micelles do not bear a cytotoxic and genotoxic potential. The concentration dependence of cytotoxicity and genotoxicity reported here, together with animal experiments [46], and the record of the long-term and extensive dietary exposure of humans to native curcumin and the absence of any reports of associated toxicity, support the notion that the natural form of curcumin and the micellar formulation do not bear harmful side effects. The normal intake quantities of curcumin and the recommended inclusion levels of supplements are about 100-fold lower than the concentration considered to be toxic in the in vitro assays, indicating a wide margin of safety.

Supplementary Materials: The following are available online at https://www.mdpi.com/article/10.3390/nu13072385/s1. Figure S1: Representative plot of cells stained with annexin V and PI. Figure S2: Effect of curcumin solubilized in ethanol (A) and micellar curcumin (B) on the viability of different primary human cell types. Figure S3: Effect of curcumin on freshly isolated human monocytes, macrophages and T cells. Figure S4: Cytotoxicity (apoptosis, necrosis) of VH10T, LN229 and A172 cells following treatment of exponentially growing populations with Cur-E or Cur-M (40 µM) for 1 h and post-incubated 48 h. Cells were harvested and measured by flow cytometry. N=3,median±SEM. Figure S5: Induction of ROS following Curcumin treatment. Figure S6: Effect of micelles filled with water in the FPG comet assay. Figure S7: Effect of Cur-M on VH10T cells in the alkaline comet assay.

Author Contributions: B.K. planned the experiments; B.K., L.B., A.F. and C.S. performed the experiments; B.K. and L.B. performed data analysis and statistics; B.K. wrote the manuscript; L.B. and A.F. provided the drawings. All authors have read and agreed to the published version of the manuscript.

Funding: This study received funding in the initial stage by German Cancer Aid.

Data Availability Statement: Data are available upon request.

Acknowledgments: We are grateful to Dariush Behnam, Aquanova AG, for the supply of micellar curcumin and his encouraging interest in the study. We thank Joana Kress for her contribution in the initial stage of the study.

Conflicts of Interest: The authors declare no conflict of interests. B.K. is scientific advisor of Aquanova AG. The company supported the work without having a role in the design of the study; in the collection, analyses or interpretation of data; in the writing of the manuscript or in the decision to publish the results.

References

1. Tan, W.; Lu, J.; Huang, M.; Li, Y.; Chen, M.; Wu, G.; Gong, J.; Zhong, Z.; Xu, Z.; Dang, Y.; et al. Anti-cancer natural products isolated from chinese medicinal herbs. *Chin. Med.* **2011**, *6*, 27. [CrossRef] [PubMed]
2. Demain, A.L.; Vaishnav, P. Natural products for cancer chemotherapy. *Microb. Biotechnol.* **2010**, *4*, 687–699. [CrossRef] [PubMed]
3. Frank, J.; Fukagawa, N.K.; Bilia, A.R.; Johnson, E.J.; Kwon, O.; Prakash, V.; Miyazawa, T.; Clifford, M.N.; Kay, C.; Crozier, A.; et al. Terms and nomenclature used for plant-derived components in nutrition and related research: Efforts toward harmonization. *Nutr. Rev.* **2019**, *78*, 451–458. [CrossRef]
4. Venturelli, S.; Burkard, M.; Biendl, M.; Lauer, U.M.; Frank, J.; Busch, C. Prenylated chalcones and flavonoids for the prevention and treatment of cancer. *Nutrition* **2016**, *32*, 1171–1178. [CrossRef] [PubMed]
5. Burkard, M.; Leischner, C.; Lauer, U.M.; Busch, C.; Venturelli, S.; Frank, J. Dietary flavonoids and modulation of natural killer cells: Implications in malignant and viral diseases. *J. Nutr. Biochem.* **2017**, *46*, 1–12. [CrossRef]
6. Aggarwal, B.B.; Harikumar, K. Potential therapeutic effects of curcumin, the anti-inflammatory agent, against neurodegenerative, cardiovascular, pulmonary, metabolic, autoimmune and neoplastic diseases. *Int. J. Biochem. Cell Biol.* **2009**, *41*, 40–59. [CrossRef]
7. Ghosh, S.; Banerjee, S.; Sil, P.C. The beneficial role of curcumin on inflammation, diabetes and neurodegenerative disease: A recent update. *Food Chem. Toxicol.* **2015**, *83*, 111–124. [CrossRef]
8. Shehzad, A.; Khan, S.; Lee, Y. Curcumin therapeutic promises and bioavailability in colorectal cancer. *Drugs Today* **2010**, *46*, 523–532. [CrossRef]
9. Shehzad, A.; Wahid, F.; Lee, Y.S. Curcumin in Cancer Chemoprevention: Molecular Targets, Pharmacokinetics, Bioavailability, and Clinical Trials. *Arch. Pharm.* **2010**, *343*, 489–499. [CrossRef]
10. Kasi, P.D.; Tamilselvam, R.; Skalicka-Woźniak, K.; Nabavi, S.M.; Daglia, M.; Bishayee, A.; Pazoki-Toroudi, H. Molecular targets of curcumin for cancer therapy: An updated review. *Tumor Biol.* **2016**, *37*, 13017–13028. [CrossRef]
11. Bortel, N.; Armeanu-Ebinger, S.; Schmid, E.; Kirchner, B.; Frank, J.; Kocher, A. Effects of curcumin in pediatric epithelial liver tumors: Inhibition of tumor growth and alpha-fetoprotein in vitro and in vivo involving the NFkappaB- and the beta-catenin pathways. *Oncotarget* **2015**, *6*, 40680–40691. [CrossRef]
12. Hsu, C.-H.; Chuang, S.-E.; Hergenhahn, M.; Kuo, M.-L.; Lin, J.-K.; Hsieh, C.-Y.; Cheng, A.-L. Pre-clinical and early-phase clinical studies of curcumin as chemopreventive agent for endemic cancers in Taiwan. *Gan Kagaku Ryoho* **2002**, *29* (Suppl. 1), 194–200.
13. Hsu, C.-H.; Cheng, A.-L. Clinical Studies with Curcumin. *Adv. Exp. Med. Biol.* **2007**, *595*, 471–480.
14. Allegra, A.; Innao, V.; Russo, S.; Gerace, D.; Alonci, A.; Musolino, C. Anticancer Activity of Curcumin and Its Analogues: Preclinical and Clinical Studies. *Cancer Investig.* **2017**, *35*, 1–22. [CrossRef]
15. Aller, L.L. What about bioavailability of oral curcumin? *Can. Med. Assoc. J.* **2019**, *191*, E427. [CrossRef] [PubMed]

16. Garg, A.X.; Moist, L.; Pannu, N.; Tobe, S.; Walsh, M.; Weir, M. Bioavailability of oral curcumin. *Can. Med. Assoc. J.* **2019**, *191*, E428. [CrossRef] [PubMed]
17. Prasad, S.; Tyagi, A.K.; Aggarwal, B.B. Recent Developments in Delivery, Bioavailability, Absorption and Metabolism of Curcumin: The Golden Pigment from Golden Spice. *Cancer Res. Treat.* **2014**, *46*, 2–18. [CrossRef] [PubMed]
18. Metzler, M.; Pfeiffer, E.; Schulz, S.I.; Dempe, J.S. Curcumin uptake and metabolism. *BioFactors* **2012**, *39*, 14–20. [CrossRef]
19. Tønnesen, H.H. Solubility, chemical and photochemical stability of curcumin in surfactant solutions. Studies of curcumin and curcuminoids, XXVIII. *Die Pharm.* **2002**, *57*, 820–824.
20. Liu, W.; Zhai, Y.; Heng, X.; Che, F.Y.; Chen, W.; Sun, D.; Zhai, G. Oral bioavailability of curcumin: Problems and advancements. *J. Drug Target.* **2016**, *24*, 694–702. [CrossRef]
21. Schiborr, C.; Kocher, A.; Behnam, D.; Jandasek, J.; Toelstede, S.; Frank, J. The oral bioavailability of curcumin from micronized powder and liquid micelles is significantly increased in healthy humans and differs between sexes. *Mol. Nutr. Food Res.* **2014**, *58*, 516–527. [CrossRef] [PubMed]
22. Khayyal, M.T.; El-Hazek, R.M.; Elsabbagh, W.; Frank, J.; Behnam, D.; Abdel-Tawab, M. Micellar solubilisation enhances the antiinflammatory activities of curcumin and boswellic acids in rats with adjuvant-induced arthritis. *Nutrition* **2018**, *54*, 189–196. [CrossRef]
23. Kocher, A.; Bohnert, L.; Schiborr, C.; Frank, J. Highly bioavailable micellar curcuminoids accumulate in blood, are safe and do not reduce blood lipids and inflammation markers in moderately hyperlipidemic individuals. *Mol. Nutr. Food Res.* **2016**, *60*, 1555–1563. [CrossRef] [PubMed]
24. Dützmann, S.; Schiborr, C.; Kocher, A.; Pilatus, U.; Hattingen, E.; Weissenberger, J.; Geßler, F.; Quick-Weller, J.; Franz, K.; Seifert, V.; et al. Intratumoral Concentrations and Effects of Orally Administered Micellar Curcuminoids in Glioblastoma Patients. *Nutr. Cancer* **2016**, *68*, 943–948. [CrossRef]
25. Damarla, S.R.; Komma, R.; Bhatnagar, U.; Rajesh, N.; Mulla, S.M.A. An Evaluation of the Genotoxicity and Subchronic Oral Toxicity of Synthetic Curcumin. *J. Toxicol.* **2018**, *2018*, 6872753. [CrossRef]
26. Decker, E.A. Phenolics: Prooxidants or antioxidants? *Nutr. Rev.* **1997**, *55*, 396–398. [CrossRef]
27. Srinivasan, M.; Prasad, N.R.; Menon, V.P. Protective effect of curcumin on gamma-radiation induced DNA damage and lipid peroxidation in cultured human lymphocytes. *Mutat. Res.* **2006**, *611*, 96–103. [CrossRef]
28. Seyithanoğlu, M.H.; Abdallah, A.; Kitiş, S.; Guler, E.M.; Koçyiğit, A.; Dündar, T.T.; Papaker, M.G. Investigation of cytotoxic, genotoxic, and apoptotic effects of curcumin on glioma cells. *Cell. Mol. Biol.* **2019**, *65*, 101–108. [CrossRef]
29. Cao, J.; Jia, L.; Zhou, H.-M.; Liu, Y.; Zhong, L.-F. Mitochondrial and Nuclear DNA Damage Induced by Curcumin in Human Hepatoma G2 Cells. *Toxicol. Sci.* **2006**, *91*, 476–483. [CrossRef] [PubMed]
30. Mendonça, L.M.; Dos Santos, G.C.; Antonucci, G.A.; Santos, A.C.; Bianchi, M.D.L.P.; Antunes, L.M.G. Evaluation of the cytotoxicity and genotoxicity of curcumin in PC12 cells. *Mutat. Res. Toxicol. Environ. Mutagen.* **2009**, *675*, 29–34. [CrossRef]
31. Hendrayani, S.-F.; Al-Khalaf, H.H.; Aboussekhra, A. Curcumin Triggers p16-Dependent Senescence in Active Breast Cancer-Associated Fibroblasts and Suppresses Their Paracrine Procarcinogenic Effects. *Neoplasia* **2013**, *15*, 631–640. [CrossRef] [PubMed]
32. Grabowska, W.; Kucharewicz, K.; Wnuk, M.; Lewinska, A.; Suszek, M.; Przybylska, D.; Mosieniak, G.; Sikora, E.; Bielak-Zmijewska, A. Curcumin induces senescence of primary human cells building the vasculature in a DNA damage and ATM-independent manner. *AGE* **2015**, *37*, 1–17. [CrossRef]
33. He, Y.; Roos, W.P.; Wu, Q.; Hofmann, T.G.; Kaina, B. The SIAH1–HIPK2–p53ser46 Damage Response Pathway is Involved in Temozolomide-Induced Glioblastoma Cell Death. *Mol. Cancer Res.* **2019**, *17*, 1129–1141. [CrossRef] [PubMed]
34. Kaina, B.; Beltzig, L.; Piee-Staffa, A.; Haas, B. Cytotoxic and Senolytic Effects of Methadone in Combination with Temozolomide in Glioblastoma Cells. *Int. J. Mol. Sci.* **2020**, *21*, 7006. [CrossRef]
35. Nikolova, T.; Marini, F.; Kaina, B. Genotoxicity testing: Comparison of the gammaH2AX focus assay with the alkaline and neutral comet assays. *Mutat. Res.* **2017**, *822*, 10–18. [CrossRef] [PubMed]
36. Bielak-Zmijewska, A.; Grabowska, W.; Ciolko, A.; Bojko, A.; Mosieniak, G.; Bijoch, Ł.; Sikora, E. The Role of Curcumin in the Modulation of Ageing. *Int. J. Mol. Sci.* **2019**, *20*, 1239. [CrossRef]
37. Knizhnik, A.V.; Roos, W.; Nikolova, T.; Quiros, S.; Tomaszowski, K.-H.; Christmann, M.; Kaina, B. Survival and Death Strategies in Glioma Cells: Autophagy, Senescence and Apoptosis Triggered by a Single Type of Temozolomide-Induced DNA Damage. *PLoS ONE* **2013**, *8*, e55665. [CrossRef]
38. He, Y.; Kaina, B. Are There Thresholds in Glioblastoma Cell Death Responses Triggered by Temozolomide? *Int. J. Mol. Sci.* **2019**, *20*, 1562. [CrossRef]
39. Yosef, R.; Pilpel, N.; Tokarsky-Amiel, R.; Biran, A.; Ovadya, Y.; Cohen, S.; Vadai, E.; Dassa, L.; Shahar, E.; Condiotti, R.; et al. Directed elimination of senescent cells by inhibition of BCL-W and BCL-XL. *Nat. Commun.* **2016**, *7*, 11190. [CrossRef]
40. Thayyullathil, F.; Chathoth, S.; Hago, A.; Patel, M.; Galadari, S. Rapid reactive oxygen species (ROS) generation induced by curcumin leads to caspase-dependent and -independent apoptosis in L929 cells. *Free Radic. Biol. Med.* **2008**, *45*, 1403–1412. [CrossRef]
41. Collins, A.R. Measuring oxidative damage to DNA and its repair with the comet assay. *Biochim. Biophys. Acta (BBA) Gen. Subj.* **2014**, *1840*, 794–800. [CrossRef] [PubMed]
42. Chen, S.; Wu, J.; Tang, Q.; Xu, C.; Huang, Y.; Huang, D. Nano-micelles based on hydroxyethyl starch-curcumin conjugates for improved stability, antioxidant and anticancer activity of curcumin. *Carbohydr. Polym.* **2020**, *228*, 115398. [CrossRef] [PubMed]

43. Sun, Y.; Li, Y.; Shen, Y.; Wang, J.; Tang, J.; Zhao, Z. Enhanced oral delivery and anti-gastroesophageal reflux activity of curcumin by binary mixed micelles. *Drug Dev. Ind. Pharm.* **2019**, *45*, 1444–1450. [CrossRef] [PubMed]
44. Karabasz, A.; Lachowicz, D.; Karewicz, A.; Mezyk-Kopec, R.; Stalińska, K.; Werner, E.; Cierniak, A.; Dyduch, G.; Bereta, J.; Bzowska, M. Analysis of toxicity and anticancer activity of micelles of sodium alginate-curcumin. *Int. J. Nanomed.* **2019**, *14*, 7249–7262. [CrossRef] [PubMed]
45. Frank, J.; Schiborr, C.; Kocher, A.; Meins, J.; Behnam, D.; Schubert-Zsilavecz, M.; Abdel-Tawab, M. Transepithelial Transport of Curcumin in Caco-2 Cells Is significantly Enhanced by Micellar Solubilisation. *Plant Foods Hum. Nutr.* **2017**, *72*, 48–53. [CrossRef]
46. Seiwert, N.; Fahrer, J.; Nagel, G.; Frank, J.; Behnam, D.; Kaina, B. Curcumin Administered as Micellar Solution Suppresses Intestinal Inflammation and Colorectal Carcinogenesis. *Nutr. Cancer* **2021**, *73*, 686–693. [CrossRef]
47. Jiang, M.; Yang-Yen, H.; Yen, J.J.; Lin, J. Curcumin induces apoptosis in immortalized NIH 3T3 and malignant cancer cell lines. *Nutr. Cancer* **1996**, *26*, 111–120. [CrossRef]
48. Kuo, M.-L.; Huang, T.-S.; Lin, J.-K. Curcumin, an antioxidant and anti-tumor promoter, induces apoptosis in human leukemia cells. *Biochim. Biophys. Acta (BBA) Mol. Basis Dis.* **1996**, *1317*, 95–100. [CrossRef]
49. Anto, R.J.; Mukhopadhyay, A.; Denning, K.; Aggarwal, B.B. Curcumin (diferuloylmethane) induces apoptosis through activation of caspase-8, BID cleavage and cytochrome c release: Its suppression by ectopic expression of Bcl-2 and Bcl-xl. *Carcinogenesis* **2002**, *23*, 143–150. [CrossRef]
50. Kizhakkayil, J.; Thayyullathil, F.; Chathoth, S.; Hago, A.; Patel, M.; Galadari, S. Modulation of curcumin-induced Akt phosphorylation and apoptosis by PI3K inhibitor in MCF-7 cells. *Biochem. Biophys. Res. Commun.* **2010**, *394*, 476–481. [CrossRef] [PubMed]
51. Grabowska, W.; Mosieniak, G.; Achtabowska, N.; Czochara, R.; Litwinienko, G.; Bojko, A.; Sikora, E.; Bielak-Zmijewska, A. Curcumin induces multiple signaling pathways leading to vascular smooth muscle cell senescence. *Biogerontology* **2019**, *20*, 783–798. [CrossRef] [PubMed]
52. Nikolova, T.; Dvorak, M.; Jung, F.; Adam, I.; Kramer, E.; Gerhold-Ay, A. The gammaH2AX assay for genotoxic and nongenotoxic agents: Comparison of H2AX phosphorylation with cell death response. *Toxicol. Sci.* **2014**, *140*, 103–117. [CrossRef] [PubMed]
53. Sebastià, N.; Soriano, J.; Barquinero, J.F.; Villaescusa, J.I.; Almonacid, M.; Cervera, J.; Such, E.; Silla, M.A.; Montoro, A. In vitro cytogenetic and genotoxic effects of curcumin on human peripheral blood lymphocytes. *Food Chem. Toxicol.* **2012**, *50*, 3229–3233. [CrossRef] [PubMed]
54. Ganiger, S.; Malleshappa, H.; Krishnappa, H.; Rajashekhar, G.; Rao, V.R.; Sullivan, F. A two generation reproductive toxicity study with curcumin, turmeric yellow, in Wistar rats. *Food Chem. Toxicol.* **2007**, *45*, 64–69. [CrossRef] [PubMed]
55. Unlu, A.; Nayir, E.; Kalenderoglu, M.D.; Kirca, O.; Ozdogan, M. Curcumin (Turmeric) and cancer. *J. BUON* **2016**, *21*, 1050–1060. [PubMed]
56. Eckert, G.P.; Schiborr, C.; Hagl, S.; Abdel-Kader, R.; Müller, W.E.; Rimbach, G.; Frank, J. Curcumin prevents mitochondrial dysfunction in the brain of the senescence-accelerated mouse-prone 8. *Neurochem. Int.* **2013**, *62*, 595–602. [CrossRef]
57. Dörsam, B.; Wu, C.-F.; Efferth, T.; Kaina, B.; Fahrer, J. The eucalyptus oil ingredient 1,8-cineol induces oxidative DNA damage. *Arch. Toxicol.* **2014**, *89*, 797–805. [CrossRef]
58. Berdelle, N.; Nikolova, T.; Quiros, S.; Efferth, T.; Kaina, B. Artesunate Induces Oxidative DNA Damage, Sustained DNA Double-Strand Breaks, and the ATM/ATR Damage Response in Cancer Cells. *Mol. Cancer Ther.* **2011**, *10*, 2224–2233. [CrossRef]
59. Radak, Z.; Ishihara, K.; Tekus, E.; Varga, C.; Posa, A.; Balogh, L. Exercise, oxidants, and antioxidants change the shape of the bell-shaped hormesis curve. *Redox Biol.* **2017**, *12*, 285–290. [CrossRef]
60. Boldogh, I.; Hajas, G.; Aguilera-Aguirre, L.; Hegde, M.L.; Radak, Z.; Bacsi, A.; Sur, S.; Hazra, T.K.; Mitra, S. Activation of Ras Signaling Pathway by 8-Oxoguanine DNA Glycosylase Bound to Its Excision Product, 8-Oxoguanine. *J. Biol. Chem.* **2012**, *287*, 20769–20773. [CrossRef]

Article

An Exploratory Analysis of Changes in Mental Wellbeing Following Curcumin and Fish Oil Supplementation in Middle-Aged and Older Adults

Julia C. Kuszewski [1], Peter R. C. Howe [1,2,3,*] and Rachel H. X. Wong [1,2]

[1] Clinical Nutrition Research Centre, School of Biomedical Sciences and Pharmacy, University of Newcastle, Callaghan 2308, Australia; Julia.kuszewski@uon.edu.au (J.C.K.); Rachel.wong@newcastle.edu.au (R.H.X.W.)
[2] Institute for Resilient Regions, University of Southern Queensland, Springfield Central 4300, Australia
[3] UniSA Allied Health & Human Performance, University of South Australia, Adelaide 5000, Australia
* Correspondence: Peter.howe@newcastle.edu.au; Tel.: +61-2-4921-7309

Received: 19 August 2020; Accepted: 21 September 2020; Published: 23 September 2020

Abstract: Curcumin has previously been shown to enhance mood in non-depressed older adults. However, observed benefits were limited to short-term supplementation (4 weeks). In a 16 week randomized, double-blind, placebo-controlled, 2 × 2 factorial design trial, we supplemented overweight or obese non-depressed adults (50–80 years) with curcumin (160 mg/day), fish oil (2000 mg docosahexaenoic acid +400 mg eicosapentaenoic acid/day), or a combination of both. Secondary outcomes included mental wellbeing measures (mood states and subjective memory complaints (SMCs)) and quality of life (QoL). Furthermore, plasma apolipoprotein E4 (APOE4) was measured to determine whether APOE4 status influences responses to fish oil. Curcumin improved vigour ($p = 0.044$) compared to placebo and reduced SMCs compared to no curcumin treatment ($p = 0.038$). Fish oil did not affect any mood states, SMCs or QoL; however, responses to fish oil were affected by APOE4 status. In APOE4 non-carriers, fish oil increased vigour ($p = 0.030$) and reduced total mood disturbances ($p = 0.048$) compared to placebo. Improvements in mental wellbeing were correlated with increased QoL. Combining curcumin with fish oil did not result in additive effects. This exploratory analysis indicates that regular supplementation with either curcumin or fish oil (limited to APOE4 non-carriers) has the potential to improve some aspects of mental wellbeing in association with better QoL.

Keywords: curcumin; fish oil; mood; subjective memory complaints; APOE4; randomized controlled trial

1. Introduction

Approximately 10–20% of older adults worldwide are affected by late-life depression, defined as a major depressive episode after the age of 60 [1,2]. Unfortunately, depression often goes undetected in the elderly due to individuals under-reporting their symptoms and symptoms being confused with other age-related issues by family members or health care workers [3]. Consequently, depression is often left untreated, which in turn can lead to poor quality of life (QoL) and an increase in morbidity, disability and dependence [1]. Poor mood has also been shown to be closely associated with subjective memory complaints (SMCs), which are considered to reflect early cognitive changes and to increase a person's risk to progress to mild cognitive impairments or dementia [4–7]. One potential preventative strategy to reduce the risk of late-life depression is to supplement the diet with mood-enhancing bioactive nutrients, such as long-chain omega-3 fatty acids (LCn-3 PUFAs) and curcumin, to improve mood in order to counteract development of depressive symptoms and poor mental wellbeing.

Curcumin, the main active polyphenolic compound of the curry spice turmeric (*Curcuma longa*), has been shown to reduce depressive symptoms in individuals suffering from depression [8] and, more recently, to improve mood in non-depressed healthy older adults [9,10]. In 2015, Cox et al. showed that curcumin supplementation (80 mg per day) for four weeks significantly reduced fatigue and attenuated negative effects of a cognitive test battery on calmness and contentedness [9]. In a partial replication study, Cox et al. extended the curcumin supplementation and measured outcomes at four and twelve weeks [10]. Again, fatigue was shown to be reduced following four weeks as well as twelve weeks of supplementation. Furthermore, curcumin significantly reduced tension, anger, confusion and total mood disturbance. However, these beneficial effects were only found following 4 weeks of supplementation. Combining curcumin with other bioactive nutrients known to counteract depressive symptoms, such as the LCn-3 PUFAs found in fish/seafood and fish oil, could be a potential strategy to extend the mood-enhancing effects over longer periods due to potential additive or synergistic effects of the combination [11].

A large body of epidemiological and observational studies shows an inverse association between fish intake and the prevalence of depression and that depressed adults have lower blood and adipose tissue levels of LCn-3 PUFAs [12,13]. This suggests that increasing one's Omega-3 Index with fish oil supplementation might help to counter depression. This has, however, proven difficult to confirm in clinical trials, as they were mostly focused on people with clinical depression. Nevertheless, the majority of these studies showed that fish oil supplementation can reduce depressive symptoms in a variety of populations, including older adults [14–16]. Only a limited number of studies has examined the potential of fish oil to prevent the risk of depression in mentally healthy older adults by enhancing their mood, with mixed results. Additionally, these studies focused on depressive symptoms only, but did not measure fish oil's effects on other mood states [17–19]. This indicates a need for further investigation, which should also take the apolipoprotein E4 (APOE4) status of participants into account, since the e4 variant of APOE has been shown to influence the effects of fish oil [20–22]. However, it is unknown whether APOE4 status influences the effects of fish oil on mental wellbeing measures. In contrast, response to curcumin supplementation seems to be unaffected by APOE4 status [23].

We recently reported the independent and combined effects of fish oil and curcumin supplementation for 16 weeks on systemic and cerebrovascular function (primary outcome cerebrovascular responsiveness (CVR) to hypercapnia) in overweight or obese middle-aged and older adults with a sedentary lifestyle [24]. In the same study, we also examined effects on mental wellbeing measures (mood states and SMCs) and general health perception (QoL) and whether the response to fish oil might be influenced by a participant's APOE4 status. The aim of this exploratory analysis was to (1) confirm the mood-enhancing benefits of curcumin reported by Cox et al. and determine whether combining curcumin with fish oil would result in additional, longer-lasting benefits on mood states, as well as improvements in SMCs and QoL; and (2) investigate the independent effects of fish oil supplementation on mental wellbeing measures and QoL and whether they are affected by APOE4 status.

2. Materials and Methods

2.1. Study Design and Population

Community-dwelling adults residing in the Hunter region of New South Wales, Australia, were recruited to participate in a 16 week randomized, double-blind, 2 × 2 factorial, placebo-controlled intervention trial. Volunteers were eligible if they were aged between 50 and 80 years, were overweight or obesity (body mass index (BMI) 25–40 kg/m^2) and had a sedentary lifestyle (<150 min of planned physical activity per week). Volunteers were excluded if they had an average fish/seafood intake above two serves per week or more than 300 mg/day of LCn-3PUFA from fish oil supplements, had suspected dementia (<82/100 points on Addenbrooke's Cognitive Examination III, determined during first screening visit), were diagnosed with major depression (current diagnosis), had a history of cardiovascular, kidney or liver disease or neurological condition, or were currently on insulin or warfarin therapy. The trial was conducted at the University of Newcastle's Clinical Nutrition Research Centre in accordance with the International Conference on Harmonization Guidelines for Good Clinical Practice. This study was approved by the University of Newcastle's Human Research Ethics Committee (H-2016-0170) and registered with the Australian and New Zealand Clinical Trials Register (ACTRN12616000732482p). Written consent was obtained from each participant prior to commencement.

2.2. Study Procedures

Eligible participants attended the research facility for a total of four visits—two at the beginning and two at the end of the intervention. The secondary outcomes described in this manuscript were obtained during the second and fourth visit. During these visits, participants had fasted overnight (at least eight h) for collection of a fasting venous blood sample (2 × 10 mL) by a trained phlebotomist at a commercial pathology centre. One sample was used for routine analysis of cardiometabolic and inflammatory markers in serum and the other was centrifuged to separate plasma from red blood cells and respective aliquots were kept for further analysis of the Omega-3 Index (eicosapentaenoic acid (EPA) and docosahexaenoic acid (DHA) concentrations in erythrocyte membranes) [24] and plasma APOE4 concentrations.

After blood collection, participants were offered water and snacks before filling in questionnaires about their mental wellbeing and QoL. Mental wellbeing measures included mood states and SMCs. Measures were repeated in the same order at the end of the intervention. Furthermore, in order to assess participants' depressive symptoms at baseline, the Centre for Epidemiologic Studies Depression Scale (CES-D) questionnaire [25] was administered. A score above 16 out of 60 points indicates a risk of depression.

2.2.1. Mood States

The Profile of Mood States (POMS) questionnaire, containing 65 descriptive words, was used to assess participant's various mood states over the last seven days before their scheduled visit [26]. It contains six mood subscales, included tension-anxiety, depression, anger-hostility, fatigue, confusion-bewilderment and vigour, which were then expressed as percentages of their maximum score for each subscale. Total mood disturbance (TMD) was calculated by averaging the percentages for all negative mood subscales and then subtracting the percentage obtained for vigour. Values ranged from −100%, indicating low mood disturbance, to +100%, indicating high mood disturbance.

2.2.2. Subjective Memory Complaints

SMCs were assessed using a 27-item 'yes' or 'no' questionnaire, which is a self-assessment of memory complaints. The first three questions were used to determine whether participants had any subjective memory complaints: "Do you perceive any memory or cognitive difficulties?", "Would you ask a doctor about these difficulties?" and "In the last two years, has your memory or cognition declined?". The remaining 24 questions were more specific, relating to difficulties in remembering conversations/appointments/names/recent news, concentration problems, difficulties starting or keeping track of conversations and difficulties keeping track with daily activities due to any decline in memory over the past two years. The positive responses ('yes') were summed and expressed as a percentage of the maximum score (24 points).

2.2.3. General Health Perception (Quality of Life)

The 36-Item Short-Form Survey (SF-36) was used to measure participants' perception of physical and mental wellbeing in the last four weeks before their scheduled visit and reflects QoL [27,28]. It includes eight subscales (physical functioning, body pain, role of physical limitation, general health perceptions (=physical wellbeing subcomponent), social functioning, general mental health, role of mental limitation and energy/vitality (=mental wellbeing subcomponent), each with a maximum score of 100, indicating no disability. An average of all subscales yielded the overall QoL score.

2.2.4. APOE4 Analysis

Only plasma aliquots from participants who had been allocated to one of the fish oil treatments (fish oil alone or in combination with curcumin, $n = 65$) were used to measure APOE4 concentrations, since APOE4 status has previously been shown to influence responses to fish oil [20] but does not appear to influence responses to curcumin supplementation [23]. APOE4 analysis was performed at the Hunter

Medical Research Institute by a trained researcher who was blind to the intervention procedures. Plasma APOE4 concentrations were measured using a commercial Apolipoprotein E4 (human) Enzyme-linked Immunosorbent Assay (ELISA) kit (Biovision, Milpitas, CA, USA). Sensitivity of the assay was 25 ng/mL with a detection range of 50–800 ng/mL, 8% intra-assay reproducibility and 12% inter-assay reproducibility, as reported by the manufacturer. Participants with plasma APOE4 concentrations that fell within the detection range were deemed to be APOE4 carriers, i.e., with an APOE2/E4, APOE3/E4 or APOE4/E4 genotype.

2.3. Investigational Product and Intervention

The intervention supplements were supplied by Blackmores Institute (Sydney, Australia) and were identical in appearance to their respective placebos, identifiable only by code. Participants were allocated to one of the four treatment groups by an independent investigator according to Altman's allocation by minimization method [29] based on their age, BMI and sex:

- FO group: active fish oil capsules (Blackmores Omega Brain™: 400 mg EPA and 2000 mg DHA/day) with placebo curcumin capsules (maltodextrin with yellow food colouring);
- CUR group: active curcumin capsules (Blackmores Brain Active™: 800 mg Longvida® containing 160 mg curcumin/day) with placebo fish oil capsules (mix of corn and olive oil with 20 mg of fish oil to match odour);
- FO + CUR group: active fish oil and active curcumin capsules;
- PL group: placebo fish oil and placebo curcumin capsules.

Participants were instructed to consume six capsules daily, two fish oil and one curcumin (or matching placebos) in the morning and again in the evening with meals, and to record their supplement intake in an assigned diary, together with any changes in medication intake. The curcumin supplement was identical to that used in the studies by Cox et al. [9,10]. However, as our participants were supplemented twice a day, their total dose was double that used previously. The twice daily supplementation schedule was an attempt to increase curcumin's efficacy by ensuring a sustained level in the blood, as curcumin has a relatively short half-life (approx. 7.5 h for Longvida® curcumin [30]). For fish oil, splitting the dose into twice per day can also help to minimize fishy burps, which are commonly reported as an unpleasant side effect. The fish oil dose was based on previous literature indicating the need for high DHA doses to see effects [18,31,32].

Participants were further instructed to maintain their habitual diet and exercise regimen. At mid-intervention, participants were followed up with a phone call to enquire about their wellbeing. At the end of the trial, participants returned any remaining supplements. Capsule counts and changes in erythrocyte Omega-3 Index (analysed as described in Kuszewski et al. [24]) were used to monitor overall compliance. Blinding was maintained until all data analysis had been completed.

2.4. Statistical Analysis

This is an analysis of secondary outcomes from our previously published clinical trial [24]. An estimated 136 participants were needed to detect a 0.7 effect size (Cohen's d) difference between treatment groups in the primary outcome (CVR to hypercapnia) at alpha = 0.05.

Using a per-protocol analysis and setting compliance to 80%, significant effects of treatment on mean changes in the variables (mood states, SMCs and QoL) between the groups were determined by a one-way MANOVA (IBM SPSS version 24, New York, NY, USA). Effect sizes are indicated by Cohen's d or, if group sizes were different, by Hedge's g. Using the 2×2 factorial design, effects of fish oil and curcumin treatment were also assessed independently with two-way ANOVA:

- Fish oil (FO and FO + CUR group) vs. no fish oil (CUR and PL group);
- Curcumin (CUR and FO + CUR group) vs. no curcumin (FO and PL group).

To examine the influence of APOE4 status on fish oil's effects on mental wellbeing and QoL, an independent t test was used to determine differences in response to fish oil between APOE4 carriers and non-carriers. Additional post-hoc analyses (one-way MANOVA) were performed to look at treatment changes in variables between groups in APOE4 non-carriers only.

Pearson's correlation analysis was used to determine whether changes in mood states and SMCs were related to changes in QoL. All results are presented as the mean ± standard error of mean (SEM). As this is an exploratory analysis, no adjustments were made for multiple comparisons.

3. Results

3.1. Participant Disposition and Baseline Characteristics

Of the 152 participants enrolled in this study between June 2017 and August 2018, 134 completed the intervention and 126 were compliant with supplementation (PL n = 32, FO n = 32, CUR n = 31, FO + CUR n = 31; for CONSORT diagram see Kuszewski et al. [24]). Four participants experienced side effects with supplementation (digestive problems: PL n = 1, CUR n = 1, FO + CUR n = 1; reflux: PL n = 1) and seven reported unrelated health issues, of which two occurred before supplementation was commenced. Of the remaining five health issues, two occurred in the fish oil group (pneumonia), two in the curcumin group (vein thrombosis, knee operation) and one in the combination group (heart attack). However, they were unlikely to be related to supplementation. For analysis of mental wellbeing measures, three participants had to be excluded: one participant was unable to complete the mental wellbeing questionnaires at week 16 (FO + CUR group), one participant experienced major life changes during the trial, affecting mental wellbeing measures (CUR group) and one participant had incomplete data (FO group), leaving 123 participants for the final analysis (PL n = 32, FO n = 31, CUR n = 30, FO + CUR n = 30).

Participants' baseline characteristics are described in Table 1. Participants were, on average, elderly, marginally obese, had low total mood disturbance and their average CES-D score was 9.0 ± 0.6—well below the cut-off score of 16/60 for suspected depression. Furthermore, 72% (n = 109) of participants indicated that they have SMCs (first question of the SMCs questionnaire), with the total score of SMCs averaging 36 ± 2%. There were no significant differences in baseline characteristics between groups, except for the POMS mood subscale of tension (PL: 16.0 ± 2.2 vs. FO + CUR: 23.7 ± 3.1, p = 0.040).

Table 1. Participants' baseline characteristics per group.

Characteristics	PL (n = 36)	FO (n = 39)	CUR (n = 38)	FO + CUR (n = 39)
Sex (female %)	50	56	55	56
Age (years)	65.4 ± 1.3	65.4 ± 1.2	65.4 ± 1.2	66.2 ± 1.3
BMI (kg/m^2)	31.0 ± 0.7	31.0 ± 0.7	30.5 ± 0.7	30.9 ± 0.6
Depressive symptoms (%)	13.4 ± 1.9	14.0 ± 1.7	17.0 ± 2.1	15.4 ± 2.4
Mood states (POMS)				
Tension (%)	16.0 ± 2.2	16.8 ± 2.2	23.1 ± 2.9	23.7 ± 3.1 *
Depression (%)	5.6 ± 1.1	7.9 ± 1.8	9.6 ± 1.9	10.6 ± 2.1
Anger (%)	7.0 ± 1.0	7.1 ± 1.4	10.7 ± 2.0	10.5 ± 1.7
Fatigue (%)	28.1 ± 4.1	28.1 ± 4.6	30.4 ± 3.9	28.3 ± 3.5
Confusion (%)	23.8 ± 2.9	23.9 ± 3.1	28.6 ± 3.3	24.5 ± 2.8
Vigour (%)	56.2 ± 2.8	50.5 ± 3.0	49.2 ± 2.8	52.5 ± 3.5
TMD [a] (%)	−40.1 ± 3.6	−33.7 ± 4.6	−28.7 ± 4.2	−33.0 ± 5.0
Subjective memory complaints (%)	37.5 ± 4.1	33.7 ± 4.1	35.7 ± 3.9	39.4 ± 4.5
Quality of Life (%)	70.4 ± 2.4	72.5 ± 1.9	67.1 ± 2.5	69.4 ± 2.4

Values are expressed as the mean ± SEM. [a] Greater negative value equals better overall mood. * Significant compared to placebo, $p < 0.05$. BMI, body mass index; CUR, curcumin alone; FO, fish oil alone; FO + CUR, fish oil and curcumin combination; PL, placebo; TMD, total mood disturbance.

3.2. Effects of Treatment on Mental Wellbeing Measures

Curcumin supplementation improved vigour compared to placebo (p = 0.044, Cohen's d = 0.55). Supplementation with fish oil, alone or in combination with curcumin, for 16 weeks did not significantly affect mood states (Table 2).

Table 2. Treatment changes in measures of mental wellbeing and quality of life (n = 123).

	PL (n = 32)	FO (n = 31)	CUR (n = 30)	FO + CUR (n = 30)
Tension (%)	0.4 ± 1.5	−0.5 ± 2.1	0.0 ± 2.1	−0.8 ± 2.5
Depression (%)	−0.6 ± 1.3	−1.6 ± 1.8	−1.8 ± 2.1	−0.1 ± 1.6
Anger (%)	−2.1 ± 1.2	0.2 ± 1.5	−2.1 ± 2.1	−2.2 ± 1.7
Fatigue (%)	−3.1 ± 4.0	0.1 ± 4.0	−1.5 ± 3.5	3.8 ± 3.9
Confusion (%)	−3.1 ± 2.4	−3.4 ± 2.8	−7.6 ± 3.1	−3.8 ± 2.4
Vigour (%)	−3.7 ± 2.6	1.5 ± 2.9	3.5 ± 2.0 *	2.5 ± 2.4
TMD [a] (%)	2.0 ± 3.8	−2.6 ± 4.2	−6.1 ± 3.3	−3.1 ± 3.7
Subjective memory complaints (%)	−1.8 ± 2.5	−2.6 ± 2.4	−6.7 ± 2.7	−8.9 ± 3.1
Quality of Life (%)	0.3 ± 1.8	2.4 ± 1.8	2.0 ± 2.1	2.1 ± 2.4

Values are expressed as the mean ± SEM. [a] Greater reduction is favourable. * Significant compared to placebo, $p < 0.05$. BMI, body mass index; CUR, curcumin alone; FO, fish oil alone; FO + CUR, fish oil and curcumin combination; PL, placebo; TMD, total mood disturbance.

SMCs were unaffected by fish oil supplementation, while curcumin and the combination of fish oil and curcumin supplementation tended to reduce SMCs. Combining these two groups in the 2 × 2 factorial analysis showed a 21% reduction in SMCs from baseline following curcumin supplementation, which was significant compared to no curcumin supplementation (CUR n = 60: −7.8 ± 2.0% vs. no CUR n = 63: −2.2 ± 1.7%, p = 0.038, Cohen's d = 0.38). This reduction was even more significant among participants who reported SMCs at baseline (CUR n = 44: −9.4 ± 2.5% vs. no CUR n = 48: −2.4 ± 2.0%; p = 0.029, Cohen's d = 0.46). The reduction in SMCs following curcumin supplementation (n = 60) was correlated with changes in confusion (R = 0.392, p = 0.002) and depression (R = 0.356, p = 0.006).

3.3. Effects of Treatment on Quality of Life

The overall score of QoL as well as the subcomponents of physical and mental wellbeing were not significantly affected by treatment (Table 2). However, the observed changes in vigour were correlated with changes in overall QoL (whole study population n = 122: R = 0.323, $p < 0.001$; curcumin group n = 59: R = 0.418, p = 0.001) and changes in SMCs were inversely correlated with changes in overall QoL (whole study population n = 122: R = −0.360, $p < 0.001$; curcumin group n = 59: R = −0.472, $p < 0.001$).

3.4. Influence of APOE4 Status

Of the 65 participants who were taking fish oil supplements, 26% were APOE4 carriers, with an average plasma APOE4 concentration of 76.7 ± 10.9 µg/mL. APOE4 carriers were slightly younger with a lower mean BMI but had the same percentage of self-reported depressive symptoms compared to APOE4 non-carriers. Baseline demographics or wellbeing measures were not significantly different between APOE4 carriers and non-carriers (Table 3).

APOE4 status significantly influenced effects of fish oil on mental wellbeing measures but not QoL. In general, while APOE4 carriers had a negative response to fish oil supplementation, i.e., greater mood disturbance and more SMCs, APOE4 non-carriers showed improvements in all mental wellbeing measures (Figure 1). Changes in tension (p = 0.014, Hedge's g = 0.74), depression (p = 0.003, Hedge's g = 0.90), anger (p = 0.043, Hedge's g = 0.60), confusion (p = 0.028, Hedge's g = 0.66), total mood disturbance (TMD) (p = 0.016, Hedge's g = 0.72) and SMCs (p = 0.015, Hedge's g = 0.75) following fish oil supplementation were significantly different between APOE4 carriers and non-carriers.

Table 3. Baseline demographics and wellbeing measures in APOE4 carriers vs. non-carriers.

	APOE4 Non-Carrier (n = 48)	APOE4 Carrier (n = 17)	p-Value
Sex (female %)	52	53	
Age (years)	66.3 ± 1.2	64.0 ± 1.8	0.298
BMI (kg/m^2)	31.5 ± 0.6	29.3 ± 0.7	0.052
Depressive symptoms (%)	15.2 ± 1.8	15.4 ± 4.2	0.953
EPA (%)	1.06 ± 0.37	1.04 ± 0.44	0.807
DHA (%)	5.49 ± 1.36	5.73 ± 1.31	0.527
Tension (%)	20.9 ± 2.3	25.5 ± 5.3	0.356
Depression (%)	10.4 ± 1.8	9.7 ± 3.4	0.855
Anger (%)	8.5 ± 1.4	12.2 ± 3.1	0.214
Fatigue (%)	30.3 ± 4.1	23.3 ± 3.9	0.219
Confusion (%)	26.8 ± 2.8	22.3 ± 4.0	0.388
Vigour (%)	49.4 ± 2.7	52.2 ± 5.7	0.619
TMD [a] (%)	−30.0 ± 4.2	−33.6 ± 8.0	0.671
Subjective memory complaints (%)	41.9 ± 3.8	30.6 ± 6.5	0.137
Quality of life (%)	69.5 ± 2.0	72.7 ± 3.4	0.430

Values are expressed as the mean ± SEM. [a] Greater negative value equals better overall mood. APOE4, apolipoprotein E4; BMI, body mass index; DHA, docosahexaenoic acid; EPA, eicosapentaenoic acid; TMD, total mood disturbance.

There were no differences between APOE4 carriers and non-carriers in erythrocyte EPA and DHA levels at baseline or in the changes in EPA (APOE4: 1.25 ± 0.32% vs. non-APOE4: 1.21 ± 0.47%, p = 0.756) and DHA (APOE4: 4.16 ± 1.33% vs. non-APOE4: 4.65 ± 1.39%, p = 0.256) levels following treatment.

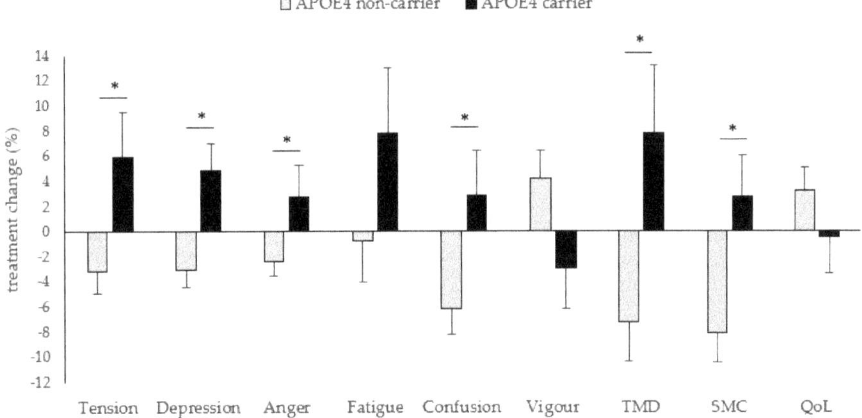

Figure 1. Differences between APOE4 carriers (n = 16) and APOE4 non-carriers (n = 44) in treatment changes of mental wellbeing measures and quality of life following fish oil supplementation. * Significant, $p < 0.05$. TMD: total mood disturbance, SMC: subjective memory complaints, QoL: quality of life.

3.5. *Subanalysis in APOE4 Non-Carriers*

Since APOE4 status influenced responses to fish oil supplementation, we re-examined the effect of fish oil supplementation in APOE4 non-carriers after excluding APOE4 carriers from the fish oil and fish oil + curcumin group (n = 17). This exploratory subanalysis showed a significant increase in vigour (p = 0.030, Hedge's g = 0.56) and decrease in TMD (p = 0.048, Hedge's g = 0.55) following fish oil supplementation compared to placebo (Table 4). The decrease in TMD following fish oil supplementation was inversely correlated with changes in vigour (R = −0.937, $p < 0.001$) and positively correlated with changes in fatigue (R = 0.590, p = 0.004). The combination of fish oil and curcumin significantly decreased SMCs (p = 0.029, Hedge's g = 0.57).

Table 4. Subanalysis of treatment changes in measures of mental wellbeing and quality of life in APOE4 non-carriers (n = 106).

	PL (n = 32)	FO (n = 22)	CUR (n = 30)	FO + CUR (n = 22)
Tension (%)	0.4 ± 1.5	−3.6 ± 1.8	0.0 ± 2.1	−2.7 ± 3.0
Depression (%)	−0.6 ± 1.3	−4.4 ± 1.9	−1.8 ± 2.1	−1.7 ± 1.9
Anger (%)	−2.1 ± 1.2	−1.8 ± 1.3	−2.1 ± 2.1	−2.9 ± 2.0
Fatigue (%)	−3.1 ± 4.0	−3.8 ± 4.4	−1.5 ± 3.5	2.3 ± 4.8
Confusion (%)	−3.1 ± 2.4	−7.9 ± 2.7	−7.6 ± 3.1	−4.4 ± 3.1
Vigour (%)	−3.7 ± 2.6	4.8 ± 3.3 *	3.5 ± 2.0 *	3.6 ± 3.1
TMD [a] (%)	2.0 ± 3.8	−9.1 ± 4.0 *	−6.1 ± 3.3	−5.4 ± 4.8
Subjective memory complaints (%)	−1.8 ± 2.5	−6.1 ± 2.6	−6.7 ± 2.7	−11.0 ± 3.9 *
Quality of life (%)	0.3 ± 1.8	4.6 ± 2.0	2.0 ± 2.1	1.8 ± 2.9

Values are expressed as the mean ± SEM. [a] greater reduction being favourable. * Significant compared to placebo, $p < 0.05$. CUR, curcumin alone; FO, fish oil alone; FO + CUR, fish oil and curcumin combination; PL, placebo; TMD, total mood disturbance.

The increase in overall QoL score following fish oil supplementation was not significant, but changes in vigour were correlated with changes in overall QoL (whole group n = 106: R = 0.287, p = 0.003; FO supplementation n = 44: R = 0.433, p = 0.003), and changes in TMD were inversely correlated with changes in overall QoL (R= −0.453, $p < 0.001$; FO supplementation n = 44: R= −0.597, $p < 0.001$).

4. Discussion

This exploratory analysis provides supportive evidence of curcumin's mood-enhancing effects and furthermore shows curcumin's potential to reduce SMCs in overweight or obese middle-aged and older adults without clinical depression. However, combining curcumin with fish oil did not result in any additional benefits on mental wellbeing. Fish oil supplementation alone did not affect mental wellbeing, although the response to fish oil was significantly affected by APOE4 status. APOE4 non-carriers showed improvements in mental wellbeing, whereas APOE4 carriers did not respond to fish oil. A subanalysis of the whole study cohort, excluding APOE4 carriers, revealed that fish oil supplementation improved vigour and decreased TMD.

QoL was not significantly improved, which might be due to the fact that baseline overall QoL scores were already relative high (average 69.9 ± 2%) and comparable to results from a large survey across Australian households [27], showing an average overall score of 71%. Nevertheless, improvements in vigour and SMCs following curcumin supplementation and improvements in vigour and TMD following fish oil supplementation were correlated with improved overall QoL, suggesting that curcumin's and fish oil's effects are clinically relevant.

4.1. Curcumin Supplementation

The observed mood-enhancing effect of curcumin, i.e., increase in vigour, is consistent with previous studies study by Cox et al., which found reductions in fatigue following 4 and 12 weeks of supplementation in older non-depressed adults [9,10]. Furthermore, Cox et al. found significant reductions in tension, anger, confusion and TMD following supplementation. However, these effects were only short-term and were not sustained after 12 weeks of supplementation [10]. In both studies, Cox et al. supplemented their participants with 400 mg Longvida curcumin (80 mg curcumin) once a day; however, we chose to give this dose twice daily to ensure a sustained level of curcumin in the blood. Consistent with findings from Cox et al. we observed reductions in confusion and TMD following 16 weeks of curcumin supplementation; however, they were not significant compared to placebo and combining curcumin with fish oil did not result in additional effects on these outcomes. Thus, it might be possible that curcumin has only short-term effects on confusion and TDM but stronger, longer-lasting effects on fatigue-vigour. Further studies are warranted to confirm the promising mood-enhancing effects of curcumin in older adults without clinical depression and identify the underlying mechanisms and how the efficacy of curcumin could be improved.

Moreover, we found that the change in vigour was positively correlated with change in quality of life; however, further investigation is needed to determine whether the mood-enhancing effects of curcumin can help to prevent the onset of depression.

Next to improvements in vigour, we found that curcumin reduced SMCs, which—to the best of our knowledge—is the first evidence of this benefit. Increasing evidence links SMCs to increased dementia risk, suggesting that SMCs reflect early, subtle cognitive changes [7]. This is supported by neuroimaging studies, which indicate changes in brain structure and function in individuals with SMCs [33]. SMCs have also been shown to be closely related to poor mood and negatively impact QoL in healthy older adults [4,34]. The observed attenuation in SMCs following curcumin might thus be partly mediated by its mood-enhancing effects, which resulted in slight, non-significant reductions in depressive mood and confusion that were, however, significantly correlated with decreased SMCs. Additionally, our observation that the reduction in SMCs following curcumin was related to increased QoL is encouraging. Targeting SMCs might thus offer an early window of opportunity to intervene prior to the development of poor mood and the detection of measurable cognitive deficits.

Nevertheless, changes in SMCs during the 16 week intervention period and correlations with changes in mood states and QoL need to be interpreted with caution. The assessment of SMCs estimates SMCs over the last two years; therefore a longer intervention time is needed to meaningfully monitor changes in SMCs. Moreover, assessment of SMCs in a clinical setting has its limitations as it is difficult to distinguish between worried and well individuals and those who have subtle cognitive impairments [35]. Lastly, a further limitation is that the SMCs questionnaire utilized in this study, although more extensive than others with 27 items, has not been validated before in cognitively unimpaired elderly.

To the best of our knowledge, only two other studies have examined the potential of diet or dietary supplements to reduce SMCs. An observational study showed an inverse correlation between adherence to a healthy diet—mix of a Mediterranean diet and Dietary Approaches to Stop Hypertension (DASH) diet—and SMCs in adults aged above 70 years who did not suffer from depression [36]. Moreover, a 12 week clinical trial of BrainPower Advanced, a supplement containing a mix of 15 ingredients (Ginkgo biloba extract, green tea extract, L-pyroglutamic acid amongst others), showed improvements in SMCs compared to placebo in older adults (average 67 years) [37].

A healthy diet and dietary supplements, such as curcumin, might thus have the potential to attenuate SMCs. However, future studies are needed to confirm these encouraging findings, further investigate the long-term effects of reducing SMCs on cognitive function, mood and QoL and explore potential underlying mechanisms. These studies might also need to incorporate neuroimaging to more accurately assess SMCs in individuals at baseline.

4.2. *Effects of Fish oil Supplementation Influenced by APOE4 Status*

In line with two previous studies conducted in healthy older adults without clinical depression, fish oil supplementation did not significantly affect mental wellbeing measures [17,19]. However, we found that the effect of fish oil on mental wellbeing was significantly influenced by APOE4 status; subanalysis in APOE4 non-carriers showed that fish oil supplementation significantly improved vigour and reduced TMD. The reduction in TMD was mostly driven by changes in vigour and fatigue. Improvements in these mood subscales might increase motivation to be more proactive in making healthy lifestyle choices such as eating a more healthy diet, which in turn can further improve mood and mental wellbeing [38,39].

Our finding that APOE4 status affects the response to fish oil supplementation is consistent with previous literature demonstrating a fish oil–APOE4 interaction [20,22]. For instance, APOE4 carriers were shown to be less sensitive to the protective effects of fish oil/consumption of fatty fish on cognitive function and all-cause dementia risk [40,41]. The mechanisms underlying the differential responses to fish oil are, however, poorly understood and need further investigation.

4.3. Combination of Fish Oil and Curcumin

The combination of fish oil and curcumin had neither an additive nor synergistic influence on mental wellbeing and QoL. Since this is the first exploratory analysis to look at the effects of this nutrient combination on mental wellbeing, further investigations are needed to identify whether there are potential synergistic effects between LCn-3PUFAs and curcumin in humans and whether they might depend on an optimal dose combination of fish oil and curcumin or might only benefit certain population groups.

4.4. Study Limitations

As this was an exploratory analysis of secondary outcomes of a large interventions study, it was not powered to detect differences in changes in mental wellbeing measures and QoL.

Another limitation is that APOE4 status was determined by measuring APOE4 in plasma, which is not as informative as APOE4 genotyping. It only indicates that a participant has the APOE4 allele, but not how many copies. Therefore, the exact genotype remains unknown. Our finding that the response to fish oil on mental wellbeing measures is dependent on APOE4 status is preliminary; further studies are needed to confirm this finding and whether it can be extrapolated to other populations.

5. Conclusions

This exploratory analysis of a 16 week dietary intervention trial in overweight or obese middle-aged and older adults without clinical depression adds further evidence that curcumin supplementation has potential beneficial effects on mood. Furthermore, our findings indicate that curcumin supplementation can reduce SMCs and that the improvements in both mood and SMCs are associated with improved QoL. These potential benefits of curcumin warrant further evaluation. Combining curcumin with fish oil did not result in additional benefits. Fish oil independently improved vigour and total mood disturbance, but only in APOE4 non-carriers. The observation that the mental wellbeing response to fish oil was influenced by APOE4 status should be followed up with studies designed to compare effects of fish oil on mental wellbeing between APOE4 carriers and non-carriers.

Author Contributions: Conceptualization, P.R.C.H. and R.H.X.W.; data curation, J.C.K.; formal analysis, J.C.K.; funding acquisition, P.R.C.H. and R.H.X.W.; investigation, J.C.K.; methodology, J.C.K.; project administration, J.C.K.; supervision, P.R.C.H. and R.H.X.W.; validation, P.R.C.H. and R.H.X.W.; writing—original draft, J.C.K.; writing—review and editing, P.R.C.H. and R.H.X.W. All authors have read and agreed to the published version of the manuscript.

Funding: This study was funded by Blackmores Institute (research and education division of Blackmores Ltd.), who also provided the test materials but had no role in the study design, data collection, analysis and interpretation or writing of the manuscript.

Acknowledgments: The authors would like to thank Hamish Evans and Natasha Baker for their assistance with this study.

Conflicts of Interest: The authors declare no conflict of interest. The funders had no role in the design of this study; in the collection, analyses, or interpretation of data; in the writing of the manuscript, or in the decision to publish the results.

References

1. Farioli-Vecchioli, S.; Sacchetti, S.; di Robilant, N.V.; Cutuli, D. The Role of Physical Exercise and Omega-3 Fatty Acids in Depressive Illness in the Elderly. *Curr. Neuropharmacol.* **2018**, *16*, 308–326. [CrossRef] [PubMed]
2. Barua, A.; Ghosh, M.K.; Kar, N.; Basilio, M.A. Prevalence of depressive disorders in the elderly. *Ann. Saudi Med.* **2011**, *31*, 620–644. [CrossRef] [PubMed]
3. Okereke, O.I.; Reynolds, C.F., 3rd; Mischoulon, D.; Chang, G.; Cook, N.R.; Copeland, T.; Friedenberg, G.; Buring, J.E.; Manson, J.E. The VITamin D and OmegA-3 TriaL-Depression Endpoint Prevention (VITAL-DEP): Rationale and design of a large-scale ancillary study evaluating vitamin D and marine omega-3 fatty acid supplements for prevention of late-life depression. *Contemp. Clin. Trials* **2018**, *68*, 133–145. [CrossRef] [PubMed]

4. Balash, Y.; Mordechovich, M.; Shabtai, H.; Giladi, N.; Gurevich, T.; Korczyn, A.D. Subjective memory complaints in elders: Depression, anxiety, or cognitive decline? *Acta Neurol. Scand.* **2013**, *127*, 344–350. [CrossRef]
5. Glodzik-Sobanska, L.; Reisberg, B.; De Santi, S.; Babb, J.S.; Pirraglia, E.; Rich, K.E.; Brys, M.; de Leon, M.J. Subjective memory complaints: Presence, severity and future outcome in normal older subjects. *Dement. Geriatr. Cogn. Disord.* **2007**, *24*, 177–184. [CrossRef]
6. Jessen, F.; Wiese, B.; Bachmann, C.; Eifflaender-Gorfer, S.; Haller, F.; Kölsch, H.; Luck, T.; Mösch, E.; van den Bussche, H.; Wagner, M.; et al. Prediction of dementia by subjective memory impairment: Effects of severity and temporal association with cognitive impairment. *Arch. Gen. Psychiatry* **2010**, *67*, 414–422. [CrossRef]
7. Kaup, A.R.; Nettiksimmons, J.; LeBlanc, E.S.; Yaffe, K. Memory complaints and risk of cognitive impairment after nearly 2 decades among older women. *Neurology* **2015**, *85*, 1852–1858. [CrossRef]
8. Fusar-Poli, L.; Vozza, L.; Gabbiadini, A.; Vanella, A.; Concas, I.; Tinacci, S.; Petralia, A.; Signorelli, M.S.; Aguglia, E. Curcumin for depression: A meta-analysis. *Crit. Rev. Food Sci. Nutr.* **2020**, *60*, 2643–2653. [CrossRef]
9. Cox, K.H.; Pipingas, A.; Scholey, A.B. Investigation of the effects of solid lipid curcumin on cognition and mood in a healthy older population. *J. Psychopharmacol.* **2015**, *29*, 642–651. [CrossRef]
10. Cox, K.H.M.; White, D.J.; Pipingas, A.; Poorun, K.; Scholey, A. Further Evidence of Benefits to Mood and Working Memory from Lipidated Curcumin in Healthy Older People: A 12-Week, Double-Blind, Placebo-Controlled, Partial Replication Study. *Nutrients* **2020**, *12*, 1678. [CrossRef]
11. Abdolahi, M.; Sarraf, P.; Javanbakht, M.H.; Honarvar, N.M.; Hatami, M.; Soveyd, N.; Tafakhori, A.; Sedighiyan, M.; Djalali, M.; Jafarieh, A.; et al. A Novel Combination of ω-3 Fatty Acids and Nano-Curcumin Modulates Interleukin-6 Gene Expression and High Sensitivity C-reactive Protein Serum Levels in Patients with Migraine: A Randomized Clinical Trial Study. *CNS Neurol. Disord. Drug Targets* **2018**, *17*, 430–438. [CrossRef] [PubMed]
12. Larrieu, T.; Layé, S. Food for Mood: Relevance of Nutritional Omega-3 Fatty Acids for Depression and Anxiety. *Front. Physiol.* **2018**, *9*, 1047. [CrossRef] [PubMed]
13. Yang, Y.; Kim, Y.; Je, Y. Fish consumption and risk of depression: Epidemiological evidence from prospective studies. *Asia Pac. Psychiatry* **2018**, *10*, e12335. [CrossRef] [PubMed]
14. Bae, J.H.; Kim, G. Systematic review and meta-analysis of omega-3-fatty acids in elderly patients with depression. *Nutr. Res.* **2018**, *50*, 1–9. [CrossRef]
15. Grenyer, B.F.; Crowe, T.; Meyer, B.; Owen, A.J.; Grigonis-Deane, E.M.; Caputi, P.; Howe, P.R. Fish oil supplementation in the treatment of major depression: A randomised double-blind placebo-controlled trial. *Prog. Neuropsychopharmacol. Biol. Psychiatry* **2007**, *31*, 1393–1396. [CrossRef]
16. Meyer, B.J.; Grenyer, B.F.; Crowe, T.; Owen, A.J.; Grigonis-Deane, E.M.; Howe, P.R. Improvement of major depression is associated with increased erythrocyte DHA. *Lipids* **2013**, *48*, 863–868. [CrossRef]
17. Giltay, E.J.; Geleijnse, J.M.; Kromhout, D. Effects of n-3 fatty acids on depressive symptoms and dispositional optimism after myocardial infarction. *Am. J. Clin. Nutr.* **2011**, *94*, 1442–1450. [CrossRef]
18. Sinn, N.; Milte, C.M.; Street, S.J.; Buckley, J.D.; Coates, A.M.; Petkov, J.; Howe, P.R. Effects of n-3 fatty acids, EPA v. DHA, on depressive symptoms, quality of life, memory and executive function in older adults with mild cognitive impairment: A 6-month randomised controlled trial. *Br. J. Nutr.* **2012**, *107*, 1682–1693. [CrossRef]
19. Van de Rest, O.; Geleijnse, J.M.; Kok, F.J.; van Staveren, W.A.; Hoefnagels, W.H.; Beekman, A.T.; de Groot, L.C. Effect of fish-oil supplementation on mental well-being in older subjects: A randomized, double-blind, placebo-controlled trial. *Am. J. Clin. Nutr.* **2008**, *88*, 706–713. [CrossRef]
20. Minihane, A.M. Impact of Genotype on EPA and DHA Status and Responsiveness to Increased Intakes. *Nutrients* **2016**, *8*, 123. [CrossRef]
21. Minihane, A.M.; Khan, S.; Leigh-Firbank, E.C.; Talmud, P.; Wright, J.W.; Murphy, M.C.; Griffin, B.A.; Williams, C.M. ApoE polymorphism and fish oil supplementation in subjects with an atherogenic lipoprotein phenotype. *Arter. Thromb. Vasc. Biol.* **2000**, *20*, 1990–1997. [CrossRef] [PubMed]
22. Pontifex, M.; Vauzour, D.; Minihane, A.M. The effect of APOE genotype on Alzheimer's disease risk is influenced by sex and docosahexaenoic acid status. *Neurobiol. Aging* **2018**, *69*, 209–220. [CrossRef] [PubMed]
23. Small, G.W.; Siddarth, P.; Li, Z.; Miller, K.J.; Ercoli, L.; Emerson, N.D.; Martinez, J.; Wong, K.P.; Liu, J.; Merrill, D.A.; et al. Memory and Brain Amyloid and Tau Effects of a Bioavailable Form of Curcumin in

Non-Demented Adults: A Double-Blind, Placebo-Controlled 18-Month Trial. *Am. J. Geriatr. Psychiatry* **2018**, *26*, 266–277. [CrossRef] [PubMed]

24. Kuszewski, J.C.; Wong, R.H.X.; Wood, L.G.; Howe, P.R.C. Effects of fish oil and curcumin supplementation on cerebrovascular function in older adults: A randomized controlled trial. *Nutr. Metab. Cardiovasc. Dis.* **2020**, *30*, 625–633. [CrossRef]
25. Radloff, L.S. The CES-D Scale: A self-report depression scale for research in the general population. *Appl. Psychol. Meas.* **1977**, *1*, 385–401. [CrossRef]
26. Wyrwich, K.W.; Yu, H. Validation of POMS questionnaire in postmenopausal women. *Qual. Life Res.* **2011**, *20*, 1111–1121. [CrossRef]
27. Butterworth, P.; Crosier, T. The validity of the SF-36 in an Australian National Household Survey: Demonstrating the applicability of the Household Income and Labour Dynamics in Australia (HILDA) Survey to examination of health inequalities. *BMC Public Health* **2004**, *4*, 44. [CrossRef]
28. Mishra, G.; Schofield, M.J. Norms for the physical and mental health component summary scores of the SF-36 for young, middle-aged and older Australian women. *Qual. Life Res.* **1998**, *7*, 215–220. [CrossRef]
29. Altman, D.G.; Bland, J.M. Treatment allocation by minimisation. *BMJ* **2005**, *330*, 843. [CrossRef]
30. Gota, V.S.; Maru, G.B.; Soni, T.G.; Gandhi, T.R.; Kochar, N.; Agarwal, M.G. Safety and pharmacokinetics of a solid lipid curcumin particle formulation in osteosarcoma patients and healthy volunteers. *J. Agric. Food Chem.* **2010**, *58*, 2095–2099. [CrossRef]
31. Cassiday, L. Sink or swim: Fish oil supplements and human health. *INFORM* **2016**, *27*, 6–13. [CrossRef]
32. Luo, X.D.; Feng, J.S.; Yang, Z.; Huang, Q.T.; Lin, J.D.; Yang, B.; Su, K.P.; Pan, J.Y. High-dose omega-3 polyunsaturated fatty acid supplementation might be more superior than low-dose for major depressive disorder in early therapy period: A network meta-analysis. *BMC Psychiatry* **2020**, *20*, 248. [CrossRef] [PubMed]
33. Hafkemeijer, A.; Altmann-Schneider, I.; Oleksik, A.M.; van de Wiel, L.; Middelkoop, H.A.; van Buchem, M.A.; van der Grond, J.; Rombouts, S.A. Increased functional connectivity and brain atrophy in elderly with subjective memory complaints. *Brain Connect.* **2013**, *3*, 353–362. [CrossRef] [PubMed]
34. Mol, M.; Carpay, M.; Ramakers, I.; Rozendaal, N.; Verhey, F.; Jolles, J. The effect of perceived forgetfulness on quality of life in older adults; a qualitative review. *Int. J. Geriatr. Psychiatry* **2007**, *22*, 393–400. [CrossRef] [PubMed]
35. Ahmed, S.; Mitchell, J.; Arnold, R.; Dawson, K.; Nestor, P.J.; Hodges, J.R. Memory complaints in mild cognitive impairment, worried well, and semantic dementia patients. *Alzheimer Dis. Assoc. Disord.* **2008**, *22*, 227–235. [CrossRef] [PubMed]
36. Adjibade, M.; Assmann, K.E.; Julia, C.; Galan, P.; Hercberg, S.; Kesse-Guyot, E. Prospective association between adherence to the MIND diet and subjective memory complaints in the French NutriNet-Santé cohort. *J. Neurol.* **2019**, *266*, 942–952. [CrossRef]
37. Zhu, J.; Shi, R.; Chen, S.; Dai, L.; Shen, T.; Feng, Y.; Gu, P.; Shariff, M.; Nguyen, T.; Ye, Y.; et al. The Relieving Effects of BrainPower Advanced, a Dietary Supplement, in Older Adults with Subjective Memory Complaints: A Randomized, Double-Blind, Placebo-Controlled Trial. *Evid. Based Complement. Altern. Med.* **2016**, *2016*, 7898093. [CrossRef]
38. Firth, J.; Gangwisch, J.E.; Borisini, A.; Wootton, R.E.; Mayer, E.A. Food and mood: How do diet and nutrition affect mental wellbeing? *BMJ* **2020**, *369*, m2382. [CrossRef]
39. Gardner, M.P.; Wansink, B.; Kim, J.; Park, S. Better moods for better eating? How mood influences food choice. *J. Consum. Psychol.* **2014**, *24*, 320–335. [CrossRef]
40. Huang, T.L.; Zandi, P.P.; Tucker, K.L.; Fitzpatrick, A.L.; Kuller, L.H.; Fried, L.P.; Burke, G.L.; Carlson, M.C. Benefits of fatty fish on dementia risk are stronger for those without APOE epsilon4. *Neurology* **2005**, *65*, 1409–1414. [CrossRef]
41. Whalley, L.J.; Deary, I.J.; Starr, J.M.; Wahle, K.W.; Rance, K.A.; Bourne, V.J.; Fox, H.C. n-3 Fatty acid erythrocyte membrane content, APOE varepsilon4, and cognitive variation: An observational follow-up study in late adulthood. *Am. J. Clin. Nutr.* **2008**, *87*, 449–454. [CrossRef] [PubMed]

© 2020 by the authors. Licensee MDPI, Basel, Switzerland. This article is an open access article distributed under the terms and conditions of the Creative Commons Attribution (CC BY) license (http://creativecommons.org/licenses/by/4.0/).

Article

Dietary Supplementation with Curcumin Reduce Circulating Levels of Glycogen Synthase Kinase-3β and Islet Amyloid Polypeptide in Adults with High Risk of Type 2 Diabetes and Alzheimer's Disease

Rohith N Thota [1,2], Jessica I Rosato [1,3], Cintia B Dias [1,4], Tracy L Burrows [1,3], Ralph N Martins [4] and Manohar L Garg [1,2,*]

1 Nutraceuticals Research Program, School of Biomedical Sciences & Pharmacy, University of Newcastle, Callaghan, NSW 2308, Australia; rohith.thota@newcastle.edu.au (R.N.T.); jessica.rosato@uon.edu.au (J.I.R.); Cintia.botelhodias@mq.edu.au (C.B.D.); tracy.burrows@newcastle.edu.au (T.L.B.)
2 Riddet Institute, Massey University, Palmerston North 4474, New Zealand
3 School of Health Sciences, University of Newcastle, Callaghan, NSW 2308, Australia
4 School of Biomedical Sciences, Macquarie University, Macquarie, NSW 2109, Australia; ralph.martins@mq.edu.au
* Correspondence: Manohar.Garg@newcastle.edu.au; Tel.: +61-2-4921-5647; Fax: +61-2-49212028

Received: 17 March 2020; Accepted: 7 April 2020; Published: 9 April 2020

Abstract: Dietary supplementation with curcumin has been previously reported to have beneficial effects in people with insulin resistance, type 2 diabetes (T2D) and Alzheimer's disease (AD). This study investigated the effects of dietary supplementation with curcumin on key peptides implicated in insulin resistance in individuals with high risk of developing T2D. Plasma samples from participants recruited for a randomised controlled trial with curcumin (180 mg/day) for 12 weeks were analysed for circulating glycogen synthase kinase-3 β (GSK-3β) and islet amyloid polypeptide (IAPP). Outcome measures were determined using ELISA kits. The homeostasis model for assessment of insulin resistance (HOMA-IR) was measured as parameters of glycaemic control. Curcumin supplementation significantly reduced circulating GSK-3β (−2.4 ± 0.4 ng/mL vs. −0.3 ± 0.6, $p = 0.0068$) and IAPP (−2.0 ± 0.7 ng/mL vs. 0.4 ± 0.6, $p = 0.0163$) levels compared with the placebo group. Curcumin supplementation significantly reduced insulin resistance (−0.3 ± 0.1 vs. 0.01 ± 0.05, $p = 0.0142$) compared with placebo group. Dietary supplementation with curcumin reduced circulating levels of IAPP and GSK-3β, thus suggesting a novel mechanism through which curcumin could potentially be used for alleviating insulin resistance related markers for reducing the risk of T2D and AD.

Keywords: curcumin; glycogen synthase kinase-3; insulin resistance; Islet amyloid polypeptide; type 2 diabetes mellitus

1. Introduction

Insulin resistance is of particular interest as defective insulin signalling in the brain contributes to the accumulation of amyloid beta (Aβ) and tau protein, presenting a pathophysiological link between Alzheimer's disease (AD) and type 2 diabetes (T2D) [1]. It has been previously described as diabetes of the brain, or type 3 diabetes [2]. The prevalence of insulin resistance is growing exponentially, with over 463 million people suffering from diabetes worldwide according to the recent edition of the International Diabetes Federation Atlas 2019, of which >90% are T2D. Recent evidence from a population-based study shows a link between T2D and AD, with an incidence 2–5 times higher in those with T2D [3] and hyperinsulinemia [4].

Islet amyloid polypeptide (IAPP) is a peptide hormone co-secreted with insulin by pancreatic β-cells [5]. Elevated serum levels of IAPP is a pathological hallmark of insulin resistance and correlates with AD diagnosis [6]. Furthermore, the amyloidosis of IAPP involving formation of Aβ-like structures and extracellular deposits of amyloid in the pancreas is a distinctive feature of T2D [5]. IAPP can also induce peripheral insulin resistance by antagonising insulin activity, further linking to the overexpression of glycogen synthase kinase-3 (GSK-3) [7]. Impaired insulin signalling and subsequent hyperactivity of GSK-3 in rodent and human models have been associated with the accumulation of Aβ and tau protein in the brain [8]. Recent research has uncovered a pathophysiological link between T2D and AD involving insulin resistance and the activation of GSK-3, a serine-threonine kinase involved in a multitude of physiological processes including glycogen metabolism and microtubule stability [9,10]. GSK-3 produces two isoforms (α and β) upon activation. Along with its pleiotropic roles in human physiology in skeletal muscle and liver, GSK-3 is linked to cognitive disorders and is thought to play an important role in the pathogenesis of AD [10]. Insulin triggers the phosphorylation (inactivation) of GSK-3 via the PI3k/Akt signalling cascade, while defective insulin signalling results in decreased phosphorylation and consequently elevated activation of GSK-3 in the brain [11]. Over-activity of GSK-3α mainly enhances plaque-associated aggregation of insoluble Aβ, while GSK-3β primarily contributes to the hyper-phosphorylation of tau [11]. The purported role of GSK-3 in the development of AD has led to it being investigated as a potential therapeutic target.

Curcumin is a bio-active curcuminoid, extracted from the rhizomes of turmeric with a wide range of pharmacological properties including the ability to reduce inflammation, oxidative stress and insulin resistance [12–14]. In vivo studies in animal models [13] and a few clinical trials [15] have shown beneficial effects of curcumin on insulin resistance. Systematic reviews have provided strong evidence for investigating curcumin efficacy for management of type 2 diabetes mellitus [15]. Substantial in vitro data are available on the anti-oxidant anti-inflammatory activities of curcumin, suggesting a possible link to its protective effect on dementia and AD [16]. In animal models, curcumin has been shown to reduce systemic inflammatory markers (cyclooxygenase (COX-2) and phospholipases; transcription factors such as nuclear factor kappa-B; pro-inflammatory cytokines such as tumour necrosis factor-α and IL-1β and C-reactive protein (CRP) concentrations) suggesting a possible link between its anti-inflammatory and cognitive protection effects [17]. Curcumin has been proven to have strong antioxidant action by the inhibition of the formation and of free radicals [18]. It decreases the oxidation of low-density lipoprotein that cause the deterioration of neurons, not only in AD but also in other neuron degenerative disorders (Huntington's and Parkinson's disease) [18]. Curcumin also increased memory function and non-spatial memory related parameters in aged rodent models with cognitive impairments [19–21]. Curcumin inhibits the activity of activator protein-1, a transcription factor implicated in the expression of Aβ [22]. Increasing evidence suggests that curcumin supplementation mitigates Aβ deposition and tauopathy whilst exerting inhibitory effects on GSK-3 activity via interactions with the PI3k/Akt cascade [23]. Epidemiologic studies also suggest a link between curcumin and cognitive benefits. Ng et al. [24] observed that subjects with higher curry (curcumin is a common ingredient) consumption had 6% higher Mini-Mental State Examination (MMSE) scores compared with subjects who never or rarely consume curry.

The widespread use of curcumin as an additive, the relatively high safety profile established in a number of short-term trials and the potency of curcumin to suppress insulin resistance could be a beneficial factor in management of both T2D and AD. The aim of this study was to determine if dietary supplementation with curcumin reduce plasma levels of peptides, GSK-3β and IAPP that are implicated in the insulin resistance in people at a high risk of developing T2D.

2. Materials and Methods

2.1. Participants

Participants were recruited for the purposes of the curcumin and omega-3 fatty acids for prevention of type 2 diabetes (COP-D) study [25] from the Hunter region in New South Wales, Australia. Interested participants were screened through telephone interviews per the inclusion and exclusion criteria. If eligible, potential participants were posted self-administered health/medical, diet, and physical activity questionnaires as well as a consent form. Inclusion criteria for the current study included: aged of 30–70 (years); body mass index (BMI) of 25–45 kg/m^2; ≥12 score in the Australian Type

2 Diabetes Risk (AUSDRISK) questionnaire (a non-invasive questionnaire for assessing the risk of developing type 2 diabetes); diagnosed with either impaired fasting glucose (IFG, fasting glucose of 6.1–6.9 mmol/L) or impaired glucose tolerance (IGT, 2-h plasma glucose ≥7.8 mmol/L and <11.1 mmol/L) or both; and glycosylated haemoglobin (HbA1c) levels of 5.7–6.4%. Participants were excluded if they were unwilling to provide blood samples at the baseline and post-intervention (12-week) site visits; diagnosed with T2D; gallbladder problems; pacemaker implants; severe neurological diseases or seizures; pregnant, planning to become pregnant or breastfeeding/lactating; taking any dietary supplements (such as fish oil, cinnamon, probiotics, vitamin D, chromium, etc.) known to influence blood glucose levels; consuming ≥2 servings of oily fish per week; or taking any medications known to have drug-nutrient interactions with curcumin (blood thinning medications such as Aspirin and warfarin). All participants gave their written informed consent. The study was conducted in accordance with the Declaration of Helsinki, and has been approved by the University of Newcastle Human Research Ethics Committee (H-2014-0385). The trial is registered with the Australia & New Zealand Clinical Trial Registry (ACTRN12615000559516).

2.2. Study Design

The detailed protocol of the current study is previously published [25]. Screened participants were randomised to placebo (2 × placebo tablets matching for curcumin) and curcumin (2 × 500 mg curcumin tablets [Meriva®] providing 180 mg of curcumin per day). Participant compliance was measured during the follow-up (6-week) and post-intervention (12-week) site visits via a capsule count-back method and capsule intake log. Any illnesses, changes in medications or medical diagnoses during the study timeframe were recorded.

2.3. Data Collection and Outcome Measures

The primary outcome of this study was to evaluate the effects of curcumin on circulating levels of GSK-3β in adults at a high risk of developing T2D. Fasting (≥10-h) blood samples were collected from participants at baseline and post-intervention (12-week), at either the Nutraceuticals Research Program clinical trial facility or John Hunter Hospital in Newcastle, NSW. These samples were analysed using GSK-3β enzyme linked immunosorbent assay (ELISA) kits, which have high specificity for human GSK-3β and no detectable cross-reactivity with other relevant proteins (manufacturer, Aviva systems biology; detection range, 0.625–40 ng/mL; mean intra-Assay CV, <10%; and mean inter-Assay CV, <12%). The secondary outcome was to evaluate the effects of curcumin supplements on IAPP, which was analysed via ELISA (manufacturer, Aviva systems biology; detection range, 0.156–10 ng/mL; mean intra-assay CV%, <4.6%; Mean inter-assay CV%, <7.4%). Serum insulin and glucose were measured by Hunter Area Pathology Services using radio immunoassay technique. HOMA2-IR was calculated using the Diabetes Trials Unit online calculator.

Questionnaires (Diet, Physical Activity and Medical History)

Medical history data collected include medication and supplement use as well as family medical history. Demographic characteristics collected from participants include age, sex and ethnicity. Participants were advised to maintain their regular dietary pattern and physical activity levels throughout the 12-week study period. To estimate their habitual dietary patterns, participants were asked to complete a 3-day (2 weekdays plus 1 weekend day) food diary prior to all three site visits. Food diaries were analysed using FoodWorks Xyris (version 8.0) to assess changes in energy intakes (kJ) during the study period. Habitual physical activity (METs, minutes/week) was assessed using the International Physical Activity Questionnaire (IPAQ) long form version (2002), which was completed by participants prior to all three site visits.

2.4. Body Composition and Anthropometric Measures

Body composition measurements included body weight (kg), muscle mass (kg), body fat mass (kg), body mass index (BMI, kg/m^2) and body fat per cent (%). Body composition was measured using direct segmental multi-frequency bioelectrical impedance (InBody 230, Biospace Co., Ltd. Seoul, Korea). Anthropometric measurements included height (cm; SE206, Seca), BMI (kg/m^2) and waist-circumference (cm).

2.5. Statistical Analysis

For the purposes of the current sub-study, n = 29 (15 allocated to placebo group and 14 allocated to curcumin treatment group) participants were analysed for GSK-3β and IAPP resulting in a study power of 97.7% for the detection of a 2.4 ng/mL reduction in serum GSK-3β level (1.3 ng/mL SD, *p* value = 0.01). Data collected at baseline were analysed for normality using histograms with a normal distribution curve overlaid the Shapiro–Wilk test, and are presented as mean ± SEM (standard error of the mean) or median (IQR, interquartile range) as appropriate. Significant changes in the baseline data between the two intervention groups were assessed through t-test or Mann–Whitney U Test when the normality assumption was not met. Post-intervention data are presented as mean ± SEM or median (IQR) of absolute change (post-intervention value minus baseline value). Changes from baseline to post-intervention within treatment groups were assessed through t-test or Wilcoxon signed-rank test.

3. Results

3.1. Baseline Characteristics

Twenty-nine serum samples were analysed from the COP-D trial for the primary outcome, GSK-3β, and for IAPP. No significant differences were observed between the two participant groups for all baseline characteristics, including demographics and serum outcome measures (Table 1). All participants were insulin resistant, with an average fasting glucose of 5.4 ± 0.1 mmol/L and median fasting serum insulin of 9.9 (4.9) mIU/L. Trial participants had a median GSK-3β of 3.0 (1.7) ng/mL and median IAPP of 4.5 (2.6) ng/mL. Likewise, there were no significant changes in dietary intake or physical activity within and between the two groups post-intervention (Table 2). Comparisons between the placebo and curcumin group showed no significant differences in the body composition and anthropometric measurements collected at baseline and post intervention (Tables 1 and 2). The average BMI of participants at baseline was within the obese category (≥30 kg/m^2) and was accompanied by high body fat per cent (34.7 ± 1.8%) and waist circumference (105.4 ± 2.4 cm). There were no significant changes in body composition and anthropometric measurements post-intervention (Table 2). According to the capsule count, mean compliance to the randomised intervention was 94.9 ± 5.80%. Curcumin was well tolerated by participants and no adverse events due to the allocated intervention were reported during the 12-week study period.

Table 1. Baseline characteristics of the trial participants.

Characteristics	Total (n = 29)	Placebo (n = 15)	Curcumin (n = 14)	*p* Value
Age (years)	52.3 ± 1.9	50.4 ± 2.6	54.5 ± 2.9	0.2998
Males/females (n/n)	12/17	6/9	6/8	-
Ethnicity—no (%)				
Caucasian	23	12 (80)	11 (78.6)	-
Asian	3	1 (6.7)	2 (14.3)	-
Others	3	2 (13.3)	1 (7.1)	-
Anthropometry measures				
Body weight (kg)	88.8 ± 3.0	90.7 ± 4.9	86.7 ± 3.5	0.5206
Muscle mass (kg)	33.3 ± 1.4	32.4 ± 1.7	34.4 ± 2.4	0.4998
Body fat mass (kg)	32.4 ± 2.2	33.7 ± 3.5	31.1 ± 3.0	0.5785
Body mass index (kg. m^{-2})	31.3 ± 1.0	32.3 ± 1.7	30.2 ± 1.1	0.3276
Waist circumference(cm)	105.4 ± 2.4	106.0 ± 3.9	104.9 ± 2.9	0.8246
Percent body fat (%)	34.7 ± 1.8	35.3 ± 2.2	34.8 ± 2.5	0.5467
Plasma outcome measures				
Fasting glucose (mmol/L)	5.4 ± 0.1	5.2 ± 0.1	5.6 ± 0.2	0.1121
Fasting serum insulin (mIU/L)	9.9 (4.9)	10.3 (7.9)	9.1 (4.6)	0.6005
HOMA2-IR	1.3 (0.6)	1.3 (1.1)	1.2 (0.6)	0.7268
IAPP (ng/mL)	4.5 (2.6)	4.1 (2.6)	3.9 (3.1)	0.8948
GSK-3β (ng/mL)	3.0 (1.7)	2.7 (1.8)	3.4 (2.7)	0.1625
Dietary intakes (kj)	9047.3 ± 424.9	8497.1 ± 599.6	9682.1 ± 573.06	0.1685
Physical Activity (METs-minutes/week)	2432 (4920)	3894 (5214)	1765 (1597)	0.1161

Data are presented as mean ± SEM or median (IQR) for continuous variables and n (%) for categorical variables. n, number of participants; METs, metabolic equivalents; SEM, standard error of the mean; IAPP, islet amyloid polypeptide; GSK-3β, glycogen synthase kinase-3 beta. *p*-values represent the significant differences between the groups.

Table 2. Changes in the study parameters of the participants in placebo and curcumin group after three-month intervention period.

Outcome Measures	Treatment Group	Mean Change	*p* Value	Mean Difference between Treatment Groups	*p* Value
Body weight (kg)	Placebo	0.64 ± 0.4	0.1731		
	Curcumin	−0.1 ± 0.4	0.8272	−0.7 ± 0.6	0.2292
Muscle mass (kg)	Placebo	0.1 (0.1)	0.8902		
	Curcumin	0.25 (0.7)	0.4440	0.1(0.8)	0.5257
Body fat mass (kg)	Placebo	0.1 (0.4)	0.7577		
	Curcumin	−0.85 (0.9)	0.8487	−0.5 (2.1)	0.3478
Body mass index (kg/m^2)	Placebo	0.20 (0.2)	0.1945		
	Curcumin	0.03 (0.2)	0.8296	0 (0.7)	0.3573
Waist circumference (cm)	Placebo	0.87 ± 0.7	0.2258		
	Curcumin	−0.10 ± 0.8	0.8940	−0.1 ± 1.0	0.3557
Percent body fat (%)	Placebo	0.5 (1.8)	0.4388		
	Curcumin	−0.6 (1.6)	0.5980	0 (2)	0.7699
Fasting glucose (mmol/L)	Placebo	−0.06 ± 0.1	0.3625		
	Curcumin	−0.07 ± 0.1	0.6041	−0.004 ± 0.1	0.9747
Fasting serum insulin (μIU/L)	Placebo	0.1 ± 0.4	0.8251		
	Curcumin	−1.9 ± 0.6	0.0076	−2.0 ± 0.4	0.0115
Dietary intakes (kj)	Placebo	−134.5 ± 479.2	0.7830		
	Curcumin	298.4 ± 487.9	0.5520	−433.0 ± 686.6	0.5338
Physical activity (Metabolic equivalent-minute/week)	Placebo	−473 (3880)	0.2202		
	Curcumin	104.5 (1508)	0.6249	0 (2618)	0.2386

Data are presented as mean ± SEM or median (IQR) as appropriate. *p*-values represent the significant differences with-in and between the groups.

3.2. *Effects of Curcumin on GSK-3β and IAPP*

After 12- weeks of curcumin supplementation, circulating GSK-3β levels were significantly lower in the curcumin group (−2.4 ± 0.4 ng/mL, *p* value = 0.00001) (Figure 1). When compared to placebo, there was also significant (*p* value = 0.0068) reduction in serum GSK-3β levels in the curcumin group. Similar observation was found with IAPP (−2.0 ± 0.7 ng/mL, *p* value 0.01) within curcumin group post-intervention (Figure 2); When compared to PL, there was also a significant (p value = 0.0163) change in mean IAPP.

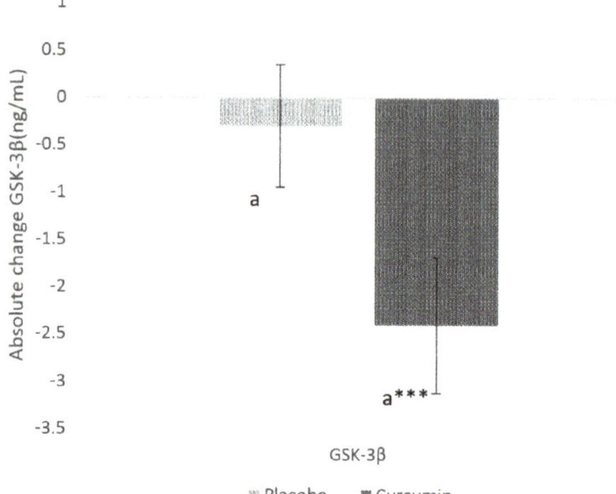

Figure 1. Absolute change in serum glycogen synthase kinase –β (GSK-3β) from baseline to post-intervention in placebo and curcumin groups for 12 weeks. *** $p < 0.001$ represents the difference within the treatment group. Small letter (a) represents the significant difference between the treatment groups.

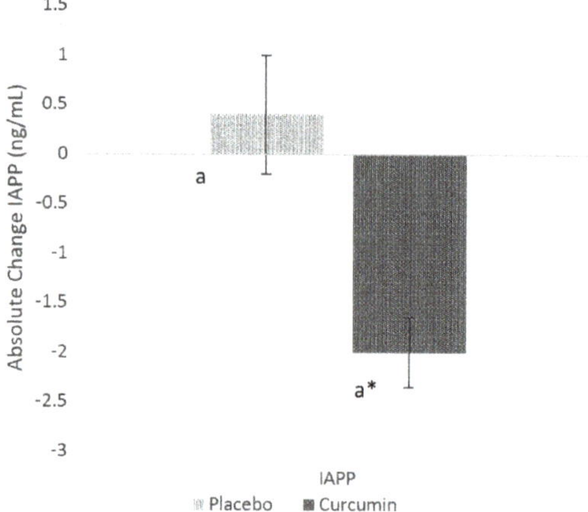

Figure 2. Absolute change in serum islet amyloid peptide (IAPP) from baseline to post-intervention in placebo and curcumin groups for 12 weeks. * $p < 0.05$ represents the difference within the treatment group. Small letter (a) represents the significant difference between the treatment groups.

3.3. Glycaemic Indices

Serum insulin was significantly reduced (p value = 0.0076) in the curcumin treatment group (−1.9 μIU) from baseline (Table 2) and was also significantly different from placebo (0.0115). Similar trends were observed with respect to HOMA2-IR (Figure 3) in curcumin group. Post-intervention, significant changes (p value = 0.0142) in HOMA2-IR were only observed in the curcumin treated group (−0.11 ± 0.05) compared to the placebo group. No significant changes were observed on blood glucose levels after supplementation with curcumin (p = 0.9747) compared to the placebo group.

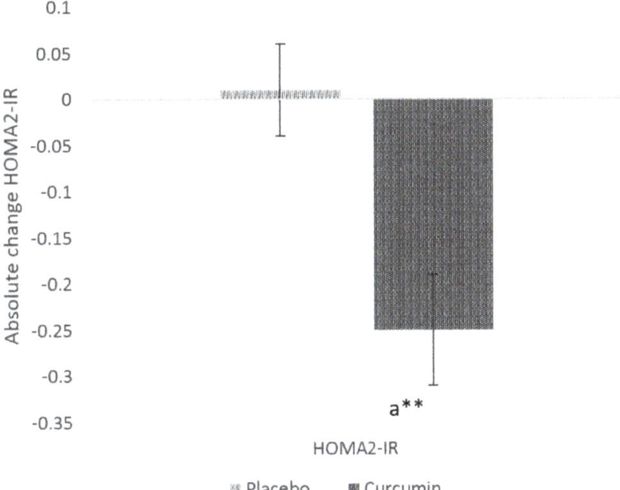

Figure 3. Absolute change in HOMA2-IR from baseline to post-intervention in placebo and curcumin groups for 12 weeks. ** $p < 0.01$ represents the difference within the treatment group. Small letter (a) represents the significant difference between the treatment groups.

4. Discussion

The primary finding of this study is that oral supplementation with curcumin (180 mg per day) for 12 weeks reduces the circulating levels of peptides that are implicated in insulin resistance, namelt GSK-3β and IAPP (Figure 4). In addition, we demonstrated that curcumin supplementation positively affects glycaemic control via reduction in insulin resistance and fasting serum insulin compared with the placebo group.

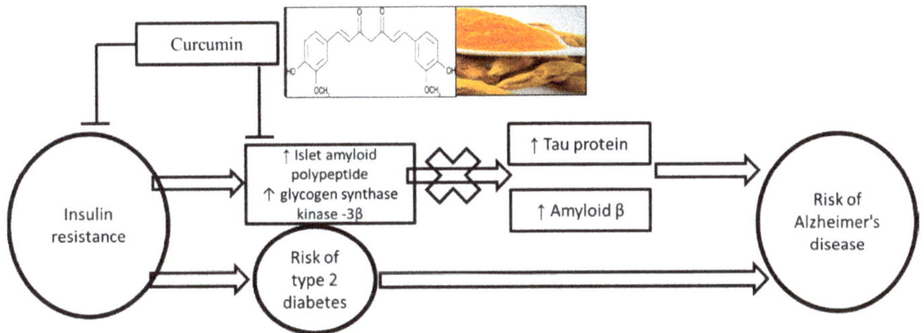

Figure 4. Summary of potential mechanism of curcumin in reducing the risk of type 2 diabetes and Alzheimer's disease.

The aetiology of insulin resistance is dependent on multiple of factors [26] and there is substantial evidence implicating the role of GSK-3 in insulin resistance [27]. In T2D, GSK-3 is an enzyme of glycogen synthesis, which plays a key role in regulating blood glucose. Its role in insulin deficiency and insulin resistance [28] is implicated via insulin/PI3-kinase/protein kinase B (insulin/PI3K/Akt) signalling pathway [29]. In AD, GSK-3β is involved in the hyperphosphorylation of microtubule-associated protein tau (tau), which is one of the pathological features in AD [30].

There are multiple potential reasons for evaluating the effects of curcumin on serum GSK-3β levels. Pre-clinical studies indicated that curcumin supplementation reduced GSK-3 activity resulting in protection against Aβ accumulation and hyper-phosphorylation of tau [23,31,32]. Simulated docking studies have shown optimal binding capacity of curcumin with GSK-3β and interactions with key amino acids resulting in the deactivation of the kinase [33]. Glycogen metabolism is a highly regulated process in which GSK-3β plays an essential role [33]. Follow-up in vitro and in vivo studies have confirmed curcumin's pharmacological activities by demonstrating that this bioactive potently inhibits GSK-3β (IC_{50} = 66.3 nM) and increases fasting liver glycogen levels [33]. A recent randomised placebo-controlled trial with curcumin (90 mg), similar to the current study dose, demonstrated reduction in Aβand tau in a certain brain areas leading to the improvement memory and attention in adults aged 51–84 years [33]. Amyloid and tau accumulation on brain were assessed by positron emission tomography (FDDNP-PET). Curcumin significantly lowered binding in the amygdala (ES = −0.41, p = 0.04) compared with a placebo [34]. In line with these observations from the pre-clinical reports, in the current study we showed that dietary supplementation with curcumin significantly reduce serum levels of GSK-3β in adults with insulin resistance. Follow-up studies are required to determine if this effect has direct implications in reducing the risk for T2D and AD.

IAPP, a pancreatic beta cell peptide, can evoke insulin resistance by antagonising insulin in a non-competitive manner [7,35]. Although IAPP has been previously shown to have no significant effect on glucose transport, it decreased insulin-stimulated glucose transport by about 30% [35]. IAPP also increased GSK-3 activity, which in turn led to increased phosphorylation of glycogen synthase and decreased glycogen synthesis de novo [7]. IAPP mediates several important brain functions via binding to its receptor in the brain, including regulating glucose metabolism, inflammatory responses, and potentially in neurogenesis [36,37]. However, amylin can aggregate when concentrations are high and become neurotoxic in cell cultures and is associated with brain amyloid burden and cognitive impairment in AD [38,39]. Recent studies have also indicated that the high plasma levels of IAPP is a pathological hallmark of insulin resistance [40] and correlates with AD diagnosis and brain structure [41]. In vitro studies revealed that curcumin significantly reduces h-IAPP fibril formation and aggregates formed in the presence of curcumin display alternative structure compared to the actual peptide [42]. Curcumin increased the time required for the conversion of IAPP monomer to assemblies that are visible in NMR, including β-sheet, suggesting that curcumin has the potential to inhibit the formation of the oligomers that are on-pathway to formation of amyloid [42]. In the current study, we reported if curcumin has a direct effect on the plasma levels of IAPP. Curcumin supplementation significantly reduced the plasma IAPP levels in the current study, providing further insights into the beneficial effects of curcumin on glycaemic homeostasis.

Insulin resistance and hyperinsulinemia is associated with the development of both T2D and AD [43]. Converging evidence from cross-sectional studies has shown significant associations of HOMA-IR with mild cognitive impairment and AD [44,45]. Moreover, presence of IR accelerates the formation of Neuritic plaques which are involved in the pathogenic process of AD [45]. Curcumin has repeatedly demonstrated efficacy in regard to improving insulin resistance [15]. Curcumin treatment reduced both serum insulin and insulin resistance (measured via HOMA2-IR). A similar effect of curcumin on insulin resistance was observed in a nine-month randomised controlled trial with curcumin extract in a pre-diabetic population [46]. Curcumin intervention reduced insulin resistance, as indicated by an increased HOMA-IR and reduced C-peptide levels [47]. Another three-month randomised controlled trial with overweight/obese T2DM patients has also indicated that the curcumin supplementation group resulted in a significant reduction of fasting glycaemia and insulin resistance [48]. However, in contrast to these, we did not observe any significant reductions in fasting glucose with curcumin supplementation. Curcumin mediated reduction in the insulin levels and insulin resistance (HOMA-IR) could reduce the risk factors for cognitive impairment such as neuritic plaques and amyloid formation. As IR and hyperinsulinemia are significantly associated and there is a common molecular pathway for T2D and AD, this study's findings on the effect of curcumin on insulin resistance implicates a potential role of curcumin in reducing the risk for AD. In this study, we provided novel insights on the efficacy of curcumin in insulin resistance by providing evidence on regulation of key peptides such as GSK-3β and IAPP, which play an important role in insulin resistance.

Participants compliance to the intervention in the current study was high. Aviva Systems Biology GSK-3β ELISA kits used to measure the primary outcome have high specificity for human GSK-3β and no detectable cross-reactivity with other relevant proteins. The link between AD and T2D is a relatively new and emerging area of research. In this study, we were able to provide novel potential adjuvants for ameliorating common risk factors underlying T2D and AD. However, the results of the current sub-study are limited by their preliminary nature, and as such a follow-up study is required to substantiate effects of curcumin supplementation on GSK-3β and IAPP and whether it can affect neurological/metabolic parameters specific to T2D and AD in high-risk adults. Furthermore, the preliminary results of this sub-study may not be generalisable or transferable to other populations as only adults with insulin resistance and high risk of T2D were studied.

Author Contributions: R.N.T. and M.L.G. designed the research; R.N.T. conducted the research; R.N.T. and J.I.R. analysed the data and drafted the manuscript; and C.B.D., T.L.B., and R.N.M. revised the manuscript. All authors have read and agreed to the published version of the manuscript.

Funding: This project was supported by a pliot grant from the Mary Castello bequest for Alzheimer s Disease Research and the Faculty of Health and Medicine, University of Newcastle.

Acknowledgments: Community members (Hunter and Newcastle, NSW, Australia) are acknowledged for their participation in the study and Indena SpA for providing curcumin and placebo tablets at no cost.

Conflicts of Interest: The authors declare no conflict of interest.

Abbreviations

AUSDRISK	Australian Type 2 Diabetes Risk
AD	Alzheimer's disease
GSK-3β	glycogen synthase kinase-3 β
HOMA-IR	homeostasis model for assessment of insulin resistance
IAPP	islet amyloid polypeptide
T2D	type 2 diabetes

References

1. Ferreira, L.S.S.; Fernandes, C.S.; Vieira, M.N.N.; De Felice, F.G. Insulin resistance in Alzheimer's disease. *Front. Neurosci.* **2018**, *12*, 830. [CrossRef]
2. de la Monte, S.M.; Wands, J.R. Alzheimer's disease is type 3 diabetes-evidence reviewed. *J. Diabetes Sci. Technol.* **2008**, *2*, 1101–1113. [CrossRef]
3. Luchsinger, J.A.; Reitz, C.; Honig, L.S.; Tang, M.X.; Shea, S.; Mayeux, R. Aggregation of vascular risk factors and risk of incident Alzheimer disease. *Neurology* **2005**, *65*, 545–551. [CrossRef]
4. Luchsinger, J.A.; Tang, M.-X.; Shea, S.; Mayeux, R. Hyperinsulinemia and risk of Alzheimer disease. *Neurology* **2004**, *63*, 1187–1192. [CrossRef]
5. Westermark, P.; Andersson, A.; Westermark, G.T. Islet amyloid polypeptide, islet amyloid, and diabetes mellitus. *Physiol. Rev.* **2011**, *91*, 795–826. [CrossRef]
6. N Fawver, J.; Ghiwot, Y.; Koola, C.; Carrera, W.; Rodriguez-Rivera, J.; Hernandez, C.; TDineley, K.; Kong, Y.; Li, J.; Jhamandas, J.; et al. Islet Amyloid Polypeptide (IAPP): A Second Amyloid in Alzheimer's Disease. *Curr. Alzheimer Res.* **2014**, *11*, 928–940. [CrossRef]
7. Abaffy, T.; Cooper, G.J. GSK3 involvement in amylin signaling in isolated rat soleus muscle. *Peptides* **2004**, *25*, 2119–2125. [CrossRef]
8. Zhang, Y.; Song, W. Islet amyloid polypeptide: Another key molecule in Alzheimer's pathogenesis? *Prog. Neurobiol.* **2017**, *153*, 100–120. [CrossRef]
9. Hooper, C.; Killick, R.; Lovestone, S. The GSK3 hypothesis of Alzheimer's disease. *J. Neurochem.* **2008**, *104*, 1433–1439. [CrossRef]
10. Gao, C.; Holscher, C.; Liu, Y.; Li, L. GSK3: A key target for the development of novel treatments for type 2 diabetes mellitus and Alzheimer disease. *Rev. Neurosci.* **2011**, *23*, 1–11. [CrossRef]
11. Beurel, E.; Grieco, S.F.; Jope, R.S. Glycogen synthase kinase-3 (GSK3): Regulation, actions, and diseases. *Pharmacol. Ther.* **2015**, *148*, 114–131. [CrossRef]
12. Thota, R.N.; Acharya, S.H.; Abbott, K.A.; Garg, M.L. Curcumin and long-chain Omega-3 polyunsaturated fatty acids for Prevention of type 2 Diabetes (COP-D): Study protocol for a randomised controlled trial. *Trials* **2016**, *17*, 565. [CrossRef]

13. Weisberg, S.P.; Leibel, R.; Tortoriello, D.V. Dietary curcumin significantly improves obesity-associated inflammation and diabetes in mouse models of diabesity. *Endocrinology* **2008**, *149*, 3549–3558. [CrossRef]
14. Shen, L.; Ji, H.F. The pharmacology of curcumin: Is it the degradation products? *Trends Mol. Med.* **2012**, *18*, 138–144. [CrossRef]
15. Pivari, F.; Mingione, A.; Brasacchio, C.; Soldati, L. Curcumin and Type 2 Diabetes Mellitus: Prevention and Treatment. *Nutrients* **2019**, *11*, 1837. [CrossRef]
16. Mishra, S.; Palanivelu, K. The effect of curcumin (turmeric) on Alzheimer's disease: An overview. *Ann. Indian Acad. Neurol.* **2008**, *11*, 13–19. [CrossRef]
17. Sarker, M.R.; Franks, S.; Sumien, N.; Thangthaeng, N.; Filipetto, F.; Forster, M. Curcumin Mimics the Neurocognitive and Anti-Inflammatory Effects of Caloric Restriction in a Mouse Model of Midlife Obesity. *PLoS ONE* **2015**, *10*, e0140431. [CrossRef]
18. Kim, G.Y.; Kim, K.H.; Lee, S.H.; Yoon, M.S.; Lee, H.J.; Moon, D.O.; Lee, C.M.; Ahn, S.C.; Park, Y.C.; Park, Y.M. Curcumin inhibits immunostimulatory function of dendritic cells: MAPKs and translocation of NF-kappa B as potential targets. *J. Immunol.* **2005**, *174*, 8116–8124. [CrossRef]
19. Dong, S.; Zeng, Q.; Mitchell, E.S.; Xiu, J.; Duan, Y.; Li, C.; Tiwari, J.K.; Hu, Y.; Cao, X.; Zhao, Z. Curcumin enhances neurogenesis and cognition in aged rats: Implications for transcriptional interactions related to growth and synaptic plasticity. *PLoS ONE* **2012**, *7*, e31211. [CrossRef]
20. Yu, S.Y.; Zhang, M.; Luo, J.; Zhang, L.; Shao, Y.; Li, G. Curcumin ameliorates memory deficits via neuronal nitric oxide synthase in aged mice. *Prog. Neuro Psychopharmacol. Biol. Psychiatry* **2013**, *45*, 47–53. [CrossRef]
21. Cheng, K.K.; Yeung, C.F.; Ho, S.W.; Chow, S.F.; Chow, A.H.; Baum, L. Highly stabilized curcumin nanoparticles tested in an in vitro blood-brain barrier model and in Alzheimer's disease Tg2576 mice. *AAPS J.* **2013**, *15*, 324–336. [CrossRef]
22. Yang, F.; Lim, G.P.; Begum, A.N.; Ubeda, O.J.; Simmons, M.R.; Ambegaokar, S.S.; Chen, P.P.; Kayed, R.; Glabe, C.G.; Frautschy, S.A.; et al. Curcumin inhibits formation of amyloid beta oligomers and fibrils, binds plaques, and reduces amyloid in vivo. *J. Biol. Chem.* **2005**, *280*, 5892–5901. [CrossRef]
23. Hoppe, J.B.; Frozza, R.L.; Pires, E.N.; Meneghetti, A.B.; Salbego, C. The curry spice curcumin attenuates beta-amyloid-induced toxicity through beta-catenin and PI3K signaling in rat organotypic hippocampal slice culture. *Neurol. Res.* **2013**, *35*, 857–866. [CrossRef]
24. Ng, T.P.; Chiam, P.C.; Lee, T.; Chua, H.C.; Lim, L.; Kua, E.H. Curry consumption and cognitive function in the elderly. *Am. J. Epidemiol.* **2006**, *164*, 898–906. [CrossRef]
25. Thota, R.N.; Acharya, S.H.; Garg, M.L. Curcumin and/or omega-3 polyunsaturated fatty acids supplementation reduces insulin resistance and blood lipids in individuals with high risk of type 2 diabetes: A randomised controlled trial. *Lipids Health Dis.* **2019**, *18*, 31. [CrossRef]
26. Sesti, G. Pathophysiology of insulin resistance. *Best Pract. Res. Clin. Endocrinol. Metab.* **2006**, *20*, 665–679. [CrossRef]
27. Zhang, Y.; Huang, N.Q.; Yan, F.; Jin, H.; Zhou, S.Y.; Shi, J.S.; Jin, F. Diabetes mellitus and Alzheimer's disease: GSK-3β as a potential link. *Behav. Brain Res.* **2018**, *339*, 57–65. [CrossRef]
28. Feng, Z.-C.; Donnelly, L.; Li, J.; Krishnamurthy, M.; Riopel, M.; Wang, R. Inhibition of Gsk3β activity improves β-cell function in c-KitWv/+ male mice. *Lab. Investig.* **2012**, *92*, 543–555. [CrossRef]
29. Gleason, J.E.; Szyleyko, E.A.; Eisenmann, D.M. Multiple redundant Wnt signaling components function in two processes during C. elegans vulval development. *Dev. Biol.* **2006**, *298*, 442–457. [CrossRef]
30. Jope, R.S.; Johnson, G.V. The glamour and gloom of glycogen synthase kinase-3. *Trends Biochem. Sci.* **2004**, *29*, 95–102. [CrossRef]
31. Xiong, Z.; Hongmei, Z.; Lu, S.; Yu, L. Curcumin mediates presenilin-1 activity to reduce β-amyloid production in a model of Alzheimer's disease. *Pharmacol. Rep.* **2011**, *63*, 1101–1108. [CrossRef]
32. Huang, H.-C.; Tang, D.; Xu, K.; Jiang, Z.-F. Curcumin attenuates amyloid-β-induced tau hyperphosphorylation in human neuroblastoma SH-SY5Y cells involving PTEN/Akt/GSK-3β signaling pathway. *J. Recept. Signal Transduct.* **2014**, *34*, 26–37. [CrossRef]
33. Bustanji, Y.; Taha, M.O.; Almasri, I.M.; Al-Ghussein, M.A.S.; Mohammad, M.K.; Alkhatib, H.S. Inhibition of glycogen synthase kinase by curcumin: Investigation by simulated molecular docking and subsequent in vitro/in vivo evaluation. *J. Enzym. Inhib. Med. Chem.* **2009**, *24*, 771–778. [CrossRef]
34. Small, G.W.; Siddarth, P.; Li, Z.; Miller, K.J.; Ercoli, L.; Emerson, N.D.; Martinez, J.; Wong, K.P.; Liu, J.; Merrill, D.A.; et al. Memory and Brain Amyloid and Tau Effects of a Bioavailable Form of Curcumin in

Non-Demented Adults: A Double-Blind, Placebo-Controlled 18-Month Trial. *Am. J. Geriatr. Psychiatry Off. J. Am. Assoc. Geriatr. Psychiatry* **2018**, *26*, 266–277. [CrossRef]

35. Leighton, B.; Cooper, G.J. Pancreatic amylin and calcitonin gene-related peptide cause resistance to insulin in skeletal muscle in vitro. *Nature* **1988**, *335*, 632–635. [CrossRef]
36. Roth, J.D. Amylin and the regulation of appetite and adiposity: Recent advances in receptor signaling, neurobiology and pharmacology. *Curr. Opin. Endocrinol. Diabetes Obes.* **2013**, *20*, 8–13. [CrossRef]
37. Edvinsson, L.; Goadsby, P.J.; Uddman, R. Amylin: Localization, effects on cerebral arteries and on local cerebral blood flow in the cat. *Sci. World J.* **2001**, *1*, 168–180. [CrossRef]
38. Lorenzo, A.; Razzaboni, B.; Weir, G.C.; Yankner, B.A. Pancreatic islet cell toxicity of amylin associated with type-2 diabetes mellitus. *Nature* **1994**, *368*, 756–760. [CrossRef]
39. May, P.C.; Boggs, L.N.; Fuson, K.S. Neurotoxicity of human amylin in rat primary hippocampal cultures: Similarity to Alzheimer's disease amyloid-beta neurotoxicity. *J. Neurochem.* **1993**, *61*, 2330–2333. [CrossRef]
40. Hull, R.L.; Westermark, G.T.; Westermark, P.; Kahn, S.E. Islet Amyloid: A Critical Entity in the Pathogenesis of Type 2 Diabetes. *J. Clin. Endocrinol. Metab.* **2004**, *89*, 3629–3643. [CrossRef]
41. Zhu, H.; Tao, Q.; Ang, T.F.A.; Massaro, J.; Gan, Q.; Salim, S.; Zhu, R.Y.; Kolachalama, V.B.; Zhang, X.; Devine, S.; et al. Association of Plasma Amylin Concentration with Alzheimer Disease and Brain Structure in Older Adults. *JAMA Netw. Open* **2019**, *2*, e199826. [CrossRef] [PubMed]
42. Daval, M.; Bedrood, S.; Gurlo, T.; Huang, C.J.; Costes, S.; Butler, P.C.; Langen, R. The effect of curcumin on human islet amyloid polypeptide misfolding and toxicity. *Amyloid* **2010**, *17*, 118–128. [CrossRef] [PubMed]
43. Zhao, W.Q.; Townsend, M. Insulin resistance and amyloidogenesis as common molecular foundation for type 2 diabetes and Alzheimer's disease. *Biochim. Biophys. Acta BBA Mol. Basis Dis.* **2009**, *1792*, 482–496. [CrossRef] [PubMed]
44. Rasgon, N.L.; Kenna, H.A.; Wroolie, T.E.; Kelley, R.; Silverman, D.; Brooks, J.; Williams, K.E.; Powers, B.N.; Hallmayer, J.; Reiss, A. Insulin resistance and hippocampal volume in women at risk for Alzheimer's disease. *Neurobiol. Aging* **2011**, *32*, 1942–1948. [CrossRef] [PubMed]
45. Willette, A.A.; Bendlin, B.B.; Starks, E.J.; Birdsill, A.C.; Johnson, S.C.; Christian, B.T.; Okonkwo, O.C.; La Rue, A.; Hermann, B.P.; Koscik, R.L.; et al. Association of Insulin Resistance With Cerebral Glucose Uptake in Late Middle-Aged Adults at Risk for Alzheimer Disease. *JAMA Neurol.* **2015**, *72*, 1013–1020. [CrossRef]
46. Matsuzaki, T.; Sasaki, K.; Tanizaki, Y.; Hata, J.; Fujimi, K.; Matsui, Y.; Sekita, A.; Suzuki, S.O.; Kanba, S.; Kiyohara, Y.; et al. Insulin resistance is associated with the pathology of Alzheimer disease. *Hisayama Study* **2010**, *75*, 764–770.
47. Chuengsamarn, S.; Rattanamongkolgul, S.; Luechapudiporn, R.; Phisalaphong, C.; Jirawatnotai, S. Curcumin extract for prevention of type 2 diabetes. *Diabetes Care* **2012**, *35*, 2121–2127. [CrossRef]
48. Na, L.X.; Li, Y.; Pan, H.Z.; Zhou, X.L.; Sun, D.J.; Meng, M.; Li, X.X.; Sun, C.H. Curcuminoids exert glucose-lowering effect in type 2 diabetes by decreasing serum free fatty acids: A double-blind, placebo-controlled trial. *Mol. Nutr. Food Res.* **2013**, *57*, 1569–1577. [CrossRef]

© 2020 by the authors. Licensee MDPI, Basel, Switzerland. This article is an open access article distributed under the terms and conditions of the Creative Commons Attribution (CC BY) license (http://creativecommons.org/licenses/by/4.0/).

Review

Curcumin and Its Potential Impact on Microbiota

Marzena Jabczyk [1], Justyna Nowak [2,*], Bartosz Hudzik [2,3] and Barbara Zubelewicz-Szkodzińska [1]

1 Department of Nutrition-Related Disease Prevention, Faculty of Health Sciences in Bytom, Medical University of Silesia, 41-900 Bytom, Poland; marzena.jabczyk@gmail.com (M.J.); bzubelewicz-szkodzinska@sum.edu.pl (B.Z.-S.)
2 Department of Cardiovascular Disease Prevention, Faculty of Health Sciences in Bytom, Medical University of Silesia, 41-900 Bytom, Poland; bartekh@mp.pl
3 Silesian Center for Heart Diseases, Third Department of Cardiology, Faculty of Medical Science in Zabrze, Medical University of Silesia, 41-800 Zabrze, Poland
* Correspondence: justyna.nowak@sum.edu.pl; Tel.: +48-323-976-541

Abstract: Curcumin is one of the most frequently researched herbal substances; however, it has been reported to have a poor bioavailability and fast metabolism, which has led to doubts about its effectiveness. Curcumin has antioxidant and anti-inflammatory effects, and has demonstrated favorable health effects. Nevertheless, well-reported in vivo pharmacological activities of curcumin are limited by its poor solubility, bioavailability, and pharmacokinetic profile. The bidirectional interactions between curcumin and gut microbiota play key roles in understanding the ambiguity between the bioavailability and biological activity of curcumin, including its wider health impact.

Keywords: curcumin; antioxidant; microbiota; health

Citation: Jabczyk, M.; Nowak, J.; Hudzik, B.; Zubelewicz-Szkodzińska, B. Curcumin and Its Potential Impact on Microbiota. *Nutrients* **2021**, *13*, 2004. https://doi.org/10.3390/nu13062004

Academic Editor: Roberta Masella

Received: 23 April 2021
Accepted: 8 June 2021
Published: 10 June 2021

Publisher's Note: MDPI stays neutral with regard to jurisdictional claims in published maps and institutional affiliations.

Copyright: © 2021 by the authors. Licensee MDPI, Basel, Switzerland. This article is an open access article distributed under the terms and conditions of the Creative Commons Attribution (CC BY) license (https://creativecommons.org/licenses/by/4.0/).

1. Introduction

Curcumin is a polyphenol substance isolated from the rhizome of Zingiberaceae and Araceae plants. It is a major active constituent of turmeric, a common Asian spice used as a dietary spice, food-coloring, as a herbal remedy, and in the beverage industries. Its bioactive components have been investigated recently [1,2]. Diferuloylmethane (1,7-bis(4-hydroxy-3-methoxyphenyl)-1,6-heptadiene-3,5-dione), which is commonly referred to as curcumin, has been shown to have activity at the cellular level, by signaling multiple molecules. In addition it exerts antioxidant and anti-inflammatory properties. It may have many therapeutic effects [2,3], having exhibited antitumor, chemosensitizing, hepatoprotective, lipid-modifying, and neuroprotective effects [4]. Nevertheless, well-reported in vivo pharmacological activities of curcumin are limited by its poor solubility, bioavailability, and pharmacokinetic profile. As many data have suggested, despite its low absorption, curcumin may demonstrate beneficial effects on health by influencing the intestinal barrier function, sustaining high concentrations in the intestinal mucosa, modulating the functioning of the intestinal barrier, and decreasing high concentrations of bacterial lipopolysaccharide (LPS) levels [5–7]. This review presents the mechanisms of action of curcumin and its potential influence for prevention and treatment by regulating the gut microbiota.

1.1. Curcumin Safety

The Joint United Nations and World Health Organization Expert Committee on Food Additives (JECFA) and the European Food Safety Authority (EFSA) have set the allowable daily intake (ADI) for curcumin at 0–3 mg/kg body weight. Some studies have reported negative side effects in healthy adults that received 500–12,000 mg in a dose response manner [2]. Some studies have showed that nausea, diarrhea, and elevated serum alkaline phosphatase and lactate dehydrogenase levels may be observed in subjects receiving

0.45–3.6 g/day [8]. On the other hand, Shabbir et al. [4] suggested that many studies have shown that it is safe to consume 8 g/day of curcumin.

1.2. Oral Bioavailability of Curcumin

Curcumin belongs to the family of polyphenol compounds, which has at least one aromatic ring structure with at least one hydroxyl group. Polyphenols are found in many vegetables and secondary metabolites, and derive from the shikimic acid pathway [9].

Factors impacting on oral curcumin bioavailability in the diet may be affected by food processing, macronutrients, and origin. In food, polyphenol composition is affected not only by heating changes but also drying, grinding, climate, and plant stress, among others. The poor availability of curcumin has resulted in the development of several curcumin formulations over the last decade, which could improve its bioavailability. Those formulations include curcumin nanoparticles, curcumin in lecithin, phosphatidylcholine carrier, and solid lipid curcumin nanoparticles [7].

As Shabbir et al. [4] emphasized, dietary lipids may affect the solubility and absorption of curcumin. The lipophilic nature of curcumin and the existence of hydroxyl groups in its structure allow it to be metabolized very rapidly in the kidneys and liver. More, it is unstable in most bodily fluids, such as water. Thus, it is recommended to be mixed with oil (or, even, milk) before consumption, in order to improve its absorption [4].

Moreover, Dei Cas and Ghidoni [10] pointed out that, to increase the dietary intake of curcuminoids, turmeric should be paired with lecithin-rich products, such as vegetable oils or eggs. In a recent study, Jardim et al. [11] demonstrated that the addition of lecithin is a promising strategy for improving the properties of nanoparticles such as curcumin, leading to its enhanced efficiency.

Piperine is a natural alkaloid that is found in black pepper (*Piper nigrum*), which is capable of increasing the bioavailability of curcumin by inhibition of biotransformation—especially glucuronidation [10]—in the liver and small intestine [12]. As Hewlings and Kalman [13] emphasized in their work, piperine has been associated with an increase in the bioavailability of curcumin by 2000%.

Interestingly, Schiborr et al. [14] revealed the existence of sex differences, with respect to the plasma levels of oral micellar curcumin. They have demonstrated that women absorbed curcumin to a larger extent than men (1.4-fold higher in women, in comparison to men). The researchers hypothesized that it may be explained by the increased expression and activity of drug efflux in the liver and some enzymes involved in curcumin biotransformation in men. It is worth mentioning that these results may also be due to differences in body weight, blood volume, and body fat. As such, this topic requires further research.

The potential beneficial effects of curcumin depend on its dietary and supplementation intake, as well as the individual's capacity for metabolism, which depends on the biodiversity of their microbiota.

Many studies have reported that high concentrations of curcumin after oral administration have been detected in the gastrointestinal tract. Scazzocchio et al. [15] noted the ambiguity between the low systemic bioavailability and broad pharmacological activity of curcumin. The hypothesis stated that polyphenol directly exerts regulatory effects on microbiota. In turn, Ng et al. [16], in their meta-analysis, suggested another thing which may contribute to the therapeutic effect of curcumin. They emphasized that poor oral bioavailability and the fact that curcumin ingested may be excreted in the feces unmetabolized lead to the point that, after digestion, curcumin reaches the intestine almost unchanged and that may demonstrate hypothetical beneficial effects on the intestinal microflora. Moreover, the potential beneficial effects depend largely on the dietary intake of curcumin, either in terms of an individual's capacity for metabolizing it, or as a consequence of the composition of the intestinal microflora of individuals [15].

1.3. Curcumin Metabolism

The primary sites of metabolism for curcumin are the liver, together with the intestine and gut microbiota. After several reactions, curcumin's double bonds are reduced in hepatocytes and enterocytes, and form dihydrocurcumin, tetrahydrocurcumin, hexahydrocurcumin, and octahydrocurcumin [10,17]. Following oral ingestion, the extensive metabolism of curcumin include reduction, sulfation, and glucuronidation in the liver, kidneys, and intestinal mucosa. 99% of plasma curcumin is present as glucuronide and sulphate conjugate metabolites which are less active [18]. Interestingly, a recent study has shown greater metabolic conjugation and reduction in the human gastrointestinal tract, in comparison to rat intestines [19].

Many other studies have revealed the presence of curcumin metabolites in the human intestinal tract [17]. Due to its resistance to low pH (i.e., stable in the range of pH 2.5 to 6.5), curcumin reaches the large intestine and undergoes extensive metabolic phases [15]. So far, two phases of curcumin metabolism have been reported [10,17,20].

Phase I includes the reduction of the four double bonds of the heptadiene-3,5-dione structure, reduction of curcumin to dihydrocurcumin (DHC), then to tetrahydrocurcumin (THC), and later to hexahydrocurcumin and, finally, octahydrocurcumin.

In phase II which takes place in the intestinal and hepatic cytosol [10], curcumin and its reduced metabolites are conjugated with monoglucuronide, a monosulfate, and then a mixed sulfate/glucuronide (conjugated curcumin), followed by conjugated DHC, conjugated THC, conjugated hexahydrocurcumin and, finally, conjugated octahydrocurcumin [20]. Hexahydrocurcumin and THC are the major products observed in most studies, whereas DHC and octahydrocurcuminol are generally not detected at all [21]. The lack of these two products may be the result of the enzymes responsible for the bioreduction which have been found to reside in the cytosol of the liver and intestine. These compounds may be easily conjugated in vivo and in vitro. For instance, hexahydrocurcumin shares the same phenolic groups or diketo moieties as curcumin, but has no olefinic double bonds. This fact makes it more stable than curcumin at a physiological pH level of 7.4 [22].

Interestingly, the intestinal microflora can deconjugate phase II metabolites and convert them back to the corresponding phase I metabolites, which may also lead to some fission products, such as ferulic acid [23]. Moreover, curcumin also may be metabolized alternatively by intestinal microflora, such as *Escherichia coli* and *Blautia* sp. *E. coli* has been reported to be active by an nicotinamide adenine dinucleotide phosphate (NADPH)-dependent reductase in a two-step reduction pathway from curcumin to DHC, then to THC. In turn, *Blautia* sp. carries out curcumin demethylation into two derivates: demethylcurcumin and bis-demethylcurcumin [10,24]. Curcumin and its reduced metabolites seem to be readily conjugated in vivo and in vitro [21]. Reductive conjugative metabolism of curcumin and an alternative metabolism by intestinal microbiota are shown in Figure 1.

Wang et al. [25] have recently developed a rapid LC-MS/MS (consisting of an LC-20 AD parallel pump, a SIL-20A autosampler, a DGU-20A3R degasser unit, a CTO-20A column oven and an MS-8040 spectrometer) method which evaluates the pharmacokinetics and tissue distribution of curcumin and its metabolites in mice. Fifty male mice were randomly divided into ten groups, and received 20 mg/kg curcumin. After drug administration, blood samples were drawn from the heart at various intervals. The study described decreased THC and DHC in plasma, while THC was present as a main metabolite of curcumin in plasma. Moreover, the results showed that curcumin and THC may be detected in the liver, and curcumin and DHC may be detected in the kidneys. Interestingly, only curcumin was detected in the brain, which (as suggested) means that curcumin may cross the blood-brain barrier [4,25].

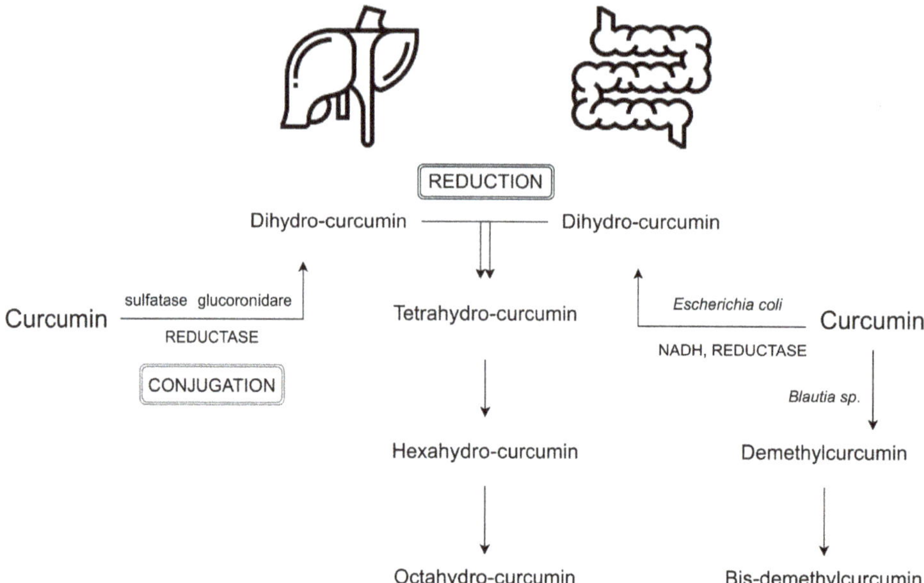

Figure 1. Reductive conjugative metabolism of curcumin and an alternative metabolism by intestinal microbiota; based on [10,20,24]. Abbreviation: NADH - nicotinamide adenine dinucleotide.

In a recent article, Shabbir et al. [4] drew attention to these metabolites and their antioxidant, anti-inflammatory, and neuroprotective effects, considering the gut-brain axis and pathways such as the demethylation, reduction, acetylation, hydroxylation, and demethoxylation of curcumin. They highlighted the results of altered microbial abundance and biodiversity through curcumin biotransformation, exerting indirect health benefits in Alzheimer disease-induced in transgenic mice [26]. This topic needs further study.

1.4. Curcumin and Microbiota

The human intestinal microbiota contains micro-organisms, estimated to consist of more than 1000 bacterial species, which are indispensable for organism physiology and metabolism and, as a consequence, play key roles in maintaining general health [1]. To date, many studies have examined alterations in the composition of microbiota and its link to spectrum of diseases, including diabetes, obesity, inflammatory bowel, liver diseases, depression, psoriasis, and neurogenerative disorders, and the field is growing at a dynamic rate.

Interestingly, despite its poor plasma and tissue bioavailability, preferential distribution and accumulation of curcumin in the intestine has been reported after oral or intraperitoneal administration [1]. Several studies have examined whether curcumin may exert regulative effects on the microbiota community. Human intestinal microbiota can transform curcumin in various metabolism pathways, including producing active metabolites which are able to exert local and systemic effects, but also by reducing the heptadienone backbone and demethylation by *Blautia* spp. [5,24]. As Pandey et al. [17] have suggested, an alternative metabolism of curcumin occurs by the intestinal microflora, especially by *Escherichia coli*. Hassaninasab et al. [20] also identified micro-organisms (strain *Escherichia coli*) capable of converting curcumin, isolated from the feces of two healthy subjects. An NADPH-dependent enzyme (CurA) in *E. coli* converts curcumin into dihydrocurcumin (DHC) and tetrahydrocurcumin (THC) [17]. In turn, the results of a study on *Bifidobacteria* reported by Jazayeri et al. [27] presented that micro-organisms such as *Bifidobacteria pseudocatenulaum, Enterococcus faecalis, Bifidobacteria longum, Lactobacillus acidophilus*, and

Lactobacillus casei are important bacterial strains which are capable of reducing the parent compound of curcumin more than 50% and, in this way, can metabolize curcumin.

Shen et al. [28] reported ananalysis of the intestinal microflora in curcumin-administered mice and controls by pyrosequencing the V3 and V4 regions of the bacterial 16S ribosomal RNA genes. The researchers investigated whether the concentration of curcumin in the intestines following oral administration may result in regulative effects on the intestinal microbiota. One group of mice (n = 6) received a standard diet enriched with natural mixtures isolated from turmeric, which contained the three major parts of curcumin (40.9%), demethoxycurcumin (33.2%), and bidemethoxycurcumin (23.3%), in a dose of 100 mg/kg body weight for 15 days. The control group (n = 6) was supplied with the same feed, but without curcumin gavage. Following the addition of curcumin, significant increases ($p < 0.05$) in several representative families in the gut, including *Prevotellaceae, Bacterioidaceae*, and *Rikenellaceae,* were recorded. The abundance of *Prevotellaceae* decreased significantly in the curcumin group, in comparison to controls (from 15.48% to 6.16%; $p = 0.01$). A significant reduction was also reported in *Prevotella* abundance (from 13.29% to 4.63%; $p = 0.00$), Conversely, *Bacteroidaceae* in the curcumin group was significantly higher in comparison to the control group (3.21% to 1.15%; $p = 0.00$). Similarly, the amount of *Rikenellaceae* increased (from 4.73% to 7.96%; $p = 0.04$) with *Alistipes* abundance (from 4.73% to 7.96%; $p = 0.04$). These results indicated that, despite no significant alteration, oral administration of curcumin tended to decrease the microbiome biodiversity. More studies are required to extend the current microbiome outcomes, in order to prove the therapeutic application of curcumin for the gut microbiome.

The action of curcumin (as well as resveratrol and simvastatin) has also been examined in animals affected by *Toxoplasma gondii*. It was found that curcumin, resveratrol, and simvastatin-treated animals had less proinflammatory *Enterobacteria* and *Enterococci* by 3.0–3.5 ($p < 0.000001$) and 1.0–1.5 ($p < 0.0005$), respectively. The potentially anti-inflammatory *Lactobacilli* or *Bifidobacteria* were slightly increased, by up to 1.0 log orders ($p < 0.05$–0.0005) [29].

The influence of curcumin on the intestinal microbiota is still not clear. However, current evidence indicates that curcumin is biotransformed not only in the gut to various metabolites via different pathways which include among others demethylation, hydroxylation, and demethoxylation. These metabolites have been reported to be even more bioactive than the parent curcumin, but also the gut microbiota may be regulated by administration of curcumin and lead to changes in the biodiversity and composition of microbes. This plays a key role in potential indirect health benefits [1]. The bi-directional interplay between curcumin and gut microflora is presented in Figure 2.

1.5. Curcumin and Microbiota in Liver Disease

Feng et al. [30] have investigated, using taxon-based analysis, the correlation between changes in gut microbiota regulation in curcumin-mediated attenuation and the development of non-alcoholic fatty liver disease in rats. Their data revealed that curcumin treatment significantly changed the gut microbial composition. The operational taxonomic units that were altered by curcumin treatment were related to hepatic steatosis parameters. Types such as *Spirochateae, Tenericutes*, and *Elusimicrobia* were decreased, conversely to *Actinobacteria*, which were markedly increased. In their discussion, they paid attention to the immunomodulatory effects of curcumin, due to its alteration in intestinal bacterial communities, which led to the relative abundance of short-chain fatty acid (SCFA) producing bacteria (including *Blautia* and *Allobaculum*). That may contribute to the alleviation of inflammation, insulin resistance, and obesity [31]. As a consequence, it may serve as a prevention target in mucosal abnormalities and hepatic steatosis. Midura-Kiela et al. [32] and Feng et al. [30] have reported that curcumin acts as an interferon gamma (IFN-γ) signaling inhibitor in colonocytes.

CURCUMIN

- promotes beneficial bacterial strains
- improves intestinal barrier functions
- positively alters microbial biodiversity and composition
- counteracts the expression of pro-inflammatory mediators

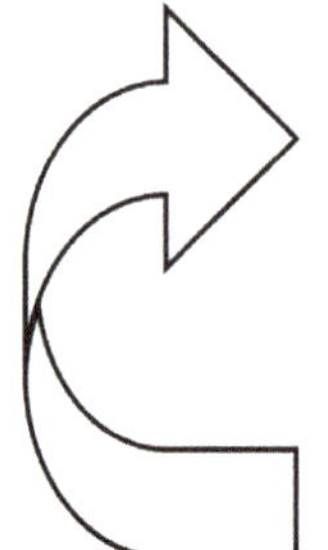

GUT MICROBIOTA

- activates biological pathways through transformation of curcumin: demethylation, hydroxylation, demethoxylation (by produced metabolites)

Figure 2. Interplay between curcumin and gut microbiota; based on [1,4,15,23].

1.6. *Curcumin and Microbiota in Colitis*

Two independent studies [33,34] have explored the modulative effects of nanoparticle curcumin administration on colonic microbiota during colitis, and suggested a modulating structure of intestine biodiversity. Interestingly, the results of randomized, double-blind, placebo-controlled study published in 2021 showed that 8 weeks administration of Curcugen™, a curcumin extract (DolCas Biotech, LLC, 9 Lenel Rd, Landing, NJ 07850, USA), was correlated with an improvement in gastrointestinal symptoms (abdominal pain, indigestion, diarrhea, and constipation) in adults [35]. The curcumin extract Curcugen™ contains 98.5% turmeric-based ingredients (50% curcuminoids, 1.5% essential oils, and other native turmeric molecules). Additionally, the potential mechanisms associated with curcumin's influence on microbial diversity have been examined. In comparison to placebo, no significant changes in the biodiversity (from the phylum to genus level) of bacteria which are regularly observed in adults with irritable bowel syndrome (IBS) were reported. These results could suggest that changes in the gut microbiota are not responsible for improving gastrointestinal symptoms in adults. In another previous study by McFadden et al., curcumin elevated microbial richness, prevented alpha diversity decrease, and increased the proportion of *Lactobacillales*, while conversely decreasing the proportion of *Coriobacterales* in mice, during colitis and colon cancer prevention [34].

1.7. *Curcumin and Microbiota and Urinary Metabolism*

Another interesting in vivo study has been carried out in healthy adults [36]. The aim of this study was to describe the potential effects of curcumin on changes in 24-hour urinary composition in healthy volunteers in response to daily consumption of a dried *Curcuma longa* extract containing a standardized amount of curcuminoid (equivalent to 100 mg of curcuminoids) for 28 days. Their results showed curcumin-induced alterations in urinary metabolites. Curcumin and two metabolic derivates (HTC and DHC) were detected in the urine, this fact points to the absorption of the major curcuminoid from the extract and its further metabolism by the liver and intestinal microflora.

Peterson et al. [5], in their pilot study, examined gut microflora profiles in healthy humans. All study participants were blinded to the treatments they received (curcumin tablets, turmeric, or placebo). The results concluded that the respondent's microbiota response was individualized over time; however, the comparison of microbiota alterations in the turmeric and curcumin supplementation groups were highly similar. The authors interpreted this observation of the turmeric response as a reflection of the catabolism of polysaccharide compounds in the root, involving a wide range of glycosyl hydrolases encoded by the species, which were elevated in the responsive subjects (*Bacteroides, Bifidobacterium, Alistipes, Parabacteroides*).

1.8. Curcumin and Microbiota in Exercise Performance

There is some evidence suggesting that exercise training has an effect on gut microbiota, through its impact on the autonomic nervous system (vagal tone) by improving gut microbiota composition [37]. Chen et al. [38] investigated the effect of nano-bubble curcumin extract (NCE-5x) supplementation on microbiota and exercise performance in mice. They found that six weeks of nano-bubble curcumin extract treatment have modulated the gut microbiota composition and led to anti-fatigue effects, which resulted in improved exercise performance in the mice. Animals following the nano-curcumin extract had a reduced amount of *Bacteroidetes* and an elevated amount of *Firmicutes*. The number of bacteria from the family *Lactobacillaceae* and *Lactobacillus* genus was higher in the NCE-5x group, in comparison to the vehicle group. However, the amount of *Clostridiales* and *Allobaculum* were lower in the NCE-5x group than in the vehicle group. Therefore, there seems to be a definite need for establishing an optimal dosage of curcumin which is safe and effective, as a result of further studies in this field, especially in humans.

1.9. Curcumin and Microbiota in Dental Disease

Interestingly, some studies have shown the effects of curcumin on oral bacteria associated with dental disease. Li et al. [39] evaluated the inhibitory properties of curcumin on the metabolism and the formation of a biofilm in planktonic *Streptococcus mutans*, which has been shown to be a main etiological factor in dental caries [40]. They exhibited its therapeutic antibacterial activity by proving that curcumin decreased the biofilm metabolism after both 5 min and 24 h, thus inhibiting the number of live and total bacteria in the short- and long-term. The in vitro biofilm models of *S. mutans* were treated with different concentrations of curcumin. To analyze the composition and extracellular polysaccharide content of *S. mutans* biofilm after curcumin treatment, the researchers used confocal laser scanning microscopy.

Moreover, their study revealed that curcumin treatment decreased not only the two-component transduction system but also metabolism adherence, carbohydrate metabolism, and the expression of genes that are linked to the synthesis of extracellular polysaccharide. Curcumin did not demonstrate any effect on the growth kinetics of *S. mutans*, so the reduced number of alive microorganisms in the study indicated a direct effect of curcumin on *S. mutans*. Despite the fact that Li et al. have explored the antibacterial effect of curcumin on *S. mutans*, it is important to take into account that the bacterial strains of *S. mutans* are heterogenic, with a variety of strains. Therefore, many more strains of *S. mutans* in the future should be assessed, in order to obtain more relevant results [39]. Table 1 lists some of the alterations in the number of gut microbes after curcumin treatment.

Table 1. Alteration in gut microbiota after curcumin treatment.

Dose of Curcumin	Alterations in Gut Microbiota	p	Author, Year, [Reference]
100 mg/kg body weight of mice (15 days)	Decrease: *Prevotellaceae* (15.48% -> 6.16%) *Prevotella* (13.29% -> 4.63%) Increase: *Bacteroidaceae* (3.21% -> 1.15%) *Rikenellaceae* (4.73% -> 7.96%) *Alistipes* (4.73% -> 7.96%) *Bacteroides* (1.15% -> 3.21%)	 $p = 0.01$ $p = 0.00$ $p = 0.00$ $p = 0.04$ $p = 0.04$ $p = 0.00$	Shen L., 2017 [28]
200 mg/kg body weight of rats (4 weeks) after HFD (high-fat diet)	Decrease: *Spirochaetae* (0.0091%) *Tenericutes* (0.013%) *Elusimicrobia* (0.0045%) Increase: *Actinobateria* (7.47%) *Collinsella* (7.18%) *Streptococcus* (0.66%) *Suterella* (0.23%) *Gemella* (0.09%) *Thalassospira* (0.26%) *Gordonibacter* (0.071%) *Actinomyces* (0.038%)	$p < 0.05$	Feng W., 2017 [30]
8–162 mg/kg body weight of mice during colitis and colon cancer prevention (the entire study lasted 30 weeks)	Decrease: *Coriobacterales* Increase: *Lactobacillales*		McFadden R.M.T., 2015 [34]
Pilot study in humans (three groups: placebo, turmeric, and curcumin) tablets 1000 mg of curcumin + 1.25 mg black pepper)	Decrease: *Blautia* spp. Increase: *Clostridium* spp. *Bacteroides* spp. *Citrobacter* spp. *Cronobacter* spp. *Enterobacter* spp. *Enterococcus* spp. *Klebsiella* spp. *Parabacteroides* spp. *Pseudosomonas* spp.		Peterson C.T., 2018 [5]
100 mg/kg body weight of rats (three groups: ovariectomized, sham operation, curcumin)	Decrease: *Anaerotruncus* *Helicobacter* Increase: *Serratia* *Shewanella* *Pseudomonas* *Papillibacter* *Exiguobacterium*	 $p = 0.004$ $p = 0.049$ $p = 0.002$ $p = 0.006$ $p = 0.014$ $p = 0.029$ $p = 0.032$	Zhang Z., 2017 [41]
Male mice divided into 3 groups: Vehicle, 0; NCE-1x, 3.075 g/kg^{-1} day^{-1}; NCE-5x, 15.375 g/kg^{-1} day^{-1}	Increased: *Firmicutes* *Lactobacillaceae* *Lactobacillus* Decreased: *Bacteroidetes* *Clostridiales* *Allobaculum*	$p < 0.05$	Chen Y-M., 2020 [38]

Abbreviations: HFD—high fat diet; NCE-1x—nano-bubble curcumin extract-1x; NCE-5x—nano-bubble curcumin extract-5x.

1.10. Curcumin and Metabolic Health

Numerous studies have shown that the changes in the microbiome are associated with many metabolic conditions which include among others obesity, diabetes, and chronic liver disease. These entities may be linked with changes in intestinal microbiome [28]. The first study reporting on the association between curcumin ingestion and the diversity of gut microbiome in a menopausal rat model was published in 2017 [41]. Their results suggested that curcumin may partially reverse the alterations in the biodiversity of rat gut microbiota by the changed distribution of intestinal microflora due to estrogen deficiency induced by ovariectomy. Curcumin promoted increases in the number of species of *Serratia, Shewanella, Pseudomonas, Papillibacter,* and *Exiguobacterium*, and decreases in *Anaerotruncus* and *Helicobacter pylori*. Moreover, Wang et al. [6] have delineated the underlying mechanism of the association of curcumin with metabolic diseases and chronic inflammation, through modulating the function of intestinal epithelial cells (IECs) and the intestinal barrier function. They used the human IEC lines immortalized cell line of human colorectal adenocarcinoma cells (Caco-2) and human colorectal adenocarcinoma cell line with epithelial morphology (HT-29), in this case examining the modulation effects of LPS on intracellular signaling as well as tight junctions. The presented results showed that curcumin attenuated the activation of IECs as well as macrophages and as a consequence led to decreased secretion of interleukin 1β (IL-1β). It is worth mentioning that IECs are activated by luminal LPS, which stimulates (through intestinal macrophages) proinflammatory cytokines. Circulating LPS levels and metabolic diseases (e.g., type 2 diabetes and atherosclerosis) have been associated. Western-type diets rich in high-fat foods induce alterations in the gut microbiota by releasing bacteria-derived products (e.g., LPS). Zam W. [42], in his review, mentioned that treatment with the nanoparticle curcumin (named Theracurmin) reduced mucosal mRNA expression of inflammatory mediators and the activation of nuclear factor kappa-light-chain-enhancer of activated B cells (NF-κB) in colonic epithelial cells. This has resulted in the elevation of the amount of butyrate-producing bacteria and fecal butyrate levels. The results of this study have been examined in experimental colitis mice [33].

Kato et al. [43] used curcumin dispersed in colloidal nanoparticles and stimulated the secretion of glucagon-like peptide 1 (GLP-1) thus leading to an increased synthesis and secretion of insulin, which resulted in a better glycemia control. Oral administration of highly dispersible and bioavailable curcumin, through the stimulation of insulin secretion in vivo, resulted in lower glycemia. Curcumin was administrated prior to IP glucose injection to confirm that the hypoglycemic and insulin secretion effects were linked to the stimulation of GLP-1 secretion. This finding suggests a potential role of curcumin in the treatment of diabetes mellitus. However, the use of native curcumin did not achieve therapeutic outcomes and did not improve glucose tolerance in mice.

Another interesting study has determined the effects of different nano-curcumin levels, and have revealed beneficial effects on growth, cholesterol levels, blood constitution, antioxidant indices, immunity, and growth of quails, as well as increased amounts of lactic acid bacteria and reduced amounts of pathogenic bacteria. In the study conducted by Reda et al. [44], liver function reflected by the alanine aminotransferase (ALT) and asparate aminotransferase (AST) levels significantly improved following the addition of dietary nano-curcumin. The addition of any amount of nano-curcumin in the quail diet also significantly improved the lipid profile. Moreover, the caecal microbiota of quail was altered, following the supplementation of nano-curcumin levels. Reductions in *Enterobacter, Salmonella,* and *E. coli* counts were observed. However, they highlighted the use of nanoparticles, such as nanocurcumin, which more easily pass through cell membranes and interplay promptly with biological systems.

2. Summary

Polyphenols, such as curcuminoids, are naturally occurring bioactive compounds that, due to their antioxidant abilities, play important roles in human nutrition. A substantial amount of promising evidence has indicated that curcumin may be capable of preventing and combating several metabolic syndromes, cancer, and obesity, and may even play a neuroprotective role. The metabolism of curcumin, which occurs in the intestine, enhances its biological activity and, as a consequence, biotransforms it into active metabolites, which may promote beneficial effects in the gut microbiota. However, there are also many limitations, such as its poor bioavailability, determining the dosage level of curcumin to achieve the optimal health results, and limitations related to animal studies. Currently, based on available data, there are several contraindications for use of cur-cumin. Curcumin should be used with caution given the dual effects of curcumin on alcoholic liver injury in people with liver disease, like cirrhosis, biliary obstruction, gallstones, biliary colic, and people abusing alcohol. In addition, curcumin may interact with nonsteroidal anti-inflammatory drugs, reserpine, and blood thinners. Therefore, the use of curcumin in people who are taking those drugs is also not recommended. Another important limitation is that curcumin cannot be tested in randomized placebo-controlled studies. Hence, there are much more difficulties to conduct research in any given clinical condition. Moreover, curcumin metabolism may be different from person to person due to the individual's variety of microorganism content. At present, extrapolation of animal studies has lead to the recommendation of administering oral curcumin in the amount of about 500 mg per day (about 4 g raw curcumin). The use of nano-curcumin is one of the possible mechanism for enhancing the bioavailability of curcumin increasing its absorption. Nanoparticles such as nanocurcumin seem to give the best effects on gut microbiota and have shown the best properties for the body.

In summary, many studies have indicated that a curcumin intervention using a high-bioavailability formulation would be more effective than treatment with standard curcumin, even considering a longer period of time. Additionally, some considerations for the effective dosage might be specific to the particular formulation in question, as well as the bioavailability of the curcumin within the compound. Future studies should focus on providing more comprehensive data concerning high-bioavailability curcumin supplementation, especially in humans. There is also a need to better understand the bi-directional metabolism of curcumin, as well as its potential beneficial effects on microbiota and overall health.

Author Contributions: Conceptualization, J.N. and M.J.; methodology, M.J.; data curation, M.J., B.H., and B.Z.-S.; writing—original draft preparation, M.J. and J.N.; writing—review and editing, J.N., B.H., and B.Z.-S.; visualization, M.J. and B.H.; supervision, B.H. and B.Z.-S.; project administration, M.J. and J.N.; funding acquisition, J.N. All authors have read and agreed to the published version of the manuscript.

Funding: This research was funded by Grant by the Medical University of Silesia (KNW-2K46/D/9/N)

Institutional Review Board Statement: Not applicable.

Informed Consent Statement: Not applicable.

Conflicts of Interest: The authors declare no conflict of interest.

References

1. Shen, L.; Ji, H.F. Bidirectional interactions between dietary curcumin and gut microbiota. *Crit. Rev. Food Sci. Nutr.* **2019**, *18*, 2896–2902. [CrossRef] [PubMed]
2. Shabbir, U.; Rubab, M.; Daliri, E.B.-M.; Chelliah, R.; Javed, A.; Oh, D.H. Settings Open Access Review Curcumin, Quercetin, Catechins and Metabolic Diseases: The Role of Gut Microbiota. *Nutrients* **2021**, *13*, 206. [CrossRef] [PubMed]
3. Nisari, M.; Yilmaz, S.; Ertekin, T.; Hanim, A.Y.; Ceylan, D.; Inanc, N.; Al, Ö.; Ülger, H. Effects of curcumin on lipid peroxidation and antioxidant enzymes in kidney, liver, brain and testis of mice bearing Ehrlich Solid Tumor. *Adv. Hyg. Exp. Med.* **2020**, *74*, 416–425. [CrossRef]
4. Shabbir, U.; Rubab, M.; Tyagi, A.; Oh, D.H. Curcumin and Its Derivatives as Theranostic Agents in Alzheimer's Disease: The Implication of Nanotechnology. *Int. J. Mol. Sci.* **2021**, *22*, 1.
5. Peterson, C.T.; Vaughn, A.R.; Sharma, V.; Chopra, D.; Mills, P.J.; Peterson, S.N.; Sivamani, R.K. Effects of Turmeric and Curcumin Dietary Supplementation on Human Gut Microbiota: A Double-Blind, Randomized, Placebo-Controlled Pilot Study. *J. Evid. Based Integr. Med.* **2018**, *23*, 2515690X18790725. [CrossRef] [PubMed]

6. Wang, J.; Siddhartha, S.; Ghosh, S. Curcumin improves intestinal barrier function: Modulation of intracellular signaling, and organization of tight junctions. *Am. J. Physiol. Cell Physiol.* **2017**, *312*, C438–C445. [CrossRef]
7. Adiwidjaja, J.; McLachlan, A.J.; Boddy, A.V. Curcumin as a clinically-promising anti-cancer agent: Pharmacokinetics and drug interactions. *Expert Opin. Drug Metab. Toxicol.* **2017**, *13*, 953–972. [CrossRef]
8. Sharma, R.A.; Euden, S.A.; Platton, S.L.; Cooke, D.N.; Shafayat, A.; Hewitt, H.R.; Marczylo, T.H.; Morgan, B.; Hemingway, D.; Plummer, S.M.; et al. Phase I clinical trial of oral curcumin: Biomarkers of systemic activity and compliance. *Clin. Cancer Res.* **2004**, *10*, 6847–6854. [CrossRef]
9. Di Meo, F.; Margarucci, S.; Galderisi, U.; Crispi, S.; Peluso, G. Curcumin, Gut Microbiota, and Neuroprotection. *Nutrients* **2019**, *11*, 2426. [CrossRef]
10. Dei Cas, M.; Ghidoni, R. Dietary Curcumin: Correlation between Bioavailability and Health Potential. *Nutrients* **2019**, *11*, 92147. [CrossRef]
11. Jardim, K.V.; Neves Siqueira, J.L.; Bao, S.N.; Sousa, M.H.; Parize, A.L. The role of the lecithin addition in the properties and cytotoxic activity of chitosan and chondroitin sulfate nanoparticles containing curcumin. *Carbohydr. Polym.* **2020**, *227*, 115351. [CrossRef]
12. Maryam, S.H.-Z.; Sarhadi, M.; Zarei, M.; Thilagavathi, R.; Selvam, C. Synergistic effects of curcumin and its analogs with other bioactive compounds: A comprehensive review. *Eur. J. Med. Chem.* **2021**, *210*, 113072.
13. Hewlings, S.J.; Kalman, D.S. Curcumin: A Review of Its Effects on Human Health. *Foods* **2017**, *6*, 92. [CrossRef]
14. Schiborr, C.; Kocher, A.; Behnam, D.; Jandasek, J.; Toelstede, S.; Frank, J. The oral bioavailability of curcumin from micronized powder and liquid micelles is significantly increased in healthy humans and differs between sexes. *Mol. Nutr. Food Res.* **2014**, *58*, 3. [CrossRef]
15. Scazzocchio, B.; Minghetti, L.; D'Archivio, M. Interaction between Gut Microbiota and Curcumin: A New Key of Understanding for the Health Effects of Curcumin. *Nutrients* **2020**, *12*, 2499. [CrossRef]
16. Ng, Q.X.; Sof, A.Y.S.; Loke, W.; Venkatanarayanan, N.; Lim, D.Y.; Yeo, W.S. A Meta-Analysis of the Clinical Use of Curcumin for Irritable Bowel Syndrome (IBS). *J. Clin. Med.* **2018**, *7*, 298. [CrossRef]
17. Pandey, A.; Chaturvedi, M.; Mishra, S.; Kumar, P.; Somvanshi, P.; Chaturvedi, R. Reductive metabolites of curcumin and their therapeutic effects. *Heliyon* **2020**, *6*, e05469. [CrossRef]
18. Paulraj, F.; Abas, F.; Lajis, N.H.; Othman, I.; Naidu, R. Molecular Pathways Modulated by Curcumin Analogue, Diarylpentanoids in Cancer. *Biomolecules* **2019**, *9*, 270. [CrossRef]
19. Ireson, C.R.; Jones, D.J.L.; Orr, S.; Coughtrie, M.W.H.; Boocock, D.J.; Williams, M.L.; Farmer, P.B.; Steward, W.P.; Gescher, A.J. Metabolism of the cancer chemopreventive agent curcumin in human and rat intestine. *Cancer Epidemiol. Biomark. Prev.* **2002**, *11*, 105–111.
20. Hassaninasab, A.; Hashimoto, Y.; Tomita-Ypokotani, K.; Koboyashi, M. Discovery of the curcumin metabolic pathway involving a unique enzyme in an intestinal microorganism. *Proc. Natl. Acad. Sci. USA* **2011**, *108*, 6615–6620. [CrossRef]
21. Hoehle, S.I.; Pfeiffer, E.; Solyom, A.M.; Metzler, M. Metabolism of Curcuminoids in Tissue Slices and Subcellular Fractions from Rat Liver. *J. Agric. Food Chem.* **2006**, *54*, 756–764. [CrossRef] [PubMed]
22. Xiao-Yu, X.; Xiao, M.; Li, S.; Ren-You, G.; Ya, L.; Hua-Bin, L. Bioactivity, Health Benefits, and Related Molecular Mechanisms of Curcumin: Current Progress, Challenges, and Perspectives. *Nutrients* **2018**, *10*, 1553.
23. Pluta, R.; Januszewski, S.; Ułamek-Kozioł, M. Mutual Two-Way Interactions of Curcumin and Gut Microbiota. *Int. J. Mol. Sci.* **2020**, *5*, 1055. [CrossRef] [PubMed]
24. Burapan, S.; Kim, M.; Han, J. Curcuminoid Demethylation as an Alternative Metabolism by Human Intestinal Microbiota. *J. Agric. Food Chem.* **2017**, *65*, 3305–3310. [CrossRef]
25. Wang, J.; Yu, X.; Zhang, L.; Wang, L.; Peng, Z.; Chen, Y. The pharmacokinetics and tissue distribution of curcumin and its metabolites in mice. *Biomed. Chromotography* **2018**, 32, 9. [CrossRef]
26. Sun, Z.Z.; Li, X.Y.; Wang, S.; Shen, L.; Ji, H.F. Bidirectional interactions between curcumin and gut microbiota in transgenic mice with Alzheimer's disease. *Appl. Microbiol. Biotechnol.* **2020**, *104*, 3507–3515. [CrossRef]
27. Jazayeri, S.D. Survival of Bifidobacteria and other selected intestinal bacteria in TPY medium supplemented with Curcumin as assessed in vitro. *Int. J. Probiotics Prebiorics* **2009**, *4*, 15–22.
28. Shen, L.; Liu, L.; Hong-Fang, J. Regulative effects of curcumin spice administration on gut microbiota and its pharmacological implications. *Food Nutr. Res.* **2017**, *61*, 1361780. [CrossRef] [PubMed]
29. Bereswill, S.; Munoz, M.; Fisher, A.; Plickert, R.; Haag, L.M.; Otto, B.; Kuhl, A.A.; Loddenkemper, C.; Gobel, U.B.; Heimesaat, M.M. Anti-Inflammatory Effects of Resveratrol, Curcumin and Simvastatin in Acute Small Intestinal Inflammation. *PLoS ONE* **2010**, *5*, e15099. [CrossRef] [PubMed]
30. Feng, W.; Wang, H.; Zhang, P.; Gao, C.; Tao, J.; Ge, Z.; Zhu, D.; Bi, Y. Modulation of gut microbiota contributes to curcumin-mediated attenuation of hepatic steatosis in rats. *Biochim. Biophys. Acta* **2017**, *1861*, 1801–1812. [CrossRef]
31. Zhao, Y.; Zhang, M.; Pang, X.; Xu, J.; Kang, C.; Li, M.; Zhang, C.; Zhang, Z.; Zhang, Y.; Li, X.; et al. Structural Changes of Gut Microbiota during Berberine-Mediated Prevention of Obesity and Insulin Resistance in High-Fat Diet-Fed Rats. *PLoS ONE* **2012**, *7*, e42529.
32. Midura-Kiela, M.T.; Radhakrishnan, V.M.; Marmonier, C.B.; Laubtiz, D.; Ghishan, F.K.; Kiela, P.R. Curcumin inhibits interferon-γ signaling in colonic epithelial cells. *Am. J. Physiol. Gastrointest. Liver Physiol.* **2012**, *302*, G86–G98. [CrossRef]

33. Ohno, M.; Nishidia, A.; Sugitani, Y.; Nishino, K.; Inatomi, O.; Sugimoto, M.; Kawahara, M.; Andoh, A. Nanoparticle curcumin ameliorates experimental colitis via modulation of gut microbiota and induction of regulatory T cells. *PLoS ONE* **2017**, *12*, e0185999. [CrossRef]
34. McFadden, R.M.T.; Larmonier, C.B.; Shehab, K.W.; Midura-Kiela, M.; Ramalingam, R.; Harrison, C.A.; Besselsen, D.G.; Chase, J.H.; Caporaso, J.G.; Jobin, C.; et al. The Role of Curcumin in Modulating Colonic Microbiota during Colitis and Colon Cancer Prevention. *Inflamm. Bowel Dis.* **2015**, *21*, 2483–2494. [CrossRef]
35. Lopresti, A.L.; Smith, S.J.; Rea, A.; Michel, S. Efficacy of a curcumin extract (Curcugen™) on gastrointestinal symptoms and intestinal microbiota in adults with self-reported digestive complaints: A randomized, double-blind, placebo-controlled study. *BMC Complementary Med. Ther.* **2021**, *21*, 4.
36. Peron, G.; Sut, S.; Dal Ben, S.; Voinovich, D.; Dall'Acqua, S. Untargeted UPLC-MS metabolomics reveals multiple changes of urine composition in healthy adult volunteers after consumption of curcuma longa L. extract. *Food Res. Int.* **2020**, *127*, 108730. [CrossRef]
37. Campbell, M.S.; Carlini, N.A.; Fleenor, B.S. Infulence of curcumin on performance and post-exercise recovery. *Crit. Rev. Food Sci. Nutr.* **2020**, *61*, 1152–1162. [CrossRef]
38. Chen, Y.-M.; Chiu, W.-C.; Chiu, Y.-S.; Li, T.; Sung, H.-S.; Hsiao, C.-Y. Supplementation of nano-bubble curcumin extract improves gut microbiota composition and exercise performance in mice. *Food Funct.* **2020**, *11*, 3574–3584. [CrossRef]
39. Li, B.; Li, X.; Lin, H.; Zhou, Y. Curcumin as a Promising Antibacterial Agent: Effects on Metabolism and Biofilm Formation in S. mutans. *BioMed Res. Int.* **2018**, *3*, 1–11. [CrossRef]
40. Bowen, W.H.; Koo, H. Biology of Streptococcus mutans-Derived Glucosyltransferases: Role in Extracellular Matrix Formation of Cariogenic Biofilms. *Caries Res.* **2011**, *45*, 69–86. [CrossRef]
41. Zhang, Z.; Chen, Y.; Xiang, L.; Wang, Z.; Xiao, G.G.; Hu, J. Effect of Curcumin on the Diversity of Gut Microbiota in Ovearectomized Rats. *Nutrients* **2017**, *9*, 1146. [CrossRef] [PubMed]
42. Zam, W. Gut Microbiota as a Prospective Therapeutic Target for Curcumin: A Review of Mutual Influence. *J. Nutr. Metab.* **2018**, *2018*, 1367984. [CrossRef] [PubMed]
43. Kato, M.; Nishikawa, S.; Ikehata, A.; Dochi, K.; Tani, T.; Takahashi, T.; Imaizumi, A.; Tsuda, T. Curcumin improves glucose tolerance via stimulation of glucagon-like peptide-1 secretion. *Mol. Nutr. Food Res.* **2016**, *61*, 3. [CrossRef] [PubMed]
44. Reda, F.M.; El-Saadony, M.T.; Elnesr, S.S.; Alagawany, M.; Tufarelli, V. Effect of Dietary Supplementation of Biological Curcumin Nanoparticles on Grwoth and Carcass Traits, Antioxidant Status, Immunity and Caecal Microbiota of Japanese Quails. *Animals* **2020**, *10*, 754. [CrossRef]

Review

Curcuma Longa, the "Golden Spice" to Counteract Neuroinflammaging and Cognitive Decline—What Have We Learned and What Needs to Be Done

Alessandra Berry [1,*,†], Barbara Collacchi [1,†], Roberta Masella [2], Rosaria Varì [2] and Francesca Cirulli [1,*]

1 Center for Behavioral Sciences and Mental Health, Istituto Superiore di Sanità, Viale Regina Elena 299, 00161 Rome, Italy; barbara.collacchi@iss.it
2 Center for Gender-specific Medicine, Istituto Superiore di Sanità, Viale Regina Elena 299, 00161 Rome, Italy; roberta.masella@iss.it (R.M.); rosaria.vari@iss.it (R.V.)
* Correspondence: alessandra.berry@iss.it (A.B.); francesca.cirulli@iss.it (F.C.)
† These authors contributed equally to this work.

Abstract: Due to the global increase in lifespan, the proportion of people showing cognitive impairment is expected to grow exponentially. As target-specific drugs capable of tackling dementia are lagging behind, the focus of preclinical and clinical research has recently shifted towards natural products. Curcumin, one of the best investigated botanical constituents in the biomedical literature, has been receiving increased interest due to its unique molecular structure, which targets inflammatory and antioxidant pathways. These pathways have been shown to be critical for neurodegenerative disorders such as Alzheimer's disease and more in general for cognitive decline. Despite the substantial preclinical literature on the potential biomedical effects of curcumin, its relatively low bioavailability, poor water solubility and rapid metabolism/excretion have hampered clinical trials, resulting in mixed and inconclusive findings. In this review, we highlight current knowledge on the potential effects of this natural compound on cognition. Furthermore, we focus on new strategies to overcome current limitations in its use and improve its efficacy, with attention also on gender-driven differences.

Keywords: turmeric; aging; brain; cognition; bioavailability; oxidative stress; inflammation

Citation: Berry, A.; Collacchi, B.; Masella, R.; Varì, R.; Cirulli, F. *Curcuma Longa*, the "Golden Spice" to Counteract Neuroinflammaging and Cognitive Decline—What Have We Learned and What Needs to Be Done. *Nutrients* **2021**, *13*, 1519. https://doi.org/10.3390/nu13051519

Academic Editor: Giuseppe Grosso

Received: 31 March 2021
Accepted: 26 April 2021
Published: 30 April 2021

Publisher's Note: MDPI stays neutral with regard to jurisdictional claims in published maps and institutional affiliations.

Copyright: © 2021 by the authors. Licensee MDPI, Basel, Switzerland. This article is an open access article distributed under the terms and conditions of the Creative Commons Attribution (CC BY) license (https://creativecommons.org/licenses/by/4.0/).

1. Introduction

Cognitive decline is a highly disabling and prevalent condition in the aging population, greatly affecting physical health and quality of life. Global average life expectancy, as observed in 2019 by the Global Health Observatory (GHO), was estimated to be 73.4 years in the WHO European Region (https://www.who.int/gho/mortality_burden_disease/life_tables/situation_trends_text/en/ accessed on 25 April 2021). In 2050, the number of people over the age of 60 is expected to reach a total of about 2.1 billion (https://www.who.int/ageing/publications/active_ageing/en/ accessed on 25 April 2021). As the ageing population is rapidly growing due to the global increase in life expectancy in westernized life-style countries, the number of people experiencing cognitive impairment is also expected to grow in parallel. Disregarding overt pathologies, the impact of age itself on cognitive abilities is so disruptive and so underestimated that it has been described as "the elephant in the room" [1,2].

The lack of effective pharmacotherapy has led researchers to seek alternative approaches in order to treat or prevent the cognitive decline accompanying ageing. Accumulating evidence suggests that conditions co-occurring in metabolic dysfunctions such as neuroinflammation, oxidative stress (OS), mitochondrial dysfunction or autophagy may all potentially act as triggers for cognitive decline. Indeed, metabolic syndrome (MetS, defined as the presence of three or more of the following five medical conditions: abdominal obesity, high blood pressure, high blood sugar, high serum triglycerides (TG) and low serum high-density lipoprotein—HDL), negatively impacts cognitive performance and brain function possibly increasing neuroinflammation, OS and brain lipid metabolism [3]. Insulin resistance (IR, defined as the inability of peripheral target tissues to respond normally to insulin) is a common condition experienced at old age and often associated with obesity. It typically precedes the onset of type 2 diabetes (T2D) by several years and is considered as a risk factor for cognitive decline in both diabetic and non-diabetic populations [4]. In fact, peripheral IR, by decreasing insulin signaling within the brain, may alter its metabolic functions, increasing OS and neuroinflammation, eventually setting the stage for dementia and neurodegeneration. Thus, neuroinflammation, OS and metabolic dysfunctions involve a strict connection between the brain and the overall metabolic regulation that occur in the periphery. Moreover, changes in microbiota composition and dysbiosis, can potentially influence a number of pathological conditions, include MetS, obesity, T2D, heart failure, and cognitive function (see [5] and references therein).

While novel target-specific drugs are currently lacking [6], some epidemiological studies indicate that natural antioxidant agents, such as polyphenols, polyunsaturated n-3 fatty acids or vitamin-rich foods may delay the occurrence of neurodegenerative disorders. Polyphenols, in particular (e.g., curcumin and resveratrol) having pleiotropic protective effects appear ideal to prevent or treat conditions (such as AD) whose origin is multifactorial [7]. A growing body of research suggests that regular consumption of natural products (vegetables, fruits, leaves, roots, seeds, berries etc.) rich in polyphenols might improve health outcomes through different mechanisms boosting the organisms' antioxidant defenses [8]. Natural compounds represent a major source for the discovery of drug targets and are ever increasingly attracting the interest of the scientific community with the main aim of validating their efficacy for the prevention and treatment of different conditions, including cognitive decline and metabolic disorders [9]. Notwithstanding the growing interest in this class of compounds, rigorous clinic trials addressing their specific effects are lacking or show biases due to the nutritional status of the subjects, genetic background, gender, treatment duration and dose–response relationship [10]. With regard to this latter point, a major drawback is related to their bioavailability, i.e., the amount of compound (or of its active principles) that reaches systemic circulation due to intestinal endothelium absorption and first-pass metabolism. Thus, the use of natural products and nutraceuticals poses important questions regarding human safety and calls for a better understanding of their therapeutic efficacy as well as their mechanisms of action.

Curcumin, one of the best investigated botanical constituents in the biomedical literature, has been receiving increased interest due to its unique molecular structure, which targets inflammatory and antioxidant pathways, and its potential to improve healthspan [11–15]. The genus Curcuma includes approximately 80 species and is regarded as one of the largest genera of the Zingiberaceae family [16]. Curcumin (1,7-bis(4-hydroxy-3-methoxyphenyl)-1,6-heptadiene-3,5-dione) is a lipophilic polyphenol active extract deriving from the rhizome of *Curcuma longa*. Curcumin (and curcuminoid analogs such as demethoxycurcumin and bisdemethoxycurcumin) provides the characteristic bright yellowish/golden pigment of turmeric widely used in traditional Indian and Chinese medicine from thousands of years because of a number of beneficial effects on human health [17,18]. Today curcumin is used all over the world as a supplement, spice and food additive. It is considered a safe compound suitable for daily dietary use by the United States Food and Drug Administration (FDA), the Joint FAO/WHO Expert Committee on Food Additives (JECFA) and the European Food Safety Authority (EFSA) who have indicated 0–3 mg/kg as an acceptable daily intake (https//www.fda.gov/food/generally-recognized-safe-gras/gras-notice-inventory accessed on 25 April 2021) [19]. Many of its medical uses have been mechanistically validated in in vitro and in vivo preclinical studies (more than 3000 investigations, see [20]) mainly focusing on its antioxidant and anti-inflammatory properties. In recent years, the positive effects of curcumin have been observed in several chronic diseases ranging from cardiovascular, gastrointestinal, neurological disorders and diabetes to several types of cancer [21–24]. The consumption of curcumin has been associated with a global improvement in the glycemic and lipid profile in patients with MetS [25]. In addition, amelioration in cognitive function in animal models has been widely documented due to its action on structure and functionality of neuronal membranes [26,27]. Despite this evidence it is worth mentioning that curcumin is characterized by poor stability, a feature resulting in an overall low oral absorption, though, once in the blood stream, curcumin appears to be stable and able to reach target tissues [28]. However, as far as the brain is concerned, its application raises the critical issue of its ability to cross the blood–brain barrier (BBB), an issue deserving further investigation [29].

This paper will focus on *Curcuma longa* as a very promising natural compound to counteract inflammaging and cognitive decline. It will review its possible mechanisms of actions and efficacy and will critically address the issue of its bioavailability and describe recent strategies aimed at improving its supplementation.

2. Curcumin, Cognitive Decline and Glucose Homeostasis (Peripheral and Central Actions)

The concept of energy homeostasis is receiving much attention nowadays due to the global spread of obesity and diabetes. Hyperglycemia is one of the conditions characterizing MetS, a main risk factor for multi-cause morbidity that includes T2D, cardiovascular disease and also dementia and Alzheimer's disease (AD). Insulin plays an important role in neuronal survival, in the protection of excitatory synapses and the formation of dendritic spines through the activation of AKT, mTOR and Ras-related pathways that are part of the insulin signaling cascade (see [4] and references therein). Moreover, it regulates levels of GABA, NMDA and AMPA-mediated mechanisms involved in brain plasticity. Insulin binding is highest in the cerebral cortex that plays a role in the control of executive functions as well as in the hippocampus, a brain area involved in learning and memory [4]. The mammalian brain is a highly demanding organ in terms of energy expenditure and shows reduced capacity for cellular regeneration and poor antioxidant defenses, making it particularly susceptible to metabolic and OS insults [30]. Indeed, exposure to hypercaloric diets and obesity has been shown to decrease insulin transport into the mammalian brain, a condition that is restored upon caloric restriction (see [31] and references therein). Recent evidence suggests that IR within the brain, a condition that may be described as the inability of brain cells to respond to insulin and its receptors, has major potential to impact cognitive functions and to contribute to the etiopathogenesis of AD. IR in peripheral tissues and organs (a conditions underlying hyperglycemia and diabetes) is often associated with IR within the brain leading to insulin deficiency and impaired glucose transport inside the neurons [32]. Such a condition may lead to neuronal death, apoptosis and degeneration, predisposing the individual to neurodegenerative diseases and the resultant cognitive decline. Thus, brain desensitization to insulin receptor due to untreated T2D, obesity or chronic consumption of hypercaloric diets may play a key role in what has now been defined as a novel form of diabetes (type 3 diabetes, T3DM) and its complications [33], indicating glucose homeostasis as key in the maintenance of cognitive function.

A growing body of clinical and preclinical data suggests that curcumin holds potential for the control of glucose homeostasis [34] since it may improve glucose uptake, insulin sensitivity and beta islet cell function. Moreover, curcumin may reduce glucose and lipid levels in addition to reducing OS and inflammation [35] by interacting with almost all the players involved in these processes, as demonstrated in in vitro studies [11,35]. A very large amount of literature is now available on curcumin effects. Thus, we will only summarize some of the main aspects.

Preclinical animal models have clearly shown a main effect of natural compounds, including curcumin, on cognitive function during ageing [26,36–39]. These effects are most likely related to the ability of curcumin to act directly on Aβ plaques as well as to its anti-inflammatory and antioxidant properties. Indeed, a number of preclinical studies have reported downregulation of biomarkers of inflammation (e.g., TNF-α, IL-1β) and OS (e.g., lipid peroxidation, reactive oxygen species—ROS—nitrite and glutathione) believed to be involved in cognitive impairments, confirming the anti-inflammatory and antioxidant properties of curcumin [40–50].

Results of published clinical studies, although in some cases not conclusive, show promise for curcumin's use as a therapeutic for cognitive decline [51]. The efficacy of curcumin supplementation in humans has been evaluated in several randomized controlled trials, suggesting its potential to reduce blood glucose, C-peptide, glycated hemoglobin (HbA1c), alanine aminotransferase (ALT) and aspartate aminotransferase (AST) in patients with T2D [52]. A recent meta-analysis provides evidence for curcumin's ability to reduce body mass index, body weight, body fat and leptin values and to increase adiponectin levels in patients with MetS and related disorders [53]. Its effectiveness in reducing TG and C-reactive protein (CRP) and increased adiponectin levels has also been reported [54]. In addition to the ability of curcumin to control glucose homeostasis, and indirectly, to improve/counteract cognitive decline, direct effects have also been observed (and are currently being studied) in the brain. In fact, curcumin has received increased interest due to its unique molecular structure that targets directly amyloid aggregation, one of the major hallmarks of AD. To this regard, Yang and colleagues have provided in vitro and in vivo evidence for curcumin to inhibit Aβ aggregation as well as to prevent oligomer formation [55]; Garcia-Alloza, in a mouse model of AD, found that curcumin reversed existing amyloid pathology and improved the associated neurotoxicity [56].

Beyond its use as a compound to prevent/counteract cognitive decline, intriguing evidence supports a possible application for in vivo diagnostics for AD and other related neurodegenerative pathologies [57]. In fact, AD diagnosis can be currently made only by means of clinical criteria supported by invasive and time consuming investigations [58]. Thus, a patient-friendly and repeatable amyloid or Aβ biomarker may alleviate the burden related to currently available diagnostic tools as well as support therapy monitoring in clinical trials [57]. In fact, den Haan and colleagues, by taking advantage of the peculiar feature of curcumin of being naturally fluorescent as well as its Aβ-binding properties, have provided evidence of a selective binding (of curcumin and its related isoforms) to fibrillar Aβ in plaques and cerebral amyloid angiopathy in post mortem AD brain tissues. However, in order to use curcumin as a feasible tool for in vivo detection of Aβ, its poor bioavailability and in vivo metabolism should be carefully considered (see below).

3. Curcumin, Oxidative Stress and Inflammation

The neuropathological features of the brain affected by dementia suggest that the oxidative and inflammatory burden plays a role in the progression of pathological signs by reducing brain plasticity, thus, being an important risk factor for cognitive disability [58]. To this regard, metabolic dysfunctions that are often associated with OS and inflammation, may greatly accelerate the onset and worsen the progression of cognitive functions by promoting brain ageing and reducing healthspan [59].

Neuroinflammatory processes are a main feature of neurodegenerative disorders in which microglia and astrocytes are over-activated, resulting in increased production of pro-inflammatory cytokines. Moreover, deficiencies in the anti-inflammatory response may also contribute to neuroinflammation. More specifically, the activated neuroglia by increasing both NF-κB, COX2 and iNOS levels may induce, in turn, the release of pro-inflammatory cytokines, such as IL-6, IL-1α and TNF-β. This pervasive inflammatory condition results in an overall increase in the OS burden leading to neuronal toxicity and the subsequent cognitive deficits characterizing neurodegenerative diseases. Numerous studies have indicated that curcumin is an effective antioxidant both in vivo and in vitro [9]. Curcumin treatment could attenuate cell apoptosis, decrease the level of lipid peroxidation, and increase the activity of various antioxidant enzymes including superoxide dismutase (MnSOD) and glutathione (GSH) [60], thus helping to break the vicious cycle sustaining neuroinflammation and containing the progression of neurodegenerative diseases [61]. The underlying mechanism is possibly associated with the function of NFE2-related factor-2 (Nrf2), a transcription factor promoting the upregulation of antioxidant defenses [61].

As far as apoptosis is concerned, many mechanisms have been proposed. Xi-Xun Du and colleagues indicated that curcumin's property of iron chelation and reduction may underlie its anti-apoptotic effects [62]. Chen and co-workers reported that curcumin may exert its cytoprotective effects against neurotoxic agents via its antiapoptotic and antioxidant properties through the Bcl-2–mitochondrion–ROS–inducible nitric oxide synthase pathway [63]. Moreover, Yu and colleagues reported that the inhibition of JNK pathway and the activation of caspase-3 cleavage might prevent neuronal death [64]. Indeed, the anti-inflammatory and antioxidant properties of curcumin are strictly related to its action on apoptotic pathways and on neuronal death. In fact, pro-inflammatory cytokines are not only involved in the so-called neuroinflammaging but may also trigger the apoptotic process. Likewise, excessive OS may directly lead to mitochondrial swelling and apoptosis. Thus, inflammation and apoptosis are related in a vicious cycle leading to neuronal death [65].

Recently, several studies have highlighted the role of inflammatory pathways mediated by the inflammasome in neurodegenerative diseases. In particular, the NOD-like receptor pyrin domain-containing-3 (NLRP3) has been suggested to play a pathogenic role in several neuroinflammatory diseases, including AD [66]. In vitro and in vivo studies have shown that Aβ peptide activates NLRP3 inflammasome in microglial cells. Furthermore, in a mouse model of AD, NLRP3 knockout (KO) mice were protected from impaired spatial memory performance and showed a decrease in the Aβ plaque load [67], similar results were obtained when a specific NLRP3 inhibitor was administered to mice [68]. This evidence points to the inflammasome as a potential therapeutic target for AD treatment [69]. Notably, recent evidence shows that curcumin, by modulating the activity of NLRP3 inflammasome, could be beneficial in reducing neuroinflammation and/or neurodegeneration in different neurological disorders, such as major depression, brain ischemia, AD and epilepsy [13,70,71].

OS is a condition characterizing aerobic biological systems, the major portion of ROS being generated as a by-product of the electron transport chain operating in the mitochondria [30]. As already mentioned, OS is a condition strongly associated to inflammation that may act both as (con)cause and effect of pathological conditions affecting brain ageing; however, a growing body of evidence suggests that ROS are not only responsible for oxidative damage to cells and macromolecules but they may also play a role as mediators in specific signaling cascades. Hydrogen peroxide (H_2O_2) in particular has been identified as a ROS able to affect the ageing process by specifically mediating insulin signaling and promoting fat accumulation, ultimately affecting the ageing process [72]. Worth to notice, the master regulator of this process is the p66Shc *gerontogene*, which, by acting within the mitochondrion, increases the generation of H_2O_2, amplifying insulin signaling [73–75]. Interestingly, deletion of p66Shc gene in mice resulted in the decreased formation of mitochondrial H_2O_2 [75], a feature that has been associated with reduced fat accumulation as well as decreased incidence of metabolic and cardiovascular pathologies [73,76]. Moreover, p66Shc KO mice were characterized by elevated resistance to OS, delayed brain ageing and improved overall healthspan, all features associated with increased brain and behavioral plasticity. In fact, the brain of p66Shc KO mice was characterized by reduced levels of inflammation and OS and increased levels of the neurotrophin brain-derived neurotrophic factor (BDNF); in addition, these mice showed decreased emotionality and improved cognitive function [77–79].

Lifestyles have the potential to modulate healthspan during ageing. For example, physical exercise and diet, as well as the consumption of nutraceutical compounds (including curcumin), may greatly contribute to reducing neuroinflammaging by targeting brain pathways related to OS and inflammation (see below, next paragraph). Physical exercise in elderly women has been shown to improve metabolic functions and this was paralleled by a decrease in the peripheral levels of p66Shc gene [80,81]. Very recently, a role for curcumin was also reported in the modulation of the p66Shc gene as its was able to downregulate the expression levels of this gene in peripheral blood mononuclear cells (PBMC), improving diabetic nephropathy in a rat model [82]. These data overall suggest that p66Shc might be exploited as a suitable biomarker of curcumin efficacy to counteract the ageing-related burden and to improve overall healthspan.

4. A Potential Mechanism of Action: Curcumin as a "Hormetin"

A large epidemiological Indo-US Cross National Dementia study showed that rural Indian populations have a low prevalence of AD and AD-associated dementia compared to the US population, and this may be linked to the high curcumin consumption [17], although such correlation does not necessarily imply causative connection.

An ever-increasing body of evidence suggests that natural compounds may alleviate the burden of chronic diseases by increasing individuals' ability to cope with OS. Notably, differently from the consumption/administration of antioxidants (of natural or synthetic origin), whose beneficial effects are still debated, most natural compounds do not act solely as free radical scavengers but rather, and most interestingly, as "antioxidant boosters" [83] In this regard, it is important to point out that ROS also function as signaling molecules underlying physiological processes and, for this reason, their generation and scavenging needs to be tightly regulated (see below).

As reviewed by Lee and colleagues [8], cellular stressors that are relevant to the pathogenesis of chronic diseases may be roughly categorized into four general types: (1) OS resulting from the unbalance between ROS production and the organism's antioxidant defenses; (2) metabolic stress deriving from impaired cellular bioenergetics and mitochondrial dysfunctions; (3) proteotoxic stress protein misfolding and aggregates accumulation, and (4) inflammatory stress that leads to the production of ROS from immune cells. All these stressors are relevant for brain ageing. Moreover, and most importantly, OS is also involved as a cause or as a consequence in all the above-mentioned categories. Cytotoxic effects of ROS contribute to the death of neurons during chronic neurodegenerative diseases such as AD and Parkinson's disease and also during the ageing process. However, ROS also function as signaling molecules underlying physiological processes including cell proliferation, migration, and survival through the regulation of neurotrophic factors [84,85]. Therefore, the generation and scavenging of ROS needs to be tightly regulated and nutraceutical compounds appear to be good candidates to play a role in this process. In this regard, the cells throughout the body and brain might trigger stress signaling pathways, eventually leading to the enhancement of their own resistance to further stressors, including OS. An intriguing hypothesis suggests that nutraceuticals might be perceived as potentially toxic by the organism (at high doses). However, exposure to low doses of these compounds might stimulate the organism's hormetic/adaptive responses aimed at counteracting such threats ("hormesis hypothesis", see [8]. Indeed, a recent paper by Calabrese and co-workers provides evidence that curcumin displays hormetic-like biphasic dose response features that are independent from the biological model used for investigation, cell type, and endpoints [86]. These findings hold major implications for study design, including selection of doses and sample size, also considering the specific context of bioavailability and pharmacokinetics, see Figure 1.

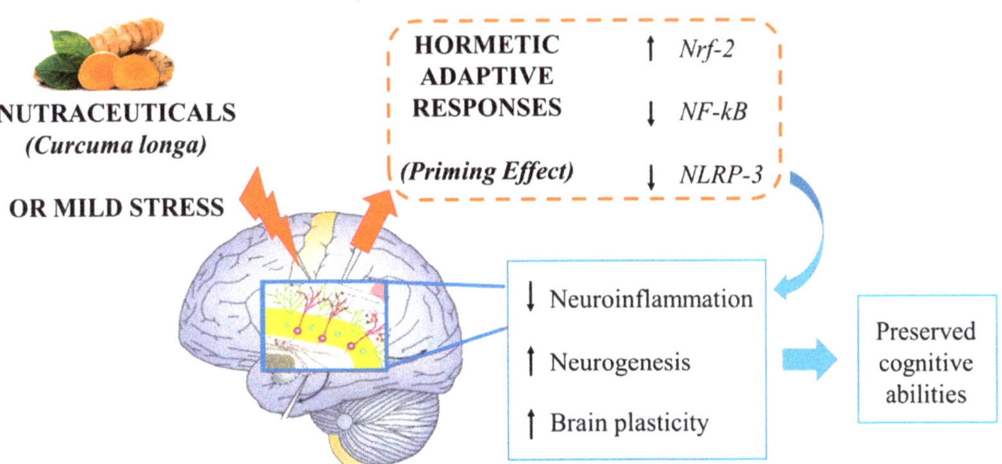

Figure 1. An intriguing hypothesis suggests that nutraceuticals might be perceived as potentially toxicant by the organism (at high doses), however, exposure to low doses of these compounds might stimulate the organism's hormetic/adaptive responses aimed at counteracting such threats. Likewise, *Curcuma longa* acting as a mild stress might trigger signaling cascades boosting antioxidant and anti-inflammatory pathways (related to Nrf2, NF-kB, NLRP3), leading to decreased neuroinflammation, increased neurogenesis and brain plasticity, finally improving cognitive abilities. Figure adapted from [87], Copyright © 2014 University of Massachusetts.

5. Criticisms to Be Considered in Curcumin Supplementation

Despite plenty of data on the positive health effects of curcumin [88,89], some problems strongly limit its effectiveness and usefulness. First of all, its low bioavailability. Curcumin is characterized by low water solubility and high instability in most body fluids; in addition, it is poorly absorbed by the gastrointestinal tract. In fact, curcumin is rapidly metabolized by the large intestine and by liver enzymes, leading ultimately to the production of sulphate and glucuronide O-conjugated metabolites [90]. Specifically, curcumin that does reach the blood flow undergoes phase I (reduction) and phase II metabolism (conjugation). Reductases reduce curcumin to dihydrocurcumin, tetrahydrocurcumin and hexahydrocurcumin (phase I) [91,92], then these phase I metabolites are conjugated to sulphates and glucuronides (phase II) [93,94].

In animal and human studies, a low concentration of curcumin in blood plasma and urine was observed after oral administration, in particular, serum curcumin levels are undetectable in humans even after high oral doses (up to 8 g/day) [95,96]. Finally, the presence of curcumin in the blood is not sufficient to ensure the delivery in the brain to exert the neuroprotective activity because several studies have demonstrated that it does not easily cross the BBB. All these factors together have pushed research towards finding new formulations or new ways of administration able to stabilize the molecule and increase its bioavailability, by reducing its metabolism and increasing the retention time in the bloodstream [97]. To this regard, the simultaneous administration of curcumin with piperine [98,99], essential oil [100] or milk [101] have been suggested to stimulate the gastrointestinal system, prevent the efflux of curcumin and to increase absorption and metabolism. Many studies have been aimed at devising and testing new drug delivery strategies using, for example, carriers such as soy lecithin phosphatidylcholine (phytosome, Meriva®) that improve both the absorption of curcumin in the intestine as well as its penetration into the cells [102–105]. In addition, nanoparticles, liposomes, micelles, phospholipid complexes, emulsions, microemulsions, nanoemulsions, solid lipid nanoparticles, nanostructured lipid carriers, biopolymer nanoparticles and microgels are able to increase curcumin bioavailability by enhancing small intestine permeation, preventing possible degradation in the microenvironment, eventually increasing plasma half-life and enhancing curcumin efficacy [106–113]. Another way to administer curcumin is by using exosomes, nanovesicles (30–100 nm) that are generated within the cell in the endosomal network. This method, which appears to be safe and non-cytotoxic [114,115], increases both the plasma concentration as well as the bioavailability of curcumin 5–10 times more than curcumin alone [116]. Other studies have confirmed that increased solubility, stability and bioavailability of curcumin can be obtained by incorporating it into exosomes. All these different curcumin formulations enhance its bioavailability and allow for greater persistence in the body, better permeability and resistance to metabolic processes and a higher efficacy [97]. Other studies have focused on changing the chemical structure of curcumin, generating curcumin derivatives that may show not only an improved pharmacological activity, but also better physicochemical and pharmacokinetic properties [117,118], see Figure 2.

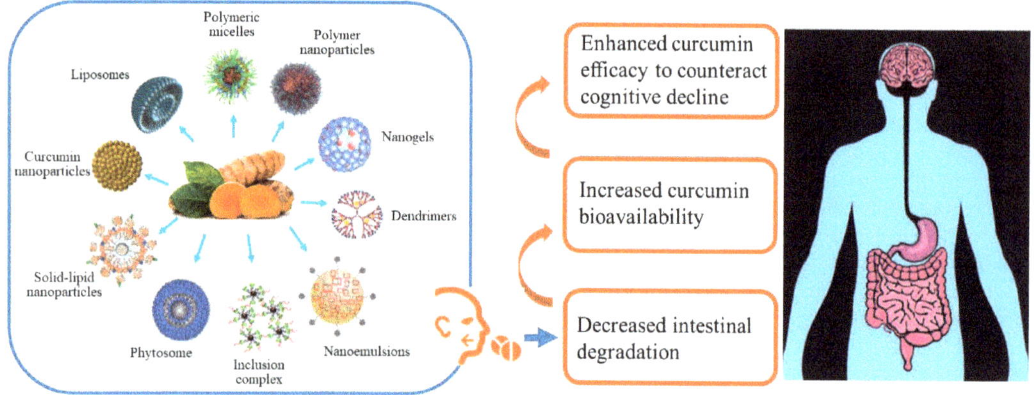

Figure 2. New delivery systems such as polymeric micelles, polymer nanoparticles, nanogels, dendrimers, nanoemul-sions, inclusion complexes, phytosomes, solid-lipid nanoparticles, curcumin nanoparticles and liposomesnanoparticles, liposomes, micelles, nanogel, dendrimers, nanoemulsions, inclusion complexes and phytosomes have the potential to reduce intestinal degradation and increase curcumin bioavailability, ultimately enhancing its efficacy throughout the body and the brain. Within this latter organ the increased curcumin bioavailability might counteract cognitive decline. Figure adapted with permission from [119]. Copyright © 2018, Said Moselhi.

It is worth noting that the paradox of the curcumin pharmacological effect, despite its poor bioavailability could be, at least partially, explained by the influence of the microbiome on curcumin metabolism. Indeed, the biological activities of curcumin are linked to the digestion by the intestinal flora, which can produce active metabolites. Thus, the beneficial effects of curcumin seem to depend on the individual ability to metabolize it, that is, to the composition of each person's intestinal microbiota [120]. Several microorganisms capable of modifying curcumin have been identified in the human microbiota, including *Bifidobacteria longum*, *Bifidobacteria pseudocatenulaum*, *Enterococcus faecalis*, *Lactobacillus acidophilus* and *Lactobacillus casei* [121]. Many of them metabolize curcumin to a large extent (more than 50%) and produce a number of metabolites (approximately 23 of them have been identified) by different metabolic pathways such as acetylation, hydroxylation, reduction, demethylation [122]. Several studies have shown that the metabolites have similar properties to curcumin [123] and many of them have been shown to exhibit neuroprotective effects [124,125], suggesting that curcumin transformed by the gut microbiota could be useful for microbiome-targeting therapies for AD. In fact, bidirectional communication exists between the central nervous system and the gut microbiota, which plays a key role in human health [5]. A growing body of evidence suggests that the gut microbiota can influence human brain and behavior, and the different metabolites secreted by the gut microbiota can affect the cognitive abilities of patients diagnosed with neurodegenerative disorders [126,127]. Changes in the gut microbiota composition, caused by dietary habits, antibiotic exposure and/or infections might result in conditions of dysbiosis (also known as dysbacteriosis) that are involved in the etiopathogenesis of different diseases in humans, including MetS, obesity, T2D and neurodegenerative disorders [127,128]. Indeed, changes in gut microbiota homeostasis leads to increased intestinal permeability that, in turn, results in the translocation of bacteria and endotoxins across the epithelial barrier, a condition that might trigger an immunological response associated with the production of pro-inflammatory cytokines. Such mechanisms have the potential to greatly contribute to neuroinflammation through the secretion of pro-inflammatory cytokines, ultimately altering brain functions [5]. Thus, curcumin supplementation may be useful to counteract/prevent cognitive decline also by mechanisms involving the preservation of individuals' microbiota homeostasis.

6. What have We Learned and What Needs to Be Done

Although only a few clinical studies have examined curcumin's effect on human cognitive functioning, the results of these trials are sometimes inconsistent, highlighting the difficulty in translating basic research to the clinic. While some studies report no cognitive enhancing effects of curcumin [129,130], other data indicate positive effects of this compound on cognitive function [131–133]. Also in preclinical studies, we found great heterogeneity and limited ability to standardize age of administration or endpoints [39,73,134,135].

Nonetheless, even in very different animal models, curcumin consistently decreases both systemic and central neuroinflammation while improving redox state [51]. As many clinical studies did not include inflammatory and oxidative biomarkers, future trials should target these measures to yield further insight into why curcumin has not shown consistent cognitive effects in humans. Among these, changes in p66Shc transcription in PBMC might be exploited as a potential biomarker of curcumin efficacy [82,136]. Bioavailability may be one main factor that increases variability between studies. Administration of any of the three constituents (curcumin, bisdemethoxycurcumin and demethoxycurcumin) separately instead of the parent curcuminoid mixture was recommended as a more efficient way of treatment [125]. Furthermore, a synergistic effect of curcumin with other dietary supplements, such as piperine, α-lipoic acid, N-acetylcysteine, B vitamins, vitamin C, and folate can improve its effects [137,138]. Nowadays, nanoparticles are mainly used since they demonstrate better penetration at the BBB and cause more evident biochemical changes than free curcumin [9,46,49,139,140].

Nevertheless, there is still room for improvement. As an example, use of soy lecithin phosphatidylcholine (phytosome, Meriva®), which improves both the absorption of curcumin in the intestine as well as its penetration into the cells, has been used and has resulted in significant effects in clinical studies assessing a number of variables such as liver function, inflammation, gastrointestinal disturbances and T2D [141–145]. Thus, using new delivery systems that increase bioavailability appears to be a *must-do* for future clinical trials [26,51].

Other drawbacks of previous clinical trials are limited power, different formulations of curcumin used [129] and differences in ethnicity, as an example Caucasians vs. Asians, making it difficult to compare results derived from different studies [146]. Genetic factors, such as polymorphisms or dietary differences, such as higher consumption of curcumin in Asia, could underlie differential effects in different ethnic groups.

A dose–response relationship should also be taken into account. The optimal dose would have maximum cognitive enhancing effects with the safest pharmacokinetic profile. We have previously shown in animal studies that, in general, beneficial effects of natural compounds on cognition are dose-dependent with the higher dosages generally being more effective compared to lower dosages, although a significant effect of nutraceutical compounds on memory retention and OS was also demonstrated at the lowest dose [39]. In addition, when selecting the most effective dose in basic or clinical studies it should be carefully taken into account that curcumin commonly displays a biphasic dose–response curve such as hormetic compounds do [86,147]. In this regard, it should be noted that problems such as optimal dose and bioavailability are common to different natural compounds. A recent review by Mazzanti and Di Giacomo pointed out that such issues may also represent a major drawback when trying to compare the efficacy of curcuma and resveratrol, two polyphenols with very similar antioxidant and anti-inflammatory properties, to counteract cognitive decline [7]. As an example, both curcumin and resveratrol are able to activate the Nrf2 and NF-kB pathways as well as to modify insulin signaling. Their effects mostly overlap and both have been shown to enhance cognitive function in animal models [8] More recently, Saleh and co-workers when trying to compare the protective effects of curcumin, resveratrol and sulphoraphane in an in vitro study, confirmed that enhancing the delivery of phytochemicals (either by designing novel nanostructures or using mixtures of natural compounds that work synergistically) is a priority in this field of research [148] Thus, further research is mandatory to establish the most effective substance to be used, in which conditions and for whom.

7. Targeting Both Genders

The male-female health-survival paradox, also known as the morbidity-mortality paradox or gender paradox, poses that women live longer, though they experience more disabilities and medical issues throughout life, compared to men. Such a paradox holds true in almost every country in the world since virtually all the primary causes of death are higher for men at all ages [149]. Thus, due to their greater resilience to stress, women live longer than men but experience higher rates of physical illness, leading to debilitating, though rarely lethal, conditions. Several hypotheses have been proposed for this phenomenon that could be interpreted as a sex-driven resilience to stressors (leading to longer lifespan), including more efficient female immune functioning, the protective role of estrogens as well as increased antioxidant capacity [149].

Many chronic conditions, including dementia and AD, though not lethal *per se*, are strongly linked to disability and loss of physiological functions. Alzheimer's disease is a multifactorial neurodegenerative disorder, the development of which depends upon both environmental as well as genetic risk factors. A recent review by Christensen and Pike put the attention specifically on two risk factors that may play a key role in the initiation and/or progression of AD and that may be particularly problematic for women: inflammation and obesity [150]. Alzheimer's disease and other dementias are highly prevalent among women [151] and the onset of menopause, which is associated with estrogen breakdown decreases women's vulnerability threshold to both metabolic and cognitive disorders that depend upon central adiposity and overall inflammation [150].

In women, the shift from adulthood to middle-age is characterized by an overall increase in the proportions of overweight and/or obese subjects [152]. Clinical and preclinical studies suggest that the age-related loss of ovarian hormones results in weight gain and contributes to changes in the distribution of adipose tissue leading to increased waist-to-hip ratio [153]. Elevated adiposity increases the risk of different pathological or sub-pathological conditions, including MetS [154], T2D [155] and AD [156]. One consequence of increased adiposity that may underlie its pathogenic role is chronic inflammation, which is observed both systemically as well as in the brain (see [150] and references therein) A large body of data supports the link between estrogens (and other sex steroid hormones) and the modulation of the individual inflammatory profile. Indeed, estrogens are powerful anti-inflammatory mediators, thus, the drop experienced by women at menopause triggers a rise in pro-inflammatory cytokines that may place tissues throughout the body at increased risk of inflammaging-associated diseases [157]. A recent clinical trial aimed at assessing the efficacy of oral curcumin (500 mg) administration twice/day for 8 weeks on anxiety and other specific symptoms that accompany menopause, has shown that curcumin significantly reduced hot flashes in postmenopausal women [158]. In this context, it is possible to foresee the use of curcumin as a possible preventive strategy to be administered in pre-menopausal women aimed at boosting anti-inflammatory and antioxidant capacity when natural defenses that women are endowed with start to be threatened by the absence of estrogens (see Figure 3). In fact, it is important to stress here that, given the current knowledge on age-associated cognitive decline, it would be unrealistic to foresee a therapeutic use of nutraceuticals in overt pathologies characterized by massive neurodegeneration, such as AD. Thus, early diagnosis of the pathology and prompt interventions should be aimed at preventing or slowing down its progression. In this regard, the identification of suitable, least-invasive biomarkers (e.g., through blood tests) should be considered a priority. Moreover, it is more and more evident that prevention through specific diets and physical exercise will represent the key in the future to counteract cluster conditions that put the individual at greater risk for cognitive and physical decline, including, e.g., MetS [39].

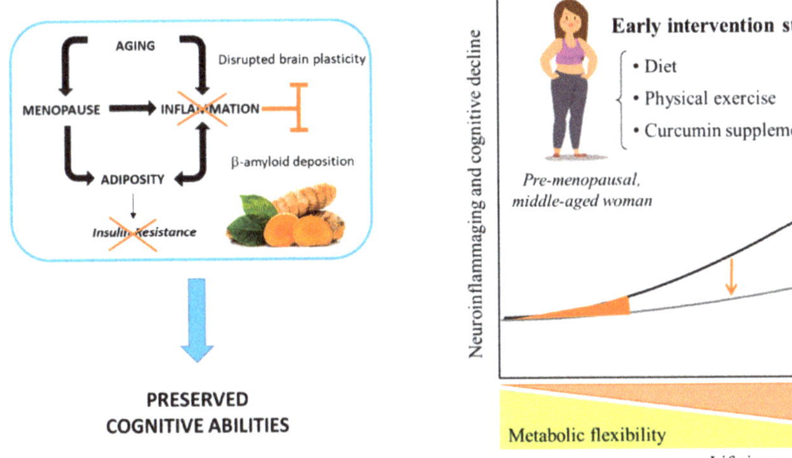

Figure 3. During middle-age, women experience menopause that is characterized by the physiological drop in the protective effects of estrogen hormones. Such a drop leads to increased inflammation and adiposity, two conditions that might reinforce one another in either decreasing brain plasticity or increasing amyloid deposition or both. Curcumin administration in middle-aged pre-menopausal women has the potential to break this vicious cycle overall boosting anti-inflammatory defenses, counteracting fat-mediated insulin resistance and preserving brain plasticity (left panel). So far, no specific sex-differences have been found with regard to curcumin dose–response efficacy. However, it is important to stress that identifying sex-specific critical time windows throughout life to start curcumin administration might be equally important and might improve the chance to protect the brain and to counteract cognitive decline. Such a time window in women might be the middle-age, right before the beginning of menopause, when the organism still retains a certain degree of metabolic flexibility. Thus, the earlier the intervention (also through specific diets and physical exercise) the greater the chance to prevent the decay in mental and physical health (**right panel**). Figure adapted with permission from [150], Copyright © 2015 Christensen and Pike, and from [136,159]: © 2007 Springer International Publishing and © 2020 Published by Elsevier Ltd on behalf of IBRO.

8. Conclusions

In this review, we have touched upon a number of critical issues that should be taken into account when designing preclinical and clinical studies aimed at assessing curcumin efficacy on cognitive functions. Among these, one important factor that clearly needs to be tackled in the future has to do with increasing its bioavailability as well as controlling the impact of nutritional status/diet and lifestyle on curcumin's effects. Diet is also likely to influence microbiota status, thus, controlling for nutritional status will be crucial for effective future studies.

Another important point relates to the need to assess the effects of this natural compound on both males and females, both in preclinical and clinical studies. As dementia and AD show a much greater prevalence in the female population, it is imperative to address this issue by targeting both sexes/genders [151].

Moreover, given the overlap between the mechanisms of action of many compounds (see, e.g., curcumin and resveratrol), it will be important to target multiple molecular pathways to maximize the effects. Traditional medicines mostly use mixtures of phytochemicals, rather than individual compounds, which suggests the need to examine, with rigorous clinical trials, the role of plant mixtures on brain health [9,160].

Ultimately, in order to assess the effects of natural compounds, such as curcuma, we need to refine our ability to measure health (and the lack of) in a life-long perspective and to characterize the conditions for the transition from health to disease [39,73,135].

Author Contributions: A.B., B.C. and F.C. conceived the review, retrieved the references and drafted the manuscript. R.V. and R.M. contributed on the issue of curcumin bioavailability, biotechnologies and inflammasome, commented on gender issues and critically read the whole manuscript.

Funding: This work was supported by the JPI HDHL-INTIMIC-085-GUTMOM project within the ERA-NET Co-fund HDHL-INTIMIC (INtesTInal MICrobiomics) programme and HDHL-INTIMIC Knowledge Platform on Food, Diet, Intestinal Microbiomics and Human Health; F.C. received funding from the Italian Ministry of Health (Ricerca Corrente).

Acknowledgments: We thank Chiara Musillo for help in retrieving and selecting bibliographic entries; Paola Rossolino for management of external funding.

Conflicts of Interest: The authors declare no conflict.

References

1. Deary, I.J.; Corley, J.; Gow, A.J.; Harris, S.E.; Houlihan, L.M.; Marioni, R.E.; Penke, L.; Rafnsson, S.B.; Starr, J.M. Age-associated cognitive decline. *Br. Med. Bull.* **2009**, *92*, 135–152. [CrossRef] [PubMed]
2. Brayne, C. The elephant in the room—healthy brains in later life, epidemiology and public health. *Nat. Rev. Neurosci.* **2007**, *8*, 233–239. [CrossRef] [PubMed]
3. Yates, K.F.; Sweat, V.; Yau, P.L.; Turchiano, M.M.; Convit, A. Impact of metabolic syndrome on cognition and brain: A selected review of the literature. *Arterioscler. Thromb. Vasc. Biol.* **2012**, *32*, 2060–2067. [CrossRef] [PubMed]
4. Nguyen, T.T.; Ta, Q.T.H.; Nguyen, T.K.O.; Nguyen, T.T.D.; Giau, V. Van Type 3 diabetes and its role implications in alzheimer's disease. *Int. J. Mol. Sci.* **2020**, *21*, 3165. [CrossRef]
5. Van Giau, V.; Wu, S.Y.; Jamerlan, A.; An, S.S.A.; Kim, S.Y.; Hulme, J. Gut microbiota and their neuroinflammatory implications in alzheimer's disease. *Nutrients* **2018**, *10*, 1765. [CrossRef]
6. Abbott, A. Dementia: A problem for our age. *Nature* **2011**, *475*, S2–S4. [CrossRef] [PubMed]
7. Mazzanti, G.; Di Giacomo, S. Curcumin and Resveratrol in the Management of Cognitive Disorders: What is the Clinical Evidence? *Molecules* **2016**, *21*, 1243. [CrossRef]
8. Lee, J.; Jo, D.G.; Park, D.; Chung, H.Y.; Mattson, M.P. Adaptive cellular stress pathways as therapeutic targets of dietary phytochemicals: Focus on the nervous system. *Pharmacol. Rev.* **2014**, *66*, 815–868. [CrossRef]
9. Tewari, D.; Stankiewicz, A.M.; Mocan, A.; Sah, A.N.; Tzvetkov, N.T.; Huminiecki, L.; Horbanczuk, J.O.; Atanasov, A.G. Ethnopharmacological Approaches for Dementia Therapy and Significance of Natural Products and Herbal Drugs. *Front. Aging Neurosci.* **2018**, *10*, 3. [CrossRef]
10. Marx, W.; Moseley, G.; Berk, M.; Jacka, F. Nutritional psychiatry: The present state of the evidence. *Proc. Nutr. Soc.* **2017**, *76*, 427–436. [CrossRef]
11. Patel, S.S.; Acharya, A.; Ray, R.S.; Agrawal, R.; Raghuwanshi, R.; Jain, P. Cellular and molecular mechanisms of curcumin in prevention and treatment of disease. *Crit. Rev. Food Sci. Nutr.* **2020**, *60*, 887–939. [CrossRef] [PubMed]
12. Tong, W.; Wang, Q.; Sun, D.; Suo, J. Curcumin suppresses colon cancer cell invasion via AMPK-induced inhibition of NF-κB, uPA activator and MMP9. *Oncol. Lett.* **2016**, *12*, 4139–4146. [CrossRef] [PubMed]
13. Hasanzadeh, S.; Read, M.I.; Bland, A.R.; Majeed, M.; Jamialahmadi, T.; Sahebkar, A. Curcumin: An inflammasome silencer. *Pharmacol. Res.* **2020**, *159*, 104921. [CrossRef] [PubMed]
14. Kahkhaie, K.R.; Mirhosseini, A.; Aliabadi, A.; Mohammadi, A.; Mousavi, M.J.; Haftcheshmeh, S.M.; Sathyapalan, T.; Sahebkar, A. Curcumin: A modulator of inflammatory signaling pathways in the immune system. *Inflammopharmacology* **2019**, *27*, 885–900. [CrossRef] [PubMed]

15. Xu, X.-Y.; Meng, X.; Li, S.; Gan, R.-Y.; Li, Y.; Li, H.-B. Bioactivity, Health Benefits, and Related Molecular Mechanisms of Curcumin: Current Progress, Challenges, and Perspectives. *Nutrients* **2018**, *10*, 1553. [CrossRef] [PubMed]
16. Sirirugsa, P.; Larsen, K.; Maknoi, C. The Genus Curcuma L. (Zingiberaceae): Distribution and Species Diversity of Curcuma in Thailand Gardens. *Bull. Singap.* **2007**, *59*, 203–220.
17. Chin, D.; Huebbe, P.; Pallauf, K.; Rimbach, G. Neuroprotective Properties of Curcumin in Alzheimer's Disease—Merits and Limitations. *Curr. Med. Chem.* **2013**, *20*, 3955–3985. [CrossRef]
18. Goel, A.; Kunnumakkara, A.B.; Aggarwal, B.B. Curcumin as "Curecumin": From kitchen to clinic. *Biochem. Pharmacol.* **2008**, *75*, 787–809. [CrossRef]
19. Sharifi-Rad, J.; El Rayess, Y.; Rizk, A.A.; Sadaka, C.; Zgheib, R.; Zam, W.; Sestito, S.; Rapposelli, S.; Neffe-Skocińska, K.; Zielińska, D.; et al. Turmeric and Its Major Compound Curcumin on Health: Bioactive Effects and Safety Profiles for Food, Pharmaceutical, Biotechnological and Medicinal Applications. *Front. Pharmacol.* **2020**, *11*, 11. [CrossRef]
20. Aggarwal, B.B.; Org, A. Targeting Inflammation-Induced Obesity and Metabolic Diseases by Curcumin and Other Nutraceuticals. *Annu. Rev. Nutr.* **2010**, *30*, 173–199. [CrossRef]
21. Masoodi, M.; Mahdiabadi, M.A.; Mokhtare, M.; Agah, S.; Kashani, A.H.F.; Rezadoost, A.M.; Sabzikarian, M.; Talebi, A.; Sahebkar, A. The efficacy of curcuminoids in improvement of ulcerative colitis symptoms and patients' self-reported well-being: A randomized double-blind controlled trial. *J. Cell. Biochem.* **2018**, *119*, 9552–9559. [CrossRef] [PubMed]
22. Mantzorou, M.; Pavlidou, E.; Vasios, G.; Tsagalioti, E.; Giaginis, C. Effects of curcumin consumption on human chronic diseases: A narrative review of the most recent clinical data. *Phyther. Res.* **2018**, *32*, 957–975. [CrossRef]
23. Kotha, R.R.; Luthria, D.L. Curcumin: Biological, Pharmaceutical, Nutraceutical, and Analytical Aspects. *Molecules* **2019**, *24*, 2930. [CrossRef]
24. Hassan, F.U.; Rehman, M.S.U.; Khan, M.S.; Ali, M.A.; Javed, A.; Nawaz, A.; Yang, C. Curcumin as an alternative epigenetic modulator: Mechanism of action and potential effects. *Front. Genet.* **2019**, *10*, 514. [CrossRef] [PubMed]
25. Tabrizi, R.; Vakili, S.; Lankarani, K.B.; Akbari, M.; Mirhosseini, N.; Ghayour-Mobarhan, M.; Ferns, G.; Karamali, F.; Karamali, M.; Taghizadeh, M.; et al. The Effects of Curcumin on Glycemic Control and Lipid Profiles Among Patients with Metabolic Syndrome and Related Disorders: A Systematic Review and Metaanalysis of Randomized Controlled Trials. *Curr. Pharm. Des.* **2018**, *24*, 3184–3199. [CrossRef] [PubMed]
26. Voulgaropoulou, S.D.; van Amelsvoort, T.A.M.J.; Prickaerts, J.; Vingerhoets, C. The effect of curcumin on cognition in Alzheimer's disease and healthy aging: A systematic review of pre-clinical and clinical studies. *Brain Res.* **2019**, *1725*, 146476. [CrossRef]
27. Albert, M.S.; Dekosky, S.T.; Dickson, D.; Dubois, B.; Feldman, H.H.; Fox, N.C.; Gamst, A.; Holtzman, D.M.; Jagust, W.J.; Petersen, R.C.; et al. The diagnosis of mild cognitive impairment due to Alzheimer's disease: Recommendations from the National Institute on Aging-Alzheimer's Association workgroups on diagnostic guidelines for Alzheimer's disease. *Alzheimers Dement* **2011**, *7*, 270–279. [CrossRef]
28. Belcaro, G.; Cesarone, M.R.; Dugall, M.; Pellegrini, L.; Ledda, A.; Grossi, M.G.; Togni, S.; Appendino, G. Efficacy and safety of Meriva®, a curcumin-phosphatidylcholine complex, during extended administration in osteoarthritis patients. *Altern. Med. Rev.* **2010**, *15*, 337–344.
29. Anand, P.; Kunnumakkara, A.B.; Newman, R.A.; Aggarwal, B.B. Bioavailability of curcumin: Problems and promises. *Mol. Pharm.* **2007**, *4*, 807–818. [CrossRef]
30. Floyd, R.A.; Hensley, K. Oxidative stress in brain aging: Implications for therapeutics of neurodegenerative diseases. *Neurobiol. Aging* **2002**, *23*, 795–807. [CrossRef]
31. Pomytkin, I.; Costa-Nunes, J.P.; Kasatkin, V.; Veniaminova, E.; Demchenko, A.; Lyundup, A.; Lesch, K.P.; Ponomarev, E.D.; Strekalova, T. Insulin receptor in the brain: Mechanisms of activation and the role in the CNS pathology and treatment. *CNS Neurosci. Ther.* **2018**, *24*, 763–774. [CrossRef]
32. Mielke, J.G.; Taghibiglou, C.; Liu, L.; Zhang, Y.; Jia, Z.; Adeli, K.; Wang, Y.T. A biochemical and functional characterization of diet-induced brain insulin resistance. *J. Neurochem.* **2005**, *93*, 1568–1578. [CrossRef]
33. Hoyer, S. The brain insulin signal transduction system and sporadic (type II) Alzheimer disease: An update. *J. Neural Transm.* **2002**, *109*, 341–360. [CrossRef]
34. Pivari, F.; Mingione, A.; Brasacchio, C.; Soldati, L. Curcumin and type 2 diabetes mellitus: Prevention and treatment. *Nutrients* **2019**, *11*, 1837. [CrossRef] [PubMed]
35. Den Hartogh, D.J.; Gabriel, A.; Tsiani, E. Antidiabetic properties of curcumin ii: Evidence from in vivo studies. *Nutrients* **2020**, *12*, 118. [CrossRef]
36. Wang, Y.; Yin, H.; Lou, J.; Han, B.; Qin, X.; Meng, F.; Geng, S.; Liu, Y. Effects of curcumin on hippocampal Bax and Bcl-2 expression and cognitive function of a rat model of Alzheimer's disease. *Neural Regen. Res.* **2011**, *6*, 1845–1849.
37. Yanagisawa, D.; Ibrahim, N.F.; Taguchi, H.; Morikawa, S.; Hirao, K.; Shirai, N.; Sogabe, T.; Tooyama, I. Curcumin derivative with the substitution at C-4 position, but not curcumin, is effective against amyloid pathology in APP/PS1 mice. *Neurobiol. Aging* **2015**, *36*, 201–210. [CrossRef]
38. Zhang, L.; Fang, Y.; Xu, Y.; Lian, Y.; Xie, N.; Wu, T.; Zhang, H.; Sun, L.; Zhang, R.; Wang, Z. Curcumin improves amyloid β-peptide (1-42) induced spatial memory deficits through BDNF-ERK signaling pathway. *PLoS ONE* **2015**, *10*, e0131525. [CrossRef] [PubMed]

39. Musillo, C.; Borgi, M.; Saul, N.; Möller, S.; Luyten, W.; Berry, A.; Cirulli, F. Natural products improve healthspan in aged mice and rats: A systematic review and meta-analysis. *Neurosci. Biobehav. Rev.* **2021**, *121*, 89–105. [CrossRef] [PubMed]
40. Agrawal, R.; Mishra, B.; Tyagi, E.; Nath, C.; Shukla, R. Effect of curcumin on brain insulin receptors and memory functions in STZ (ICV) induced dementia model of rat. *Pharmacol. Res.* **2010**, *61*, 247–252. [CrossRef] [PubMed]
41. Banji, O.J.F.; Banji, D.; Ch, K. Curcumin and hesperidin improve cognition by suppressing mitochondrial dysfunction and apoptosis induced by D-galactose in rat brain. *Food Chem. Toxicol.* **2014**, *74*, 51–59. [CrossRef] [PubMed]
42. Sundaram, J.R.; Poore, C.P.; Sulaimee, N.H.B.; Pareek, T.; Cheong, W.F.; Wenk, M.R.; Pant, H.C.; Frautschy, S.A.; Low, C.M. Kesavapany, S. Curcumin Ameliorates Neuroinflammation, Neurodegeneration, and Memory Deficits in p25 Transgenic Mouse Model that Bears Hallmarks of Alzheimer's Disease. *J. Alzheimers. Dis.* **2017**, *60*, 1429–1442. [CrossRef] [PubMed]
43. Banji, D.; Banji, O.J.F.; Dasaroju, S.; Kumar Ch, K. Curcumin and piperine abrogate lipid and protein oxidation induced by d-galactose in rat brain. *Brain Res.* **2013**, *1515*, 1–11. [CrossRef]
44. Banji, D.; Banji, O.J.F.; Dasaroju, S.; Annamalai, A.R. Piperine and curcumin exhibit synergism in attenuating D-galactose induced senescence in rats. *Eur. J. Pharmacol.* **2013**, *703*, 91–99. [CrossRef] [PubMed]
45. Bassani, T.B.; Turnes, J.M.; Moura, E.L.R.; Bonato, J.M.; Cóppola-Segovia, V.; Zanata, S.M.; Oliveira, R.M.M.W.; Vital, M.A.B.F. Effects of curcumin on short-term spatial and recognition memory, adult neurogenesis and neuroinflammation in a streptozotocin-induced rat model of dementia of Alzheimer's type. *Behav. Brain Res.* **2017**, *335*, 41–54. [CrossRef]
46. Hoppe, J.B.; Coradini, K.; Frozza, R.L.; Oliveira, C.M.; Meneghetti, A.B.; Bernardi, A.; Pires, E.S.; Beck, R.C.R.; Salbego, C.G. Free and nanoencapsulated curcumin suppress β-amyloid-induced cognitive impairments in rats: Involvement of BDNF and Akt/GSK-3β signaling pathway. *Neurobiol. Learn. Mem.* **2013**, *106*, 134–144. [CrossRef]
47. Ishrat, T.; Hoda, M.N.; Khan, M.B.; Yousuf, S.; Ahmad, M.; Khan, M.M.; Ahmad, A.; Islam, F. Amelioration of cognitive deficits and neurodegeneration by curcumin in rat model of sporadic dementia of Alzheimer's type (SDAT). *Eur. Neuropsychopharmacol* **2009**, *19*, 636–647. [CrossRef]
48. Kumar, A.; Prakash, A.; Dogra, S. Protective effect of curcumin (*Curcuma longa*) against d-galactose-induced senescence in mice. *J Asian Nat. Prod. Res.* **2011**, *13*, 42–55. [CrossRef]
49. Sandhir, R.; Yadav, A.; Mehrotra, A.; Sunkaria, A.; Singh, A.; Sharma, S. Curcumin nanoparticles attenuate neurochemical and neurobehavioral deficits in experimental model of Huntington's disease. *NeuroMol. Med.* **2014**, *16*, 106–118. [CrossRef]
50. Singh, S.; Kumar, P. Neuroprotective potential of curcumin in combination with piperine against 6-hydroxy dopamine induced motor deficit and neurochemical alterations in rats. *Inflammopharmacology* **2017**, *25*, 69–79. [CrossRef]
51. Sarker, M.R.; Franks, S.F. Efficacy of curcumin for age-associated cognitive decline: A narrative review of preclinical and clinical studies. *GeroScience* **2018**, *40*, 73–95. [CrossRef] [PubMed]
52. Panahi, Y.; Khalili, N.; Sahebi, E.; Namazi, S.; Simental-Mendía, L.E.; Majeed, M.; Sahebkar, A. Effects of Curcuminoids Plus Piperine on Glycemic, Hepatic and Inflammatory Biomarkers in Patients with Type 2 Diabetes Mellitus: A Randomized Double-Blind Placebo-Controlled Trial. *Drug Res.* **2018**, *68*, 403–409. [CrossRef] [PubMed]
53. Akbari, M.; Lankarani, K.B.; Tabrizi, R.; Ghayour-Mobarhan, M.; Peymani, P.; Ferns, G.; Ghaderi, A.; Asemi, Z. The Effects of Curcumin on Weight Loss Among Patients With Metabolic Syndrome and Related Disorders: A Systematic Review and Meta-Analysis of Randomized Controlled Trials. *Front. Pharmacol.* **2019**, *10*, 649. [CrossRef]
54. Adibian, M.; Hodaei, H.; Nikpayam, O.; Sohrab, G.; Hekmatdoost, A.; Hedayati, M. The effects of curcumin supplementation on high-sensitivity C-reactive protein, serum adiponectin, and lipid profile in patients with type 2 diabetes: A randomized, double-blind, placebo-controlled trial. *Phyther. Res.* **2019**, *33*, 1374–1383. [CrossRef] [PubMed]
55. Yang, F.; Lim, G.P.; Begum, A.N.; Ubeda, O.J.; Simmons, M.R.; Ambegaokar, S.S.; Chen, P.; Kayed, R.; Glabe, C.G.; Frautschy, S.A. et al. Curcumin inhibits formation of amyloid β oligomers and fibrils, binds plaques, and reduces amyloid in vivo. *J. Biol. Chem* **2005**, *280*, 5892–5901. [CrossRef]
56. Garcia-Alloza, M.; Borrelli, L.A.; Rozkalne, A.; Hyman, B.T.; Bacskai, B.J. Curcumin labels amyloid pathology in vivo, disrupts existing plaques, and partially restores distorted neurites in an Alzheimer mouse model. *J. Neurochem.* **2007**, *102*, 1095–1104 [CrossRef]
57. den Haan, J.; Morrema, T.H.J.; Rozemuller, A.J.; Bouwman, F.H.; Hoozemans, J.J.M. Different curcumin forms selectively bind fibrillar amyloid beta in post mortem Alzheimer's disease brains: Implications for in-vivo diagnostics. *Acta Neuropathol. Commun* **2018**, *6*, 75. [CrossRef]
58. McKhann, G.M.; Knopman, D.S.; Chertkow, H.; Hyman, B.T.; Jack, C.R.; Kawas, C.H.; Klunk, W.E.; Koroshetz, W.J.; Manly, J.J.; Mayeux, R.; et al. The diagnosis of dementia due to Alzheimer's disease: Recommendations from the National Institute on Aging-Alzheimer's Association workgroups on diagnostic guidelines for Alzheimer's disease. *Alzheimer's Dement.* **2011**, *7* 263–269. [CrossRef]
59. Mrak, R.E.; Griffin, W.S.T. Potential Inflammatory biomarkers in Alzheimer's disease. *J. Alzheimer's Dis.* **2006**, *8*, 369–375 [CrossRef]
60. Lin, X.; Bai, D.; Wei, Z.; Zhang, Y.; Huang, Y.; Deng, H.; Huang, X. Curcumin attenuates oxidative stress in RAW264.7 cells by increasing the activity of antioxidant enzymes and activating the Nrf2-Keap1 pathway. *PLoS ONE* **2019**, *14*, e0216711. [CrossRef]
61. Scuto, M.C.; Mancuso, C.; Tomasello, B.; Ontario, M.L.; Cavallaro, A.; Frasca, F.; Maiolino, L.; Salinaro, A.T.; Calabrese, E.J. Calabrese, V. Curcumin, Hormesis and the Nervous System. *Nutrients* **2019**, *11*, 2417. [CrossRef] [PubMed]

62. Du, X.X.; Xu, H.M.; Jiang, H.; Song, N.; Wang, J.; Xie, J.X. Curcumin protects nigral dopaminergic neurons by iron-chelation in the 6-hydroxydopamine rat model of Parkinson's disease. *Neurosci. Bull.* **2012**, *28*, 253–258. [CrossRef] [PubMed]
63. Chen, J.; Tang, X.Q.; Zhi, J.L.; Cui, Y.; Yu, H.M.; Tang, E.H.; Sun, S.N.; Feng, J.Q.; Chen, P.X. Curcumin protects PC12 cells against 1-methyl-4-phenylpyridinium ion-induced apoptosis by bcl-2-mitochondria-ROS-iNOS pathway. *Apoptosis* **2006**, *11*, 943–953. [CrossRef] [PubMed]
64. Yu, S.; Zheng, W.; Xin, N.; Chi, Z.H.; Wang, N.Q.; Nie, Y.X.; Feng, W.Y.; Wang, Z.Y. Curcumin prevents dopaminergic neuronal death through inhibition of the c-Jun N-terminal kinase pathway. *Rejuvenation Res.* **2010**, *13*, 55–64. [CrossRef]
65. Guo, J.; Cao, X.; Hu, X.; Li, S.; Wang, J. The anti-apoptotic, antioxidant and anti-inflammatory effects of curcumin on acrylamide-induced neurotoxicity in rats. *BMC Pharmacol. Toxicol.* **2020**, *21*, 62. [CrossRef]
66. Pennisi, M.; Crupi, R.; Di Paola, R.; Ontario, M.L.; Bella, R.; Calabrese, E.J.; Crea, R.; Cuzzocrea, S.; Calabrese, V. Inflammasomes, hormesis, and antioxidants in neuroinflammation: Role of NRLP3 in Alzheimer disease. *J. Neurosci. Res.* **2017**, *95*, 1360–1372. [CrossRef]
67. Heneka, M.T.; Kummer, M.P.; Stutz, A.; Delekate, A.; Schwartz, S.; Vieira-Saecker, A.; Griep, A.; Axt, D.; Remus, A.; Tzeng, T.C.; et al. NLRP3 is activated in Alzheimer's disease and contributes to pathology in APP/PS1 mice. *Nature* **2013**, *493*, 674–678. [CrossRef]
68. Lonnemann, N.; Hosseini, S.; Marchetti, C.; Skouras, D.B.; Stefanoni, D.; D'Alessandro, A.; Dinarello, C.A.; Korte, M. The NLRP3 inflammasome inhibitor OLT1177 rescues cognitive impairment in a mouse model of Alzheimer's disease. *Proc. Natl. Acad. Sci. USA* **2020**, *117*, 32145–32154. [CrossRef]
69. White, C.S.; Lawrence, C.B.; Brough, D.; Rivers-Auty, J. Inflammasomes as therapeutic targets for Alzheimer's disease. *Brain Pathol.* **2017**, *27*, 223–234. [CrossRef]
70. Ghosh, S.; Banerjee, S.; Sil, P.C. The beneficial role of curcumin on inflammation, diabetes and neurodegenerative disease: A recent update. *Food Chem. Toxicol.* **2015**, *83*, 111–124. [CrossRef]
71. He, Q.; Jiang, L.; Man, S.; Wu, L.; Hu, Y.; Chen, W. Curcumin Reduces Neuronal Loss and Inhibits the NLRP3 Inflammasome Activation in an Epileptic Rat Model. *Curr. Neurovasc. Res.* **2018**, *15*, 186–192. [CrossRef]
72. Giorgio, M.; Trinei, M.; Migliaccio, E.; Pelicci, P.G. Hydrogen peroxide: A metabolic by-product or a common mediator of ageing signals? *Nat. Rev. Mol. Cell Biol.* **2007**, *8*, 722–728. [CrossRef]
73. Berry, A.; Cirulli, F. The p66Shc gene paves the way for healthspan: Evolutionary and mechanistic perspectives. *Neurosci. Biobehav. Rev.* **2013**, *37*, 790–802. [CrossRef]
74. Berniakovich, I.; Trinei, M.; Stendardo, M.; Migliaccio, E.; Minucci, S.; Bernardi, P.; Pelicci, P.G.; Giorgio, M. p66Shc-generated oxidative signal promotes fat accumulation. *J. Biol. Chem.* **2008**, *283*, 34283–34293. [CrossRef] [PubMed]
75. Trinei, M.; Berniakovich, I.; Beltrami, E.; Migliaccio, E.; Fassina, A.; Pelicci, P.G.; Giorgio, M. P66Shc signals to age. *Aging* **2009**, *1*, 503–510. [CrossRef] [PubMed]
76. Bellisario, V.; Berry, A.; Capoccia, S.; Raggi, C.; Panetta, P.; Branchi, I.; Piccaro, G.; Giorgio, M.; Pelicci, P.G.; Cirulli, F. Gender-dependent resiliency to stressful and metabolic challenges following prenatal exposure to high-fat diet in the p66Shc-/- mouse. *Front. Behav. Neurosci.* **2014**, *8*, 285. [CrossRef] [PubMed]
77. Berry, A.; Greco, A.; Giorgio, M.; Pelicci, P.G.; de Kloet, R.; Alleva, E.; Minghetti, L.; Cirulli, F. Deletion of the lifespan determinant p66Shc improves performance in a spatial memory task, decreases levels of oxidative stress markers in the hippocampus and increases levels of the neurotrophin BDNF in adult mice. *Exp. Gerontol.* **2008**, *43*, 200–208. [CrossRef]
78. Berry, A.; Capone, F.; Giorgio, M.; Pelicci, P.G.; de Kloet, E.R.; Alleva, E.; Minghetti, L.; Cirulli, F. Deletion of the life span determinant p66Shc prevents age-dependent increases in emotionality and pain sensitivity in mice. *Exp. Gerontol.* **2007**, *42*, 37–45. [CrossRef] [PubMed]
79. Berry, A.; Carnevale, D.; Giorgio, M.; Pelicci, P.G.; de Kloet, E.R.; Alleva, E.; Minghetti, L.; Cirulli, F. Greater resistance to inflammation at adulthood could contribute to extended life span of p66Shc-/- mice. *Exp. Gerontol.* **2010**, *45*, 343–350. [CrossRef]
80. Bucci, M.; Huovinen, V.; Guzzardi, M.A.; Koskinen, S.; Raiko, J.R.; Lipponen, H.; Ahsan, S.; Badeau, R.M.; Honka, M.J.; Koffert, J.; et al. Resistance training improves skeletal muscle insulin sensitivity in elderly offspring of overweight and obese mothers. *Diabetologia* **2016**, *59*, 77–86. [CrossRef]
81. Berry, A.; Bucci, M.; Raggi, C.; Eriksson, J.G.; Guzzardi, M.A.; Nuutila, P.; Huovinen, V.; Iozzo, P.; Cirulli, F. Dynamic changes in p66Shc mRNA expression in peripheral blood mononuclear cells following resistance training intervention in old frail women born to obese mothers: A pilot study. *Aging Clin. Exp. Res.* **2018**, *30*, 871–876. [CrossRef] [PubMed]
82. ALTamimi, J.Z.; AlFaris, N.A.; AL-Farga, A.M.; Alshammari, G.M.; BinMowyna, M.N.; Yahya, M.A. Curcumin reverses diabetic nephropathy in streptozotocin-induced diabetes in rats by inhibition of PKCβ/p66Shc axis and activation of FOXO-3a. *J. Nutr. Biochem.* **2021**, *87*, 108515. [CrossRef]
83. Scudellari, M. The science myths that will not die. *Nature* **2015**, *528*, 322–325. [CrossRef] [PubMed]
84. Berry, A.; Bellisario, V.; Panetta, P.; Raggi, C.; Magnifico, M.C.; Arese, M.; Cirulli, F. Administration of the antioxidant n-acetyl-cysteine in pregnant mice has long-term positive effects on metabolic and behavioral endpoints of male and female offspring prenatally exposed to a high-fat diet. *Front. Behav. Neurosci.* **2018**, *12*, 48. [CrossRef] [PubMed]
85. Huang, Y.Z.; McNamara, J.O. Neuroprotective effects of reactive oxygen species mediated by BDNF-independent activation of TrkB. *J. Neurosci.* **2012**, *32*, 15521–15532. [CrossRef] [PubMed]

86. Calabrese, E.J.; Dhawan, G.; Kapoor, R.; Mattson, M.P.; Rattan, S.I. Curcumin and hormesis with particular emphasis on neural cells. *Food Chem. Toxicol.* **2019**, *129*, 399–404. [CrossRef] [PubMed]
87. Mattson, M.P. Challenging oneself intermittently to improve health. *Dose-Response* **2014**, *12*, 600–618. [CrossRef]
88. Kocaadam, B.; Şanlier, N. Curcumin, an active component of turmeric (*Curcuma longa*), and its effects on health. *Crit. Rev. Food Sci. Nutr.* **2017**, *57*, 2889–2895. [CrossRef]
89. Hewlings, S.; Kalman, D. Curcumin: A Review of Its Effects on Human Health. *Foods* **2017**, *6*, 92. [CrossRef]
90. Jäger, R.; Lowery, R.P.; Calvanese, A.V.; Joy, J.M.; Purpura, M.; Wilson, J.M. Comparative absorption of curcumin formulations. *Nutr. J.* **2014**, *13*, 11. [CrossRef]
91. Asai, A.; Miyazawa, T. Occurrence of orally administered curcuminoid as glucuronide and glucuronide/sulfate conjugates in rat plasma. *Life Sci.* **2000**, *67*, 2785–2793. [CrossRef]
92. Ireson, C.R.; Jones, D.J.L.; Orr, S.; Coughtrie, M.W.H.; Boocock, D.J.; Williams, M.L.; Farmer, P.B.; Steward, W.P.; Gescher, A.J. Metabolism of the Cancer Chemopreventive Agent Curcumin in Human and Rat Intestine. *Cancer Epidemiol. Prev. Biomark.* **2002**, *11*, 105–111.
93. Pan, M.H.; Huang, T.M.; Lin, J.K. Biotransformation of curcumin through reduction and glucuronidation in mice. *Drug Metab. Dispos.* **1999**, *27*, 486–494. [PubMed]
94. Ireson, C.; Orr, S.; Jones, D.J.; Verschoyle, R.; Lim, C.K.; Luo, J.L.; Howells, L.; Plummer, S.; Jukes, R.; Williams, M.; et al. Characterization of metabolites of the chemopreventive agent curcumin in human and rat hepatocytes and in the rat in vivo, and evaluation of their ability to inhibit phorbol ester-induced prostaglandin E2 production. *Cancer Res.* **2001**, *61*, 61.
95. Lao, C.D.; Ruffin IV, M.T.; Normolle, D.; Heath, D.D.; Murray, S.I.; Bailey, J.M.; Boggs, M.E.; Crowell, J.; Rock, C.L.; Brenner, D.E. Dose escalation of a curcuminoid formulation. *BMC Complementary Altern. Med.* **2006**, *6*, 10. [CrossRef] [PubMed]
96. Dhillon, N.; Aggarwal, B.B.; Newman, R.A.; Wolff, R.A.; Kunnumakkara, A.B.; Abbruzzese, J.L.; Ng, C.S.; Badmaev, V.; Kurzrock, R. Phase II trial of curcumin in patients with advanced pancreatic cancer. *Clin. Cancer Res.* **2008**, *14*, 4491–4499. [CrossRef]
97. Stohs, S.J.; Chen, O.; Ray, S.D.; Ji, J.; Bucci, L.R.; Preuss, H.G. Highly Bioavailable Forms of Curcumin and Promising Avenues for Curcumin-Based Research and Application: A Review. *Molecules* **2020**, *25*, 1397. [CrossRef] [PubMed]
98. Suresh, D.; Srinivasan, K. Tissue distribution & elimination of capsaicin, piperine & curcumin following oral intake in rats. *Indian J. Med. Res.* **2010**, *131*, 5.
99. Di Meo, F.; Filosa, S.; Madonna, M.; Giello, G.; Di Pardo, A.; Maglione, V.; Baldi, A.; Crispi, S. Curcumin C3 complex®/Bioperine®has antineoplastic activity in mesothelioma: An in vitro and in vivo analysis. *J. Exp. Clin. Cancer Res.* **2019**, *38*, 360. [CrossRef] [PubMed]
100. Banji, D.; Banji, O.J.F.; Srinivas, K. Neuroprotective Effect of Turmeric Extract in Combination with Its Essential Oil and Enhanced Brain Bioavailability in an Animal Model. *Biomed. Res. Int.* **2021**, *2021*, 1–12. [CrossRef]
101. Sumeet, G.; Rachna, K.; Samrat, C.; Ipshita, C.; Vikas, J.; Manu, S. Anti Inflammatory and Anti Arthritic Activity of Different Milk Based Formulation of Curcumin in Rat Model. *Curr. Drug Deliv.* **2018**, *15*, 205–214. [CrossRef]
102. Cuomo, J.; Appendino, G.; Dern, A.S.; Schneider, E.; McKinnon, T.P.; Brown, M.J.; Togni, S.; Dixon, B.M. Comparative absorption of a standardized curcuminoid mixture and its lecithin formulation. *J. Nat. Prod.* **2011**, *74*, 664–669. [CrossRef] [PubMed]
103. Asher, G.N.; Xie, Y.; Moaddel, R.; Sanghvi, M.; Dossou, K.S.S.; Kashuba, A.D.M.; Sandler, R.S.; Hawke, R.L. Randomized Pharmacokinetic Crossover Study Comparing 2 Curcumin Preparations in Plasma and Rectal Tissue of Healthy Human Volunteers. *J. Clin. Pharmacol.* **2017**, *57*, 185–193. [CrossRef] [PubMed]
104. Purpura, M.; Lowery, R.P.; Wilson, J.M.; Mannan, H.; Münch, G.; Razmovski-Naumovski, V. Analysis of different innovative formulations of curcumin for improved relative oral bioavailability in human subjects. *Eur. J. Nutr.* **2018**, *57*, 929–938. [CrossRef]
105. Mirzaei, H.; Shakeri, A.; Rashidi, B.; Jalili, A.; Banikazemi, Z.; Sahebkar, A. Phytosomal curcumin: A review of pharmacokinetic, experimental and clinical studies. *Biomed. Pharmacother.* **2017**, *85*, 102–112. [CrossRef]
106. Gera, M.; Sharma, N.; Ghosh, M.; Huynh, D.L.; Lee, S.J.; Min, T.; Kwon, T.; Jeong, D.K. Nanoformulations of curcumin: An emerging paradigm for improved remedial application. *Oncotarget* **2017**, *8*, 66680–66698. [CrossRef] [PubMed]
107. Rahimi, H.R.; Nedaeinia, R.; Sepehri Shamloo, A.; Nikdoust, S.; Kazemi Oskuee, R. Novel delivery system for natural products: Nano-curcumin formulations. *Avicenna J. Phytomed.* **2016**, *6*, 383–398. [CrossRef]
108. Ban, C.; Jo, M.; Park, Y.H.; Kim, J.H.; Han, J.Y.; Lee, K.W.; Kweon, D.H.; Choi, Y.J. Enhancing the oral bioavailability of curcumin using solid lipid nanoparticles. *Food Chem.* **2020**, *302*, 125328. [CrossRef]
109. Moballegh Nasery, M.; Abadi, B.; Poormoghadam, D.; Zarrabi, A.; Keyhanvar, P.; Khanbabaei, H.; Ashrafizadeh, M.; Mohammadinejad, R.; Tavakol, S.; Sethi, G. Curcumin Delivery Mediated by Bio-Based Nanoparticles: A Review. *Molecules* **2020**, *25*, 689. [CrossRef]
110. Del Prado-Audelo, M.L.; Caballero-Florán, I.H.; Meza-Toledo, J.A.; Mendoza-Muñoz, N.; González-Torres, M.; Florán, B.; Cortés, H.; Leyva-Gómez, G. Formulations of curcumin nanoparticles for brain diseases. *Biomolecules* **2019**, *9*, 56. [CrossRef]
111. Hu, B.; Liu, X.; Zhang, C.; Zeng, X. Food macromolecule based nanodelivery systems for enhancing the bioavailability of polyphenols. *J. Food Drug Anal.* **2017**, *25*, 3–15. [CrossRef]
112. Yavarpour-Bali, H.; Pirzadeh, M.; Ghasemi-Kasman, M. Curcumin-loaded nanoparticles: A novel therapeutic strategy in treatment of central nervous system disorders. *Int. J. Nanomed.* **2019**, *14*, 4449–4460. [CrossRef] [PubMed]
113. Yallapu, M.M.; Nagesh, P.K.B.; Jaggi, M.; Chauhan, S.C. Therapeutic Applications of Curcumin Nanoformulations. *AAPS J.* **2015**, *17*, 1341–1356. [CrossRef]

114. Kalani, A.; Tyagi, A.; Tyagi, N. Exosomes: Mediators of neurodegeneration, neuroprotection and therapeutics. *Mol. Neurobiol.* **2014**, *49*, 590–600. [CrossRef] [PubMed]
115. Aqil, F.; Munagala, R.; Jeyabalan, J.; Agrawal, A.K.; Gupta, R. Exosomes for the Enhanced Tissue Bioavailability and Efficacy of Curcumin. *AAPS J.* **2017**, *19*, 1691–1702. [CrossRef] [PubMed]
116. Oskouie, M.N.; Aghili Moghaddam, N.S.; Butler, A.E.; Zamani, P.; Sahebkar, A. Therapeutic use of curcumin-encapsulated and curcumin-primed exosomes. *J. Cell. Physiol.* **2019**, *234*, 8182–8191. [CrossRef] [PubMed]
117. Borik, R.M.; Fawzy, N.M.; Abu-Bakr, S.M.; Aly, M.S. Design, synthesis, anticancer evaluation and docking studies of novel heterocyclic derivatives obtained via reactions involving curcumin. *Molecules* **2018**, *23*, 1398. [CrossRef] [PubMed]
118. Chainoglou, E.; Hadjipavlou-Litina, D. Curcumin in Health and Diseases: Alzheimer's Disease and Curcumin Analogues, Derivatives, and Hybrids. *Int. J. Mol. Sci.* **2020**, *21*, 1975. [CrossRef]
119. Moselhy, S.S.; Razvi, S.; Hasan, N.; Balamash, K.S.; Abulnaja, K.O.; Yaghmoor, S.S.; Youssri, M.A.; Kumosani, T.A.; Al-Malki, A.L. Multifaceted role of a marvel golden molecule, curcumin: A review. *Indian J. Pharm. Sci.* **2018**, *80*, 400–411.
120. Scazzocchio, B.; Minghetti, L.; D'archivio, M. Interaction between gut microbiota and curcumin: A new key of understanding for the health effects of curcumin. *Nutrients* **2020**, *12*, 1–18. [CrossRef]
121. Hassaninasab, A.; Hashimoto, Y.; Tomita-Yokotani, K.; Kobayashi, M. Discovery of the curcumin metabolic pathway involving a unique enzyme in an intestinal microorganism. *Proc. Natl. Acad. Sci. USA* **2011**, *108*, 6615–6620. [CrossRef]
122. Burapan, S.; Kim, M.; Han, J. Curcuminoid Demethylation as an Alternative Metabolism by Human Intestinal Microbiota. *J. Agric. Food Chem.* **2017**, *65*, 3305–3310. [CrossRef] [PubMed]
123. Wu, J.C.; Tsai, M.L.; Lai, C.S.; Wang, Y.J.; Ho, C.T.; Pan, M.H. Chemopreventative effects of tetrahydrocurcumin on human diseases. *Food Funct.* **2014**, *5*, 12–17. [CrossRef] [PubMed]
124. Pinkaew, D.; Changtam, C.; Tocharus, C.; Govitrapong, P.; Jumnongprakhon, P.; Suksamrarn, A.; Tocharus, J. Association of Neuroprotective Effect of Di-O-Demethylcurcumin on Aβ25–35-Induced Neurotoxicity with Suppression of NF-κB and Activation of Nrf2. *Neurotox. Res.* **2016**, *29*, 80–91. [CrossRef]
125. Ahmed, T.; Enam, S.A.; Gilani, A.H. Curcuminoids enhance memory in an amyloid-infused rat model of Alzheimer's disease. *Neuroscience* **2010**, *169*, 1296–1306. [CrossRef]
126. Bostancıklioğlu, M. Intestinal Bacterial Flora and Alzheimer's Disease. *Neurophysiology* **2018**, *50*, 140–148. [CrossRef]
127. Rogers, G.B.; Keating, D.J.; Young, R.L.; Wong, M.L.; Licinio, J.; Wesselingh, S. From gut dysbiosis to altered brain function and mental illness: Mechanisms and pathways. *Mol. Psychiatry* **2016**, *21*, 738–748. [CrossRef] [PubMed]
128. Mulders, R.J.; de Git, K.C.G.; Schéle, E.; Dickson, S.L.; Sanz, Y.; Adan, R.A.H. Microbiota in obesity: Interactions with enteroendocrine, immune and central nervous systems. *Obes. Rev.* **2018**, *19*, 435–451. [CrossRef]
129. Baum, L.; Lam, C.W.K.; Cheung, S.K.K.; Kwok, T.; Lui, V.; Tsoh, J.; Lam, L.; Leung, V.; Hui, E.; Ng, C.; et al. Six-month randomized, placebo-controlled, double-blind, pilot clinical trial of curcumin in patients with Alzheimer disease. *J. Clin. Psychopharmacol.* **2008**, *28*, 110–113. [CrossRef]
130. Ringman, J.M.; Frautschy, S.A.; Teng, E.; Begum, A.N.; Bardens, J.; Beigi, M.; Gylys, K.H.; Badmaev, V.; Heath, D.D.; Apostolova, L.G.; et al. Oral curcumin for Alzheimer's disease: Tolerability and efficacy in a 24-week randomized, double blind, placebo-controlled study. *Alzheimer's Res. Ther.* **2012**, *4*, 43. [CrossRef]
131. Cox, K.H.M.; Pipingas, A.; Scholey, A.B. Investigation of the effects of solid lipid curcumin on cognition and mood in a healthy older population. *J. Psychopharmacol.* **2015**, *29*, 642–651. [CrossRef] [PubMed]
132. Rainey-Smith, S.R.; Brown, B.M.; Sohrabi, H.R.; Shah, T.; Goozee, K.G.; Gupta, V.B.; Martins, R.N. Curcumin and cognition: A randomised, placebo-controlled, double-blind study of community-dwelling older adults. *Br. J. Nutr.* **2016**, *115*, 2106–2113. [CrossRef]
133. Small, G.W.; Siddarth, P.; Li, Z.; Miller, K.J.; Ercoli, L.; Emerson, N.D.; Martinez, J.; Wong, K.P.; Liu, J.; Merrill, D.A.; et al. Memory and Brain Amyloid and Tau Effects of a Bioavailable Form of Curcumin in Non-Demented Adults: A Double-Blind, Placebo-Controlled 18-Month Trial. *Am. J. Geriatr. Psychiatry* **2018**, *26*, 266–277. [CrossRef]
134. Saul, N.; Möller, S.; Cirulli, F.; Berry, A.; Luyten, W.; Fuellen, G. Health and longevity studies in C. elegans: The "healthy worm database" reveals strengths, weaknesses and gaps of test compound-based studies. *Biogerontology* **2021**, *22*, 215–236. [CrossRef]
135. Cohen, A.A.; Luyten, W.; Gogol, M.; Simm, A.; Saul, N.; Cirulli, F.; Berry, A.; Antal, P.; Köhling, R.; Wouters, B.; et al. Health and aging: Unifying concepts, scores, biomarkers and pathways. *Aging Dis.* **2019**, *10*, 883–900. [CrossRef]
136. Berry, A.; Cirulli, F. High-Fat Diet and Foetal Programming: Use of P66Shc Knockouts and Implications for Human Kind. In *Diet, Nutrition, and Fetal Programming*; Springer International Publishing: Berlin/Heidelberg, Germany, 2017; pp. 557–568.
137. Rinwa, P.; Kumar, A. Piperine potentiates the protective effects of curcumin against chronic unpredictable stress-induced cognitive impairment and oxidative damage in mice. *Brain Res.* **2012**, *1488*, 38–50. [CrossRef] [PubMed]
138. Parachikova, A.; Green, K.N.; Hendrix, C.; Laferla, F.M. Formulation of a medical food cocktail for Alzheimer's disease: Beneficial effects on cognition and neuropathology in a mouse model of the disease. *PLoS ONE* **2010**, *5*, e14015. [CrossRef]
139. Kundu, P.; Das, M.; Tripathy, K.; Sahoo, S.K. Delivery of Dual Drug Loaded Lipid Based Nanoparticles across the Blood-Brain Barrier Impart Enhanced Neuroprotection in a Rotenone Induced Mouse Model of Parkinson's Disease. *ACS Chem. Neurosci.* **2016**, *7*, 1658–1670. [CrossRef]

140. Ma, Q.L.; Zuo, X.; Yang, F.; Ubeda, O.J.; Gant, D.J.; Alaverdyan, M.; Teng, E.; Hu, S.; Chen, P.P.; Maiti, P.; et al. Curcumin suppresses soluble Tau dinners and corrects molecular chaperone, synaptic, and behavioral deficits in aged human Tau transgenic mice. *J. Biol. Chem.* **2013**, *288*, 4056–4065. [CrossRef] [PubMed]
141. Appendino, G.; Belcaro, G.; Cornelli, U.; Luzzi, R.; Togni, S.; Dugall, M.; Cesarone, M.R.; Feragalli, I.B.; Ippolito, E.; Errichi, B.M.; et al. Potential role of curcumin phytosome (Meriva) in controlling the evolution of diabetic microangiopathy. A pilot study. *Panminerva Med.* **2011**, *53*, 43–49.
142. Panahi, Y.; Kianpour, P.; Mohtashami, R.; Jafari, R.; Simental-Mendiá, L.E.; Sahebkar, A. Curcumin Lowers Serum Lipids and Uric Acid in Subjects with Nonalcoholic Fatty Liver Disease: A Randomized Controlled Trial. *J. Cardiovasc. Pharmacol.* **2016**, *68*, 223–229. [CrossRef]
143. Franceschi, F.; Feregalli, B.; Togni, S.; Cornelli, U.; Giacomelli, L.; Eggenhoffner, R.; Belcaro, G. A novel phospholipid delivery system of curcumin (Meriva®) preserves muscular mass in healthy aging subjects. *Eur. Rev. Med. Pharmacol. Sci.* **2016**, *20*, 762–766. [PubMed]
144. Thota, R.N.; Dias, C.B.; Abbott, K.A.; Acharya, S.H.; Garg, M.L. Curcumin alleviates postprandial glycaemic response in healthy subjects: A cross-over, randomized controlled study. *Sci. Rep.* **2018**, *8*, 1–8. [CrossRef] [PubMed]
145. Steigerwalt, R.; Nebbioso, M.; Appendino, G.; Belcaro, G.; Ciammaichella, G.; Cornelli, U.; Luzzi, R.; Togni, S.; Dugall, M.; Cesarone, M.R.; et al. Meriva®, a lecithinized curcumin delivery system, in diabetic microangiopathy and retinopathy. *Panminerva Med.* **2012**, *54*, 11–16.
146. Burroughs, V.J.; Maxey, R.W.; Levy, R.A. Racial and ethnic differences in response to medicines: Towards individualized pharmaceutical treatment. *J. Natl. Med. Assoc.* **2002**, *94*, 1–26.
147. Ristow, M.; Schmeisser, K. Mitohormesis: Promoting health and lifespan by increased levels of reactive oxygen species (ROS). *Dose-Response* **2014**, *12*, 288–341. [CrossRef]
148. Saleh, H.A.; Ramdan, E.; Elmazar, M.M.; Azzazy, H.M.E.; Abdelnaser, A. Comparing the protective effects of resveratrol, curcumin and sulforaphane against LPS/IFN-γ-mediated inflammation in doxorubicin-treated macrophages. *Sci. Rep.* **2021**, *11*, 545. [CrossRef]
149. Austad, S.N. Why women live longer than men: Sex differences in longevity. *Gend. Med.* **2006**, *3*, 79–92. [CrossRef]
150. Christensen, A.; Pike, C.J. Menopause, obesity and inflammation: Interactive risk factors for Alzheimer's disease. *Front. Aging Neurosci.* **2015**, *7*, 130. [CrossRef]
151. Alzheimer's Association. 2020 Alzheimer's disease facts and figures. *Alzheimer's Dement.* **2020**, *16*, 391–460. [CrossRef]
152. Ogden, C.L.; Carroll, M.D.; Kit, B.K.; Flegal, K.M. Prevalence of childhood and adult obesity in the United States, 2011-2012. *JAMA J. Am. Med. Assoc.* **2014**, *311*, 806–814. [CrossRef] [PubMed]
153. Kawamoto, R.; Kikuchi, A.; Akase, T.; Ninomiya, D.; Kumagi, T. Usefulness of waist-to-height ratio in screening incident metabolic syndrome among Japanese community-dwelling elderly individuals. *PLoS ONE* **2019**, *14*, e0216069. [CrossRef] [PubMed]
154. McGill, A.T. Past and future corollaries of theories on causes of metabolic syndrome and obesity related co-morbidities part 2: A composite unifying theory review of human-specific co-adaptations to brain energy consumption. *Arch. Public Health* **2014**, *72*, 31. [CrossRef]
155. Wajchenberg, B.L. Subcutaneous and Visceral Adipose Tissue: Their Relation to the Metabolic Syndrome. *Endocr. Rev.* **2000**, *21*, 697–738. [CrossRef]
156. Jayaraman, A.; Pike, C.J. Alzheimer's disease and type 2 diabetes: Multiple mechanisms contribute to interactions topical collection on pathogenesis of type 2 diabetes and insulin Resistance. *Curr. Diab. Rep.* **2014**, *14*, 476. [CrossRef]
157. Pfeilschifter, J.; Köditz, R.; Pfohl, M.; Schatz, H. Changes in proinflammatory cytokine activity after menopause. *Endocr. Rev.* **2002**, *23*, 90–119. [CrossRef]
158. Ataei-Almanghadim, K.; Farshbaf-Khalili, A.; Ostadrahimi, A.R.; Shaseb, E.; Mirghafourvand, M. The effect of oral capsule of curcumin and vitamin E on the hot flashes and anxiety in postmenopausal women: A triple blind randomised controlled trial. *Complementary Ther. Med.* **2020**, *48*, 102267. [CrossRef] [PubMed]
159. Cirulli, F.; Musillo, C.; Berry, A. Maternal Obesity as a Risk Factor for Brain Development and Mental Health in the Offspring. *Neuroscience* **2020**, *447*, 122–135. [CrossRef]
160. Long, F.; Yang, H.; Xu, Y.; Hao, H.; Li, P. A strategy for the identification of combinatorial bioactive compounds contributing to the holistic effect of herbal medicines. *Sci. Rep.* **2015**, *5*, 12361. [CrossRef]

Review

Obesity-Associated Inflammation: Does Curcumin Exert a Beneficial Role?

Rosaria Varì, Beatrice Scazzocchio, Annalisa Silenzi, Claudio Giovannini and Roberta Masella *

Center for Gender-Specific Medicine, Gender Specific Prevention and Health Unit, Istituto Superiore di Sanità, Viale Regina Elena 299, 00161 Rome, Italy; rosaria.vari@iss.it (R.V.); beatrice.scazzocchio@iss.it (B.S.); annalisa.silenzi@iss.it (A.S.); claudio.giovannini@iss.it (C.G.)

* Correspondence: roberta.masella@iss.it

Abstract: Curcumin is a lipophilic polyphenol, isolated from the plant turmeric of Curcuma longa. Curcuma longa has always been used in traditional medicine in Asian countries because it is believed to have numerous health benefits. Nowadays it is widely used as spice component and in emerging nutraceutical food worldwide. Numerous studies have shown that curcumin possesses, among others, potential anti-inflammatory properties. Obesity represents a main risk factor for several chronic diseases, including type 2 diabetes, cardiovascular disease, and some types of cancer. The establishment of a low-grade chronic inflammation, both systemically and locally in adipose tissue, occurring in obesity most likely represents a main factor in the pathogenesis of chronic diseases. The molecular mechanisms responsible for the onset of the obesity-associated inflammation are different from those involved in the classic inflammatory response caused by infections and involves different signaling pathways. The inflammatory process in obese people is triggered by an inadequate intake of nutrients that produces quantitative and qualitative alterations of adipose tissue lipid content, as well as of various molecules that act as endogenous ligands to activate immune cells. In particular, dysfunctional adipocytes secrete inflammatory cytokines and chemokines, the adipocytokines, able to recruit immune cells into adipose tissue, amplifying the inflammatory response also at systemic level. This review summarizes the most recent studies focused at elucidating the molecular targets of curcumin activity responsible for its anti-inflammatory properties in obesity-associated inflammation and related pathologies.

Keywords: curcumin; obesity; inflammation; adipose tissue

Citation: Varì, R.; Scazzocchio, B.; Silenzi, A.; Giovannini, C.; Masella, R. Obesity-Associated Inflammation: Does Curcumin Exert a Beneficial Role? *Nutrients* **2021**, *13*, 1021. https://doi.org/10.3390/nu13031021

Academic Editor: Giuseppe Grosso

Received: 25 February 2021
Accepted: 18 March 2021
Published: 22 March 2021

Publisher's Note: MDPI stays neutral with regard to jurisdictional claims in published maps and institutional affiliations.

Copyright: © 2021 by the authors. Licensee MDPI, Basel, Switzerland. This article is an open access article distributed under the terms and conditions of the Creative Commons Attribution (CC BY) license (https://creativecommons.org/licenses/by/4.0/).

1. Introduction

Curcumin, the main natural polyphenol found in the rhizome of Curcuma longa (turmeric) [1], has been recognized for thousands of years because of its medicinal properties and potential health benefits [2]. It is used worldwide in different forms: as spice, antiseptic, anti-inflammatory, preservative or coloring agent, as well as supplement in capsules or powder form [3]. It has been reported the beneficial effect of curcumin in various diseases, including inflammatory and degenerative conditions, cancer, dyslipidemia, metabolic syndrome (MetS), and obesity [4–7]. Several studies have also shown that most of the benefits are due to its antioxidant and anti-inflammatory activities [5]. Overweight and obesity are a major public health problem all over the world [8]. Obesity is caused by the imbalance between energy intake and energy expenditure, culminating in the excess of fat accumulation in the adipose tissue (AT) [9]. It is associated with a chronic low-grade inflammation that might represent the main factor linking obesity and the development and progression of various diseases including type 2 diabetes (T2D), dyslipidemia, heart diseases, stroke, and cancer [10,11]. AT, indeed, is recognized as an endocrine organ that secretes a number of cytokines and chemokines with regulatory and immune functions [12]. Dysfunctions of the secretory activity of AT, thus, most likely play a pathogenic role in the occurrence of the obesity-related pathologies [13]. In consideration of this, the current review examines specifically the possible role of curcumin in counteracting the activation of inflammatory pathways in AT. To this purpose, we conducted a comprehensive literature search until December 2020 in PubMed, using "obesity", "inflammation", "adipose tissue", "adipocyte" as key words in combination with "curcumin" and "dietary curcumin".

2. Obesity, AT Dysfunction and Inflammation

Obesity is characterized by an excessive AT expansion due to hyperplasia (increase in number) and/or hypertrophy (increase in size) of adipocytes, the major cellular component of AT. Although the main function of adipocytes is the storage and release of lipids, they secrete also active molecules that are used for intracellular signaling and to communicate with every organ system in the body. The second largest AT cellular component beyond adipocytes are resident immune cells that, in turn, play important roles in the maintenance of AT homeostasis. The intensity and complexity of these signal networks are highly regulated, differ in each fat pad, and are dramatically affected by various disease states. In conclusion, AT is an active endocrine organ, secreting a variety of hormones and metabolites that regulate systemic metabolism. When the imbalance in the storage of lipids by fat cells is established, alterations in secretive function occur and systemic metabolic dysfunctions might happen, such as in T2D, cardiovascular and liver diseases, and cancers. Through these cellular derangement and metabolic dysfunction, an excessive caloric intake contributes to a chronic low-grade inflammation, also known as 'metainflammation'. In particular, visceral AT accumulated in the abdominal seat, shows a disrupted balance between secreted pro- and anti-inflammatory factors, with increased levels of pro-inflammatory adipocytokines, including leptin [14], and decrease of anti-inflammatory adipokines, such as adiponectin [15]. These events all together cause local alterations of the AT environment and alter the normal cross-talk with other organs, such as liver, muscle, brain, and pancreas [16,17], which leads, as a further result, to metabolic dysfunctions, such as hyperinsulinemia and insulin resistance (IR) [11]. Furthermore, the polarization profile of the resident immune cells depends on the health status of the adipocytes [18]. Changes in the adipocyte secretion profile, in fact, trigger the recruitment and activation of immune cells [19]. In particular, in obese subjects, AT macrophages shift from an anti-inflammatory profile (such as that found in normal weight people) towards a pro-inflammatory phenotype [20,21] producing themselves an alteration in the production/activation of key factors that exacerbate local and systemic inflammation [22], such as tumor necrosis factor (TNF)α, interleukin (IL)-6, IL-1β, toll-like receptor (TLR) 4, and nuclear factor (NF)-κB, that may amplify the inflammatory state and favor the onset of pathologies [23–27]. The pro-inflammatory profile in obese individuals is evidenced by elevated serum levels of TNFα and IL-6, simultaneously with adiponectin and anti-inflammatory cytokines decrease [28,29]. The expression of the pro-inflammatory cytokines is regulated by the activation of the transcription factor NF-κB. This factor is stored in the cytoplasm as inactive form bound to the inhibitor IκBα that is, in turn, regulated by the inhibitor of κB kinase (IKK) complex consisting of 2 subunits, IKKα and IKKβ. Different stimuli, including growth factors, cytokines and foreign pathogens or molecules, such as lipopolysaccharides (LPS) and free fatty acids (FFA) [30], activate the IKK kinase complex inducing proteasomal degradation of IκBα and leading to the translocation of NF-κB in the nucleus, where it induces the expression of genes of various inflammatory mediators. Obese people show an increased activation of NF-κB pathway, most likely responsible for the increased pro-inflammatory cytokine release [31]. Among the pro-inflammatory compounds, it should be mentioned leptin; it is primarily produced by AT, its level increases in obese people and participates in the control of body weight by regulating food intake and energy expenditure [32]. On the other hand, adiponectin, produced almost exclusively by AT, circulates in high concentration in plasma and has anti-inflammatory properties probably related to the inhibition of NF-κB activation and, consequently, to the reduced synthesis of pro-inflammatory cytokine [33,34]. The secretion of anti-inflammatory adipocytokines is inhibited in visceral AT from obese patients and subjects with MetS leading to a significant reduction in their plasma levels [35,36]. Despite the intense experimental work carried out, the exact molecular mechanisms responsible for the chronic low-grade metabolic inflammation in obesity are not completely clarified yet. However, with the identification of the nod-like receptor pyrin domain-containing (NLRP)3 inflammasome in AT, a new hypothesis has been formulated suggesting that it might be relevant for regulating obesity-associated inflammation and insulin sensitivity [37]. The NLRP3 inflammasome is a cytosolic molecular complex whose expression in AT directly correlates with body weight and aging, while its inactivation significantly mitigates metabolic disorders [38,39]. A number of exogenous and endogenous signals might act as NLRP3 inflammasome activator in AT leading to the production of pro-inflammatory cytokines [40,41] a potential mechanism linking an elevated intake of saturated fatty acids (SFAs) to the progression of metabolic diseases. In line with this, increased gene expressions of NLRP3 and its key effectors IL-1β and IL-18 have been observed in visceral fat of metabolically unhealthy individuals compared to those from lean healthy control or metabolically healthy obese individuals [42]. Furthermore, these inflammatory effects were suppressed, after weight loss, in the subcutaneous fat of patients with obesity and T2D, with consequent improvement in insulin sensitivity [38]. Studies have hypothesized a causal nexus between systemic inflammation and an increased release of FFA from AT in obese and insulin-resistant subjects [43,44]. Indeed, the direct drainage of free FA and adipokines from visceral AT to the liver can activate immune responses leading to the secretion of inflammatory compounds [17]. A potential mechanism of action through which FFA, and mainly dietary SFAs, can mediate AT dysfunctions contributing to the onset of inflammation involves the TLR4 [45,46]. TLR4 belongs to the TLR family and is expressed not only on leukocytes but also on many non-immune cells, including adipocytes, hepatocytes, and muscle cells. It has been hypothesized that FFA can bind and stimulate TLR4; thus, the elevated plasma level of FFA observed in obesity could activate TLR4. A recent research showed that the TLR4 activation can mediate inflammatory processes also through the impairment of adipogenesis which, in turn, elicit adipocyte and resident immune cell dysfunctions [47]. In conclusion, the onset of inflammatory processes linked to obesity and metabolic dysfunction in AT involves a number of different factors closely intertwined. The inflammation associated with obesity has been shown to derive from changes in the delicate crosstalk between adipocytes and macrophages due to an increased infiltration of macrophages into AT, the activation of a number of pro-inflammatory pathways, the alterations of adipokine production and increased expression and release of a panel of inflammatory cytokines. Understanding the molecular and metabolic switches that, starting from AT, lead to immune cells polarization towards inflammatory phenotypes may allow the definition of interventions capable of leading to the resolution of inflammation and blocking the sequence of events responsible for the occurrence of clinical complications in obesity. Targeting the key intracellular pathways underlying AT dysfunctions might represent a useful tool in counteracting obesity-related pathologies. From this point of view, the identification of potential protective activity of curcumin in positively modulating AT pro/anti-inflammatory balance has been gaining significant interest.

3. Curcumin and Inflammation in Obesity

Several studies carried out in humans have shown that curcumin attenuates inflammation in obesity and obesity-related diseases by rebalancing the equilibrium between anti- and pro-inflammatory factors via different mechanisms due to the interactions of curcumin with a wide range of biomolecules, such as transcription factors, cellular receptors, growth factors, enzymes, cytokines, and chemokines [48,49]. Moreover, some reports have suggested that curcumin can enhance weight loss induced by diet and lifestyle intervention on overweight subjects with MetS ([50,51]. However, it should be considered that a main problem in the use of curcumin is its poor bioavailability. To increase curcumin bioavailability, different delivery systems including micelles, liposomes, phospholipid complexes, nanostructured lipid carriers, and biopolymer nanoparticles have been developed, as well as the addition of piperine, a bioactive alkaloid extracted from the Piper species, which has been shown to effectively enhance the bioavailability of several nutritional supplements including curcumin [52].

4. Curcumin Decreases Circulating Inflammatory Markers in Overweight/Obese Subjects

There is an increasing evidence that curcumin treatment could be able to alleviate the altered pro-inflammatory mediator secretions present in obesity and related pathologies. In this section, data from human studies carried out on overweight and obese subjects with curcumin supplementation are collected and summarized. A research performed on 84 overweight or obese patients with non-alcoholic fatty liver disease (NAFLD) demonstrated that, curcumin supplementation with two 40 mg capsules/day after meals for 3 months, induces a decrease in many serum inflammatory markers, such as TNFα, high-sensitive C-reactive protein (hs-CRP), and IL-6 [53]. The same conclusions were reached by other studies carried out in obese/overweight people; specifically, curcumin administration (1 g/day) for 8 weeks reduces serum concentrations of TNFα, IL-6, and monocyte chemoattractant protein 1 (MCP-1) in males and females with diagnosis of MetS with respect to the placebo group [7]. In a randomized placebo-controlled clinical trial carried out on 60 adolescent girls undergoing to a slight weight-loss diet for 10 weeks, curcumin consumption (500 mg/day) was able to induce a significant decrease in hs-CRP and IL-6 compared to placebo supplementation [54]. In addition, it has been demonstrated that curcumin modulates circulating levels of IL-1β in thirty subjects randomized to receive curcumin (1 g/day) or a matched placebo for 4 weeks. Serum IL-1β was found to be significantly reduced by curcumin treatment. In contrast, no significant difference was observed in the concentrations of IL-6, and MCP-1 [55]. Finally, curcuminoids supplementation (300 mg/day) for 3 months in T2D patients led to a significant decrease in circulating FFA levels [56], that are considered a major factor linking obesity and inflammation [57–59].

5. Curcumin Modulates Adipokines

Adiponectin and leptin are two important adipokines released by adipocytes that have [18] several target organs including brain, liver, pancreas, muscle, immune system, and AT itself. They are involved in inflammation and immune response, showing, as stated above, adiponectin anti-inflammatory properties, leptin, on the contrary, pro-inflammatory ones [60]. Obese subjects are characterized by an imbalance of the two adipokines showing a low concentration of adiponectin and high levels of leptin in plasma [61]. Curcumin has been shown to increase the production of adiponectin [62]. To this regard, a systematic review [63] showed that curcuminoid administration significantly increased plasma adiponectin concentrations in randomized controlled trials. Specifically, in a double-blind randomized trial carried out over a 12-week period on 118 patients with T2D the effects of the daily administration of 1 g curcumin added with 10 mg piperine were compared to placebo. The treatment with curcumin plus piperine reduced serum levels of TNFα and increased serum level of adiponectin [64]. In another study, curcumin supplementation (1 g/day) for 6 weeks increased serum adiponectin concentrations compared to both curcumin-phospholipid complex (1 g/day) and placebo groups in 120 men and women with MetS [65]. In a randomized double-blind study 44 men and women with T2D were treated with curcumin 1500 mg/day or placebo for 10 weeks. At the end of the study, a significant increase in serum adiponectin concentration together with a decrease in the mean weight were observed in the curcumin group [66]. Conversely, no effect on adiponectin was seen in 22 young men randomly assigned to receive curcumin (500 mg/day) or placebo for 12 weeks. This finding might be determined by the low dose of curcumin used for the treatment [67]. However, the same amount of curcumin (500 mg/day) for 4 weeks reduced serum leptin and resistin and increased adiponectin content in 15 children and 15 adults [68,69]. Accordingly, elevated levels of adiponectin and decreased leptin levels were reported in diabetic men and women after 6-month intervention with a high dose of curcumin (1500 mg/day) [70]. Similar effects on serum levels of leptin were observed in males and females with NAFLD treated for 12 weeks with even higher doses of curcumin (3000 mg/day) [71]. In conclusion, all the studies discussed show that curcumin supplementation contributes to rebalance pro- and anti-inflammatory factor production significantly increasing the levels of anti-inflammatory adipocytokines, such as adiponectin, and decreasing the pro-inflammatory ones, such as TNFα, IL-6, IL-1β, and MCP-1, counteracting the chronic inflammatory condition in overweight/obese subjects (Table 1).

Table 1. Effects of curcumin on inflammation in obesity: human studies.

Study Design	Subjects	Treatment	Duration	Outcomes	References
Randomized double-blind, placebo-controlled	Overweight/obese with NAFLD (males and females, $n = 84$)	42 curcumin (40 mg/day) 42 placebo	3 months	↓TNF-alpha and IL-6	[53]
Randomized, double-blind, placebo-controlled	Overweight/obese with MetS (males and females, $n = 117$)	59 curcumin (1 g/day) 58 placebo	8 weeks	↓TNF-α, IL-6, and MCP-1	[7]
Randomized, double-blind, placebo-controlled	Overweight/obese (adolescent girls, $n = 60$)	30 curcumin (500 mg/day) 30 placebo	10 weeks	↓IL-6	[54]
Randomized, double blind, crossover	Obese (males and females, $n = 30$)	15 curcumin (1g/day + 5 mg bioperine) 15 placebo	4 weeks each treatment + 2 weeks wash-out between the regimens.	↓IL-1β no changes IL-6, and MCP-1	[55]
Randomized, double-blind, placebo-controlled	Overweight/obese with T2D (males and females, $n = 100$)	50 curcumin (300 mg/day) 50 placebo	3 months	↓FFA	[56]
Randomized, double-blind, placebo-controlled	T2D (unspecified gender $n= 100$)	50 curcumin (1 g + 10 mg piperine/day) 50 placebo	12 weeks	↓TNF-α and Leptin ↑ Adiponectin	[64]
Randomized, double-blind, placebo-controlled	Overweight with T2D (males and females, $n = 44$)	21 curcumin (1500 mg/day) 23 placebo	10 weeks	↑ Adiponectin ↓weight	[66]
Randomized, double-blind, placebo-controlled	Obese (males and females, 29 adults, 29 children)	15 children curcumin (500 mg/day) 14 children placebo 15 adults curcumin (500 mg/day) 14 adults placebo	4 weeks	↓Leptin ↓Resistin ↑Adiponectin	[68,69]
Randomized, double-blind, placebo-controlled	Obese with MetS (males and females, $n = 120$)	40 curcumin (1 g/day) 40 placebo 40 phospholipidated curcumin (1 g/day)	6 weeks	↑ Adiponectin	[65]
Randomized, double-blind, placebo-controlled	Overweight T2D (males and females, $n = 210$)	107 curcumin (1.5 g/day) 103 placebo	6 months	↓ Leptin ↑ Adiponectin	[70]
Randomized double-blind, placebo-controlled	Overweight\obese with NAFLD (males and females, $n = 46$)	23 curcumin (3 g/day) 23 placebo	12 weeks	↓Leptin	[71]

Table 1. *Cont.*

Study Design	Subjects	Treatment	Duration	Outcomes	References
Randomized double-blind, placebo-controlled	Obese (males, $n = 22$)	11 curcumin (500 mg/day) 11 placebo	12 weeks	no change Adiponectin	[67]

Abbreviations: ↑ Increases; ↓ Decreases; IL-6, interleukin-6; IL-1β, interleukin-1β; MCP-1, monocyte chemoattractant protein-1; TNFα, tumor necrosis factor α; FFA, free fatty acids; T2D, type 2 diabetes; MetS, metabolic syndrome; NAFLD; nonalcoholic fatty liver disease.

6. Effects of Curcumin on Inflammatory Signaling Pathways

Most of the potential molecular mechanisms responsible for the health effects of curcumin have been studied in animals and in vitro models using stabilized cell lines or human primary cells. Although the results obtained in this way cannot be completely extrapolated to humans, they allowed to suggest possible mechanisms of curcumin action that could explain the phenotypic effects evidenced by the studies carried out in humans (Figure 1). As regards animal studies, the anti-inflammatory effect of curcumin was first demonstrated in acute and chronic models of inflammation in rats and mice [72]. Specifically, in obese mice, curcumin treatment (3% by weight for 6 weeks) decreases NF-κB activity in liver tissue, associated with decreased hepatic expression of inflammatory molecules, such as TNFα and MCP-1. Curcumin-treated obese mice also show a decreased macrophage infiltration and an increased expression of forkhead transcription factor (Foxo)1 and adiponectin into AT, and higher circulating adiponectin levels [72]. In line with these results, an in vivo study performed in male rats, T2D insulin resistant because of high-fat diet (HFD) consumption, demonstrated oral administration of curcumin (80 mg/kg body weight) was able to improve insulin sensitivity by attenuating TNFα serum levels [73]. Moreover, dietary curcumin (4 g/kg diet added 2 days/week) attenuated the inflammatory response induced by HFD in mice through inhibiting NF-κB expression and JNK signaling pathway in epididymal AT [74]. Interestingly, administration of 0.1% curcumin associated with white pepper (0.01%) (Curcuma-P®) significantly down-regulated the proinflammatory cytokines IL-6 and TNFα, but did not modify IL-1β and MCP-1, in the subcutaneous AT of mice after 4 weeks of HFD. This effect was relatively tissue-specific and independent on macrophage infiltration. Indeed, the inflammatory cell infiltration in AT was not modified by Curcuma-P® supplementation [75]. Another interesting activity of curcumin has been demonstrated on endoplasmic reticulum (ER). The involvement of ER stress (with accumulation of misfolded proteins) in the release of FFA has been observed in several studies [76–78]. Curcumin has been demonstrated to mitigate ER stress in mice fed HFD and in primary adipocytes. Specifically, short-term HFD feeding (10 days) increased ER stress in mouse AT by increasing the expression of phospho-inositol-requiring kinase 1(p-IRE1) and phospho-eukaryotic Initiation Factor 2 (p-eIF2), two important indicators of ER stress. Oral administration of curcumin (50 mg/kg) counteracted the activation of IRE1 and eIF2 by reducing the phosphorylation, and consequently inhibiting the ER stress in vivo. Similarly, curcumin (0.1, 1, 10 μM) treatment inhibited IRE1 and eIF2 activation in mouse AT treated with 100 μM of palmitate (inductor of ER stress). Furthermore, curcumin administration reduced glycerol and FFA released from AT of HFD-fed mice blocking cAMP/PKA signaling via regulation of AMP-activated protein kinase (AMPK) [79], as well as significantly decreased plasma FFA levels in HF-induced obese rats [80]. Several in vitro models have been used to collect more detailed information about the potential molecular mechanisms of action through which curcumin exerts its effects. Curcumin has been shown to inhibit the activation of the pro-inflammatory NF-κB signaling pathway in several cell types, including human adipocytes and macrophages [48]. In adipocytes treated with TNFα to induce inflammatory processes, the contemporary treatment with 20 μM curcumin suppressed the degradation of IκBα, the NF-κB inhibitor, reducing, consequently, NF-κB translocation to the nucleus and significantly inhibiting the expression of TNFα, IL-1β, IL-6 and COX2 genes and IL-6 secretion [81]. In the same type of cells, curcumin also exerts a protective effect on hypoxia in a dose-dependent manner (5, 10, and 20 μM) reducing the secretion of the inflammatory cytokines and protecting mitochondrial functions [82]. Besides a direct effect on adipocytes, curcumin has been shown to exert anti-inflammatory effects by counteracting the increased recruitment of macrophages in AT from obese mice [72,83]. Several studies have evidenced that curcumin treatment reduces macrophage invasion of AT in mouse models of obesity [62,72]. It has been shown that the cross-talk between adipocytes and macrophages in AT triggers and increases inflammatory responses in obesity including the increased production of MCP-1 and other inflammatory cytokines [84,85]. Curcumin treatment (0.1–10 μM) of Raw 264.7 macrophages incubated with the culture medium of mesenteric AT taken from obese mice, potentially able to induce an inflammatory response, significantly inhibited the production of TNFα, MCP-1, and nitric oxide, as well as the migration capacity of the macrophages with respect to the cells not treated with curcumin. Furthermore, 10 μM curcumin treatment significantly inhibited MCP-1 release from 3T3-L1 adipocytes [86]. Studies carried out in different cell systems strongly suggest that the anti-inflammatory activity of curcumin occurs by modulating NLRP3 inflammasome. In THP-1 macrophages treated with phorbol 12-myristate 13-acetate (PMA), an activator of NLRP3 inflammasome, curcumin (6.25, 12.5, and 25 μM) reduced NLRP3 inflammasome level, the activation of caspase-1 and the secretion of IL-1β in a dose-dependent manner, most likely down-regulating TLR4/NF-κB signal transduction pathway that is involved in NLRP3 inflammasome activation [87]. In mouse bone marrow-derived macrophages (BMDM) treated with nigericin, another NLRP3 inflammasome activator, the pre-treatment with curcumin (30–50 μM) inhibited caspase-1 cleavage and IL-1β secretion. The same results were observed in human macrophages. Specifically, differentiated THP-1 cells pretreated with curcumin showed reduced caspase-1 activation and IL-1β secretion after treatment with LPS and nigericin. The inhibition of NLRP3 activation by curcumin appears to be due to the suppression of K^+ efflux [88].

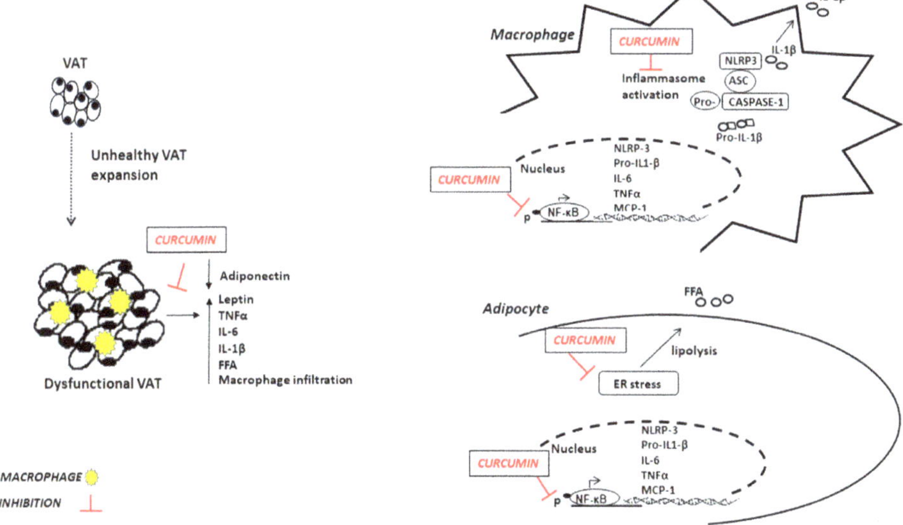

Figure 1. Potential anti-inflammatory mechanisms of curcumin in obesity. VAT: visceral adipose tissue; FFA: free fatty acids; TNFα: tumor necrosis factor α; IL-6: interleukin 6; IL-1β: interleukin β; ER: endoplasmic reticulum; NF-κB: Nuclear transcriptor factor kappa B; MCP-1: Monocyte chemoattractant protein-1; NLRP-3: nod-like receptor pyrin domain-containing 3.

In conclusion, a growing body of experimental data supports the hypothesis that the beneficial effects of curcumin on obesity-related pathologies may be related to the suppression of IL-6, TNFα, IL-1β, and MCP-1 expression from adipocytes, the inhibition of macrophage recruitment in AT, and the inhibition of the inflammatory activity of the NLRP3 inflammasome [89] (Table 2).

Table 2. Effects of curcumin on inflammation in obesity: in vivo and in vitro studies.

Animal Model	Diet	Duration	Outcome	References
Male C57BL/6 mice Wild-type and ob/ob	Standard diet (4% fat) ± curcumin 3% by weight HFD (35% fat) ± curcumin 3% by weight (*n* = 5/group)	6-weeks	in adipose tissue ↑Foxo1 and adiponectin expression ↓infiltration of macrophages ↑circulating Adiponectin levels ↓MCP-1. ↓TNFα, MCP-1 expression, and NF-κB activity in liver	[72]
Male Sprague Dawley rats	Standard diet (control) HFD HFD + curcumin (80 mg/kg/day) (*n* = 11/group)	60 days 75 days	↓FFA and TNFα serum levels in all group compared to non-treated HFD groups	[73]
Male C57BL/6J mice	Low-fat diet (10% Kcal from fat) HFD (45% Kcal from fat) HFD + curcumin (4 g/kg diet) (*n* = 12/group)	28 weeks	in adipose tissue ↓macrophage infiltration ↓NF-κB expression and JNK signaling pathway activation	[74]
Male C57BL/6J mice	Standard Diet (control) HFD HFD + Curcuma-P® (0.1% curcumin + 0.01% white pepper) (*n* = 8/group)	4 weeks	↓IL-6 and TNFα, no changes in MCP1, IL-1β, CD68, and F4/80 in adipose tissue	[75]
Male C57BL/6 mice	HFD HFD + curcumin (50 mg/kg) (*n* = 6/group)	10 days	↓ER stress in adipose tissue ↓FFA release from adipose tissue	[79]
Male Wistar rats	Standard diet (control) HFD HFD + curcuminoid (30, 60, 90 mg\Kg body weigth\day) (*n* = 12/group)	12 weeks	↓FFA plasma levels	[80]
Cell type				
Raw 264.7 macrophages treated with conditioned medium by mesenteric adipose tissue 3T3-L1 adipocytes	0.1–1–10 μM curcumin 10 μM curcumin	24 h	↓TNFα and MCP-1 release ↓MCP-1 release	[86]
3T3-L1 adipocytes treated with TNF-α	2–20 μM curcumin	62 h	↓NF-κB activation. ↓TNFα, IL-1β, IL-6, expression ↓IL-6 secretion	[81]
3T3-L1 adipocytes 24-h hypoxia	5, 10, 20 mM curcumin	24 h	↓TNFα, IL-1β, IL-6 release	[82]

Table 2. *Cont.*

Animal Model	Diet	Duration	Outcome	References
THP-1 macrophages treated with PMA	6.25, 12.5, 25 μM curcumin	24 h	↓NLRP3 inflammasome expression, caspase-1 activation, IL-1β secretion, TLR4 expression, and NF-κB activation	[87]
Mouse, bone marrow-derived macrophages (BMDM) treated with nigericin (10 mM) THP-1 cells treated with LPS	30–50 μM curcumin 30–50 μM curcumin	1 h	↓caspase-1 cleavage ↓IL-1β secretion ↓caspase-1 activation ↓IL-1β secretion	[88]

Abbreviations: ↑ Increases; ↓ Decreases; IL-6, interleukin-6; IL-1β, interleukin-1β; MCP-1, monocyte chemoattractant protein-1; TNFα, tumor necrosis factor α; FFA, free fatty acids; HFD, high-fat diet; ER, endoplasmic reticulum; LPS, lipopolysaccharides; PMA, phorbol 12-myristate 13-acetate; Foxo1, forkhead transcription factor 1; NF-κB, nuclear transcriptor factor kappa B; JNK, jun N-terminal kinase; TLR4, toll-like receptor 4; NLRP3, nod-like receptor pyrin domain-containing 3.

7. Conclusions

The inflammation present in AT is involved in the development of various obesity-related pathologies. The studies reported in this review clearly show that curcumin supplementation significantly decreases inflammatory cytokine production and increases adiponectin level in plasma of obese and overweight subjects. Furthermore, curcumin can regulate several molecular targets including transcription factors (NF-kB, NLP3), signaling pathways and other complex regulatory systems in AT resulting in the suppression/attenuation of the chronic low-grade inflammation. However, since curcumin is widely used as a supplement around the world because of its health promoting properties, further studies, both in vitro to better define the mechanisms of action, and in humans by controlled gender-based clinical trials to evaluate the real effectiveness, are mandatory. It should be reached, thus, that ultimate evidence on curcumin effects and highlighted possible differences in the response to curcumin treatment between women and men, allowing the definition of personalized advice about curcumin consumption.

Author Contributions: R.V., B.S., and R.M. provided substantial contributions to the conception of the work, as well as manuscript writing; C.G., A.S. contributed to the final discussion and revision of the manuscript. All authors have read and agreed to the published version of the manuscript.

Funding: This review received no external funding.

Institutional Review Board Statement: Not applicable.

Informed Consent Statement: Not applicable.

Data Availability Statement: Data sharing is not applicable to this article.

Conflicts of Interest: The authors declare no conflict of interest.

References

1. Jin, T.R. Curcumin and dietary polyphenol research: Beyond drug discovery. *Acta Pharmacol. Sin.* **2018**, *39*, 779–786. [CrossRef]
2. Kocaadam, B.; Sanlier, N. Curcumin, an active component of turmeric (Curcuma longa), and its effects on health. *Crit. Rev. Food Sci. Nutr.* **2017**, *57*, 2889–2895. [CrossRef] [PubMed]
3. Gupta, S.C.; Patchva, S.; Aggarwal, B.B. Therapeutic roles of curcumin: Lessons learned from clinical trials. *AAPS J.* **2013**, *15*, 195–218. [CrossRef] [PubMed]
4. Bradford, P.G. Curcumin and obesity. *BioFactors* **2013**, *39*, 78–87. [CrossRef]
5. Aggarwal, B.B.; Harikumar, K.B. Potential therapeutic effects of curcumin, the anti-inflammatory agent, against neurodegenerative, cardiovascular, pulmonary, metabolic, autoimmune and neoplastic diseases. *Int. J. Biochem. Cell Biol.* **2009**, *41*, 40–59. [CrossRef]

6. Sohaei, S.; Amani, R.; Tarrahi, M.J.; Ghasemi-Tehrani, H. The effects of curcumin supplementation on glycemic status, lipid profile and hs-CRP levels in overweight/obese women with polycystic ovary syndrome: A randomized, double-blind, placebo-controlled clinical trial. *Complementary Ther. Med.* **2019**, *47*, 102201. [CrossRef] [PubMed]
7. Panahi, Y.; Hosseini, M.S.; Khalili, N.; Naimi, E.; Simental-Mendia, L.E.; Majeed, M.; Sahebkar, A. Effects of curcumin on serum cytokine concentrations in subjects with metabolic syndrome: A post-hoc analysis of a randomized controlled trial. *Biomed. Pharmacother.* **2016**, *82*, 578–582. [CrossRef]
8. Chooi, Y.C.; Ding, C.; Magkos, F. The epidemiology of obesity. *Metab. Clin. Exp.* **2019**, *92*, 6–10. [CrossRef]
9. Thyagarajan, B.; Foster, M.T. Beiging of white adipose tissue as a therapeutic strategy for weight loss in humans. *Horm. Mol. Biol. Clin. Investig.* **2017**, *31*, 1–13. [CrossRef]
10. Emanuela, F.; Grazia, M.; Marco de, R.; Maria Paola, L.; Giorgio, F.; Marco, B. Inflammation as a link between obesity and metabolic syndrome. *J. Nutr. Metab.* **2012**, *2012*, 476380. [CrossRef]
11. Schwartz, B.; Yehuda-Shnaidman, E. Putative role of adipose tissue in growth and metabolism of colon cancer cells. *Front. Oncol.* **2014**, *4*, 164. [CrossRef]
12. Galic, S.; Oakhill, J.S.; Steinberg, G.R. Adipose tissue as an endocrine organ. *Mol. Cell. Endocrinol.* **2010**, *316*, 129–139. [CrossRef]
13. Esser, N.; Legrand-Poels, S.; Piette, J.; Scheen, A.J.; Paquot, N. Inflammation as a link between obesity, metabolic syndrome and type 2 diabetes. *Diabetes Res. Clin. Pract.* **2014**, *105*, 141–150. [CrossRef]
14. Wellen, K.E.; Hotamisligil, G.S. Obesity-induced inflammatory changes in adipose tissue. *J. Clin. Investig.* **2003**, *112*, 1785–1788. [CrossRef]
15. Lee, B.C.; Lee, J. Cellular and molecular players in adipose tissue inflammation in the development of obesity-induced insulin resistance. *Biochim. Biophys. Acta* **2014**, *1842*, 446–462. [CrossRef] [PubMed]
16. Romacho, T.; Elsen, M.; Rohrborn, D.; Eckel, J. Adipose tissue and its role in organ crosstalk. *Acta Physiol.* **2014**, *210*, 733–753. [CrossRef]
17. Stern, J.H.; Rutkowski, J.M.; Scherer, P.E. Adiponectin, leptin, and fatty acids in the maintenance of metabolic homeostasis through adipose tissue crosstalk. *Cell Metab.* **2016**, *23*, 770–784. [CrossRef]
18. Kershaw, E.E.; Flier, J.S. Adipose tissue as an endocrine organ. *J. Clin. Endocrinol. Metab.* **2004**, *89*, 2548–2556. [CrossRef]
19. Weisberg, S.P.; McCann, D.; Desai, M.; Rosenbaum, M.; Leibel, R.L.; Ferrante, A.W., Jr. Obesity is associated with macrophage accumulation in adipose tissue. *J. Clin. Investig.* **2003**, *112*, 1796–1808. [CrossRef] [PubMed]
20. Castoldi, A.; Naffah de Souza, C.; Camara, N.O.; Moraes-Vieira, P.M. The macrophage switch in obesity development. *Front. Immunol.* **2015**, *6*, 637. [CrossRef]
21. Boulenouar, S.; Michelet, X.; Duquette, D.; Alvarez, D.; Hogan, A.E.; Dold, C.; O'Connor, D.; Stutte, S.; Tavakkoli, A.; Winters, D.; et al. Adipose type one innate lymphoid cells regulate macrophage homeostasis through targeted cytotoxicity. *Immunity* **2017**, *46*, 273–286. [CrossRef]
22. Suganami, T.; Tanimoto-Koyama, K.; Nishida, J.; Itoh, M.; Yuan, X.; Mizuarai, S.; Kotani, H.; Yamaoka, S.; Miyake, K.; Aoe, S.; et al. Role of the Toll-like receptor 4/NF-kappaB pathway in saturated fatty acid-induced inflammatory changes in the interaction between adipocytes and macrophages. *Arterioscler. Thromb. Vasc. Biol.* **2007**, *27*, 84–91. [CrossRef]
23. Castellano-Castillo, D.; Morcillo, S.; Clemente-Postigo, M.; Crujeiras, A.B.; Fernandez-Garcia, J.C.; Torres, E.; Tinahones, F.J.; Macias-Gonzalez, M. Adipose tissue inflammation and VDR expression and methylation in colorectal cancer. *Clin. Epigenetics* **2018**, *10*, 60. [CrossRef]
24. Griffin, C.; Eter, L.; Lanzetta, N.; Abrishami, S.; Varghese, M.; McKernan, K.; Muir, L.; Lane, J.; Lumeng, C.N.; Singer, K. TLR4, TRIF, and MyD88 are essential for myelopoiesis and CD11c(+) adipose tissue macrophage production in obese mice. *J. Biol. Chem.* **2018**, *293*, 8775–8786. [CrossRef] [PubMed]
25. Luna-Vital, D.; Luzardo-Ocampo, I.; Cuellar-Nunez, M.L.; Loarca-Pina, G.; Gonzalez de Mejia, E. Maize extract rich in ferulic acid and anthocyanins prevents high-fat-induced obesity in mice by modulating SIRT1, AMPK and IL-6 associated metabolic and inflammatory pathways. *J. Nutr. Biochem.* **2020**, *79*, 108343. [CrossRef]
26. Chawla, A.; Nguyen, K.D.; Goh, Y.P. Macrophage-mediated inflammation in metabolic disease. Nature reviews. *Immunology* **2011**, *11*, 738–749. [CrossRef]
27. Reilly, S.M.; Saltiel, A.R. Adapting to obesity with adipose tissue inflammation. *Nat. Rev. Endocrinol.* **2017**, *13*, 633–643. [CrossRef] [PubMed]
28. Yamashita, A.S.; Belchior, T.; Lira, F.S.; Bishop, N.C.; Wessner, B.; Rosa, J.C.; Festuccia, W.T. Regulation of metabolic disease-associated inflammation by nutrient sensors. *Mediat. Inflamm.* **2018**, *2018*, 8261432. [CrossRef]
29. Hocking, S.; Samocha-Bonet, D.; Milner, K.L.; Greenfield, J.R.; Chisholm, D.J. Adiposity and insulin resistance in humans: The role of the different tissue and cellular lipid depots. *Endocr. Rev.* **2013**, *34*, 463–500. [CrossRef] [PubMed]
30. Shi, H.; Kokoeva, M.V.; Inouye, K.; Tzameli, I.; Yin, H.; Flier, J.S. TLR4 links innate immunity and fatty acid-induced insulin resistance. *J. Clin. Investig.* **2006**, *116*, 3015–3025. [CrossRef]
31. Shoelson, S.E.; Lee, J.; Yuan, M. Inflammation and the IKK beta/I kappa B/NF-kappa B axis in obesity- and diet-induced insulin resistance. International journal of obesity and related metabolic disorders. *J. Int. Assoc. Study Obes.* **2003**, *27*, S49–S52. [CrossRef]
32. Muruzabal, F.J.; Fruhbeck, G.; Gomez-Ambrosi, J.; Archanco, M.; Burrell, M.A. Immunocytochemical detection of leptin in non-mammalian vertebrate stomach. *Gen. Comp. Endocrinol.* **2002**, *128*, 149–152. [CrossRef]

33. Ajuwon, K.M.; Spurlock, M.E. Adiponectin inhibits LPS-induced NF-kappaB activation and IL-6 production and increases PPARgamma2 expression in adipocytes. *Am. J. Physiol. Regul. Integr. Comp. Physiol.* **2005**, *288*, R1220–R1225. [CrossRef] [PubMed]
34. Ouchi, N.; Kihara, S.; Arita, Y.; Okamoto, Y.; Maeda, K.; Kuriyama, H.; Hotta, K.; Nishida, M.; Takahashi, M.; Muraguchi, M.; et al. Adiponectin, an adipocyte-derived plasma protein, inhibits endothelial NF-kappaB signaling through a cAMP-dependent pathway. *Circulation* **2000**, *102*, 1296–1301. [CrossRef]
35. Schmidt, F.M.; Weschenfelder, J.; Sander, C.; Minkwitz, J.; Thormann, J.; Chittka, T.; Mergl, R.; Kirkby, K.C.; Fasshauer, M.; Stumvoll, M.; et al. Inflammatory cytokines in general and central obesity and modulating effects of physical activity. *PLoS ONE* **2015**, *10*, e0121971. [CrossRef] [PubMed]
36. Ghadge, A.A.; Khaire, A.A.; Kuvalekar, A.A. Adiponectin: A potential therapeutic target for metabolic syndrome. *Cytokine Growth Factor Rev.* **2018**, *39*, 151–158. [CrossRef]
37. Horng, T.; Hotamisligil, G.S. Linking the inflammasome to obesity-related disease. *Nat. Med.* **2011**, *17*, 164–165. [CrossRef] [PubMed]
38. Vandanmagsar, B.; Youm, Y.H.; Ravussin, A.; Galgani, J.E.; Stadler, K.; Mynatt, R.L.; Ravussin, E.; Stephens, J.M.; Dixit, V.D. The NLRP3 inflammasome instigates obesity-induced inflammation and insulin resistance. *Nat. Med.* **2011**, *17*, 179–188. [CrossRef] [PubMed]
39. Strowig, T.; Henao-Mejia, J.; Elinav, E.; Flavell, R. Inflammasomes in health and disease. *Nature* **2012**, *481*, 278–286. [CrossRef]
40. Finucane, O.M.; Sugrue, J.; Rubio-Araiz, A.; Guillot-Sestier, M.V.; Lynch, M.A. The NLRP3 inflammasome modulates glycolysis by increasing PFKFB3 in an IL-1beta-dependent manner in macrophages. *Sci. Rep.* **2019**, *9*, 4034. [CrossRef]
41. Gianfrancesco, M.A.; Dehairs, J.; L'Homme, L.; Herinckx, G.; Esser, N.; Jansen, O.; Habraken, Y.; Lassence, C.; Swinnen, J.V.; Rider, M.H.; et al. Saturated fatty acids induce NLRP3 activation in human macrophages through K(+) efflux resulting from phospholipid saturation and Na, K-ATPase disruption. *Biochim. Biophys. Acta Mol. Cell Biol. Lipids* **2019**, *1864*, 1017–1030. [CrossRef]
42. Esser, N.; L'Homme, L.; De Roover, A.; Kohnen, L.; Scheen, A.J.; Moutschen, M.; Piette, J.; Legrand-Poels, S.; Paquot, N. Obesity phenotype is related to NLRP3 inflammasome activity and immunological profile of visceral adipose tissue. *Diabetologia* **2013**, *56*, 2487–2497. [CrossRef] [PubMed]
43. Keane, K.N.; Cruzat, V.F.; Carlessi, R.; de Bittencourt, P.I., Jr.; Newsholme, P. Molecular events linking oxidative stress and inflammation to Insulin resistance and beta-cell dysfunction. *Oxid. Med. Cell. Longev.* **2015**, *2015*, 181643. [CrossRef] [PubMed]
44. Xin, Y.; Wang, Y.; Chi, J.; Zhu, X.; Zhao, H.; Zhao, S.; Wang, Y. Elevated free fatty acid level is associated with insulin-resistant state in nondiabetic Chinese people. *Diabetes Metab. Syndr. Obes. Targets Ther.* **2019**, *12*, 139–147. [CrossRef] [PubMed]
45. Saberi, M.; Woods, N.B.; de Luca, C.; Schenk, S.; Lu, J.C.; Bandyopadhyay, G.; Verma, I.M.; Olefsky, J.M. Hematopoietic cell-specific deletion of toll-like receptor 4 ameliorates hepatic and adipose tissue insulin resistance in high-fat-fed mice. *Cell Metab.* **2009**, *10*, 419–429. [CrossRef]
46. Nguyen, M.T.; Favelyukis, S.; Nguyen, A.K.; Reichart, D.; Scott, P.A.; Jenn, A.; Liu-Bryan, R.; Glass, C.K.; Neels, J.G.; Olefsky, J.M. A subpopulation of macrophages infiltrates hypertrophic adipose tissue and is activated by free fatty acids via Toll-like receptors 2 and 4 and JNK-dependent pathways. *J. Biol. Chem.* **2007**, *282*, 35279–35292. [CrossRef]
47. McKernan, K.; Varghese, M.; Patel, R.; Singer, K. Role of TLR4 in the induction of inflammatory changes in adipocytes and macrophages. *Adipocyte* **2020**, *9*, 212–222. [CrossRef]
48. Shehzad, A.; Ha, T.; Subhan, F.; Lee, Y.S. New mechanisms and the anti-inflammatory role of curcumin in obesity and obesity-related metabolic diseases. *Eur. J. Nutr.* **2011**, *50*, 151–161. [CrossRef]
49. Shimizu, K.; Funamoto, M.; Sunagawa, Y.; Shimizu, S.; Katanasaka, Y.; Miyazaki, Y.; Wada, H.; Hasegawa, K.; Morimoto, T. Anti-inflammatory action of curcumin and its use in the treatment of lifestyle-related diseases. *Eur. Cardiol.* **2019**, *14*, 117–122. [CrossRef] [PubMed]
50. Di Pierro, F.; Bressan, A.; Ranaldi, D.; Rapacioli, G.; Giacomelli, L.; Bertuccioli, A. Potential role of bioavailable curcumin in weight loss and omental adipose tissue decrease: Preliminary data of a randomized, controlled trial in overweight people with metabolic syndrome. Preliminary study. *Eur. Rev. Med. Pharmacol. Sci.* **2015**, *19*, 4195–4202.
51. Akbari, M.; Lankarani, K.B.; Tabrizi, R.; Ghayour-Mobarhan, M.; Peymani, P.; Ferns, G.; Ghaderi, A.; Asemi, Z. The effects of curcumin on weight loss among patients with metabolic syndrome and related disorders: A systematic review and meta-analysis of randomized controlled trials. *Front. Pharmacol.* **2019**, *10*, 649. [CrossRef]
52. Srinivasan, K. Black pepper and its pungent principle-piperine: A review of diverse physiological effects. *Crit. Rev. Food Sci. Nutr.* **2007**, *47*, 735–748. [CrossRef]
53. Jazayeri-Tehrani, S.A.; Rezayat, S.M.; Mansouri, S.; Qorbani, M.; Alavian, S.M.; Daneshi-Maskooni, M.; Hosseinzadeh-Attar, M.J. Nano-curcumin improves glucose indices, lipids, inflammation, and Nesfatin in overweight and obese patients with non-alcoholic fatty liver disease (NAFLD): A double-blind randomized placebo-controlled clinical trial. *Nutr. Metab.* **2019**, *16*, 8. [CrossRef]
54. Saraf-Bank, S.; Ahmadi, A.; Paknahad, Z.; Maracy, M.; Nourian, M. Effects of curcumin supplementation on markers of inflammation and oxidative stress among healthy overweight and obese girl adolescents: A randomized placebo-controlled clinical trial. *Phytother. Res. PTR* **2019**, *33*, 2015–2022. [CrossRef]
55. Ganjali, S.; Sahebkar, A.; Mahdipour, E.; Jamialahmadi, K.; Torabi, S.; Akhlaghi, S.; Ferns, G.; Parizadeh, S.M.; Ghayour-Mobarhan, M. Investigation of the effects of curcumin on serum cytokines in obese individuals: A randomized controlled trial. *Sci. World J.* **2014**, *2014*, 898361. [CrossRef]

56. Na, L.X.; Li, Y.; Pan, H.Z.; Zhou, X.L.; Sun, D.J.; Meng, M.; Li, X.X.; Sun, C.H. Curcuminoids exert glucose-lowering effect in type 2 diabetes by decreasing serum free fatty acids: A double-blind, placebo-controlled trial. *Mol. Nutr. Food Res.* **2013**, *57*, 1569–1577. [CrossRef] [PubMed]
57. Arner, P.; Ryden, M. Fatty acids, obesity and insulin resistance. *Obes. Facts* **2015**, *8*, 147–155. [CrossRef]
58. Boden, G. Obesity and free fatty acids. *Endocrinol. Metab. Clin. N. Am.* **2008**, *37*, 635–646. [CrossRef] [PubMed]
59. Glass, C.K.; Olefsky, J.M. Inflammation and lipid signaling in the etiology of insulin resistance. *Cell Metab.* **2012**, *15*, 635–645. [CrossRef] [PubMed]
60. Fruhbeck, G.; Catalan, V.; Rodriguez, A.; Ramirez, B.; Becerril, S.; Salvador, J.; Colina, I.; Gomez-Ambrosi, J. Adiponectin-leptin ratio is a functional biomarker of adipose tissue inflammation. *Nutrients* **2019**, *11*, 454. [CrossRef]
61. Liu, W.; Zhou, X.; Li, Y.; Zhang, S.; Cai, X.; Zhang, R.; Gong, S.; Han, X.; Ji, L. Serum leptin, resistin, and adiponectin levels in obese and non-obese patients with newly diagnosed type 2 diabetes mellitus: A population-based study. *Medicine* **2020**, *99*, e19052. [CrossRef]
62. Zhao, Y.; Chen, B.; Shen, J.; Wan, L.; Zhu, Y.; Yi, T.; Xiao, Z. The beneficial effects of quercetin, curcumin, and resveratrol in obesity. *Oxidative Med. Cell. Longev.* **2017**, *2017*, 1459497. [CrossRef]
63. Simental-Mendia, L.E.; Cicero, A.F.G.; Atkin, S.L.; Majeed, M.; Sahebkar, A. A systematic review and meta-analysis of the effect of curcuminoids on adiponectin levels. *Obes. Res. Clin. Pract.* **2019**, *13*, 340–344. [CrossRef] [PubMed]
64. Panahi, Y.; Khalili, N.; Sahebi, E.; Namazi, S.; Atkin, S.L.; Majeed, M.; Sahebkar, A. Curcuminoids plus piperine modulate adipokines in type 2 diabetes mellitus. *Curr. Clin. Pharmacol.* **2017**, *12*, 253–258. [CrossRef] [PubMed]
65. Salahshooh, M.M.; Parizadeh, S.M.R.; Pasdar, A.; Karimian, M.S.; Safarian, H.; Javandoost, A.; Gordon, A.; Mobarhan, M.G.; Sahebkar, A. The effect of curcumin (Curcuma longa L.) on circulating levels of adiponectin in patients with metabolic syndrome. *Comp. Clin. Pathol.* **2017**, *26*, 17–23. [CrossRef]
66. Adibian, M.; Hodaei, H.; Nikpayam, O.; Sohrab, G.; Hekmatdoost, A.; Hedayati, M. The effects of curcumin supplementation on high-sensitivity C-reactive protein, serum adiponectin, and lipid profile in patients with type 2 diabetes: A randomized, double-blind, placebo-controlled trial. *Phytother. Res. PTR* **2019**, *33*, 1374–1383. [CrossRef] [PubMed]
67. Campbell, M.S.; Ouyang, A.; Krishnakumar, I.M.; Charnigo, R.J.; Westgate, P.M.; Fleenor, B.S. Influence of enhanced bioavailable curcumin on obesity-associated cardiovascular disease risk factors and arterial function: A double-blinded, randomized, controlled trial. *Nutrition* **2019**, *62*, 135–139. [CrossRef]
68. Ismail, N.A.; Abd El Dayem, S.M.; Salama, E.; Ragab, S.; Abd El Baky, A.N.; Ezzat, W.M. Impact of Curcumin Intake on Gluco-Insulin Homeostasis, Leptin and Adiponectin in Obese Subjects. *Res. J. Pharm. Biol. Chem.* **2016**, *7*, 1891–1897.
69. Ismail, N.A.; Ragab, S.; El-Baky, A.N.E.A.; Hamed, M.; Ibrahim, A.A. Effect of oral curcumin administration on insulin resistance, serum resistin and fetuin-A in obese children: Randomized placebo-controlled study. *Res. J. Pharm. Biol. Chem. Sci.* **2014**, *5*, 887–896.
70. Chuengsamarn, S.; Rattanamongkolgul, S.; Phonrat, B.; Tungtrongchitr, R.; Jirawatnotai, S. Reduction of atherogenic risk in patients with type 2 diabetes by curcuminoid extract: A randomized controlled trial. *J. Nutr. Biochem.* **2014**, *25*, 144–150. [CrossRef]
71. Navekar, R.; Rafraf, M.; Ghaffari, A.; Asghari-Jafarabadi, M.; Khoshbaten, M. Turmeric supplementation improves serum glucose indices and leptin levels in patients with nonalcoholic fatty liver diseases. *J. Am. Coll. Nutr.* **2017**, *36*, 261–267. [CrossRef] [PubMed]
72. Weisberg, S.P.; Leibel, R.; Tortoriello, D.V. Dietary curcumin significantly improves obesity-associated inflammation and diabetes in mouse models of diabesity. *Endocrinology* **2008**, *149*, 3549–3558. [CrossRef]
73. El-Moselhy, M.A.; Taye, A.; Sharkawi, S.S.; El-Sisi, S.F.; Ahmed, A.F. The antihyperglycemic effect of curcumin in high fat diet fed rats. Role of TNF-alpha and free fatty acids. *Food Chem. Toxicol.* **2011**, *49*, 1129–1140. [CrossRef]
74. Shao, W.; Yu, Z.; Chiang, Y.; Yang, Y.; Chai, T.; Foltz, W.; Lu, H.; Fantus, I.G.; Jin, T. Curcumin prevents high fat diet induced insulin resistance and obesity via attenuating lipogenesis in liver and inflammatory pathway in adipocytes. *PLoS ONE* **2012**, *7*, e28784. [CrossRef]
75. Neyrinck, A.M.; Alligier, M.; Memvanga, P.B.; Nevraumont, E.; Larondelle, Y.; Preat, V.; Cani, P.D.; Delzenne, N.M. Curcuma longa extract associated with white pepper lessens high fat diet-induced inflammation in subcutaneous adipose tissue. *PLoS ONE* **2013**, *8*, e81252. [CrossRef]
76. Arruda, A.P.; Pers, B.M.; Parlakgul, G.; Guney, E.; Inouye, K.; Hotamisligil, G.S. Chronic enrichment of hepatic endoplasmic reticulum-mitochondria contact leads to mitochondrial dysfunction in obesity. *Nat. Med.* **2014**, *20*, 1427–1435. [CrossRef] [PubMed]
77. Deng, J.; Liu, S.; Zou, L.; Xu, C.; Geng, B.; Xu, G. Lipolysis response to endoplasmic reticulum stress in adipose cells. *J. Chem.* **2012**, *287*, 6240–6249. [CrossRef] [PubMed]
78. Carmen, G.Y.; Victor, S.M. Signaling mechanisms regulating lipolysis. *Cell. Signal.* **2006**, *18*, 401–408. [CrossRef]
79. Wang, L.; Zhang, B.; Huang, F.; Liu, B.; Xie, Y. Curcumin inhibits lipolysis via suppression of ER stress in adipose tissue and prevents hepatic insulin resistance. *J. Lipid Res.* **2016**, *57*, 1243–1255. [CrossRef]
80. Pongchaidecha, A.; Lailerd, N.; Boonprasert, W.; Chattipakorn, N. Effects of curcuminoid supplement on cardiac autonomic status in high-fat-induced obese rats. *Nutrition* **2009**, *25*, 870–878. [CrossRef] [PubMed]
81. Gonzales, A.M.; Orlando, R.A. Curcumin and resveratrol inhibit nuclear factor-kappaB-mediated cytokine expression in adipocytes. *Nutr. Metab.* **2008**, *5*, 17. [CrossRef]

82. Priyanka, A.; Anusree, S.S.; Nisha, V.M.; Raghu, K.G. Curcumin improves hypoxia induced dysfunctions in 3T3-L1 adipocytes by protecting mitochondria and down regulating inflammation. *BioFactors* **2014**, *40*, 513–523. [CrossRef]
83. Meydani, M.; Hasan, S.T. Dietary polyphenols and obesity. *Nutrients* **2010**, *2*, 737–751. [CrossRef] [PubMed]
84. Bruun, J.M.; Lihn, A.S.; Pedersen, S.B.; Richelsen, B. Monocyte chemoattractant protein-1 release is higher in visceral than subcutaneous human adipose tissue (AT): Implication of macrophages resident in the AT. *J. Clin. Endocrinol. Metab.* **2005**, *90*, 2282–2289. [CrossRef]
85. Fain, J.N.; Madan, A.K. Regulation of monocyte chemoattractant protein 1 (MCP-1) release by explants of human visceral adipose tissue. *Int. J. Obes.* **2005**, *29*, 1299–1307. [CrossRef] [PubMed]
86. Woo, H.M.; Kang, J.H.; Kawada, T.; Yoo, H.; Sung, M.K.; Yu, R. Active spice-derived components can inhibit inflammatory responses of adipose tissue in obesity by suppressing inflammatory actions of macrophages and release of monocyte chemoattractant protein-1 from adipocytes. *Life Sci.* **2007**, *80*, 926–931. [CrossRef]
87. Kong, F.; Ye, B.; Cao, J.; Cai, X.; Lin, L.; Huang, S.; Huang, W.; Huang, Z. Curcumin represses NLRP3 inflammasome activation via TLR4/MyD88/NF-kappaB and P2X7R signaling in PMA-induced macrophages. *Front. Pharmacol.* **2016**, *7*, 369. [CrossRef] [PubMed]
88. Yin, H.; Guo, Q.; Li, X.; Tang, T.; Li, C.; Wang, H.; Sun, Y.; Feng, Q.; Ma, C.; Gao, C.; et al. Curcumin suppresses IL-1beta Secretion and prevents inflammation through inhibition of the NLRP3 inflammasome. *J. Immunol.* **2018**, *200*, 2835–2846. [CrossRef]
89. Hasanzadeh, S.; Read, M.I.; Bland, A.R.; Majeed, M.; Jamialahmadi, T.; Sahebkar, A. Curcumin: An inflammasome silencer. *Pharmacol. Res.* **2020**, *159*, 104921. [CrossRef] [PubMed]

MDPI

Review

Anticancer Mechanism of Curcumin on Human Glioblastoma

Shu Chyi Wong [1], Muhamad Noor Alfarizal Kamarudin [2] and Rakesh Naidu [1,*]

1 Jeffrey Cheah School of Medicine and Health Sciences, Monash University Malaysia, Jalan Lagoon Selatan, Bandar Sunway, Selangor Darul Ehsan 47500, Malaysia; wsc_1997@hotmail.com

2 Brain Research Institute Monash Sunway (BRIMS), Jeffrey Cheah School of Medicine and Health Sciences, Monash University Malaysia, Jalan Lagoon Selatan, Bandar Sunway, Selangor Darul Ehsan 47500, Malaysia; muhamadnoor.alfarizal@monash.edu

* Correspondence: kdrakeshna@hotmail.com; Tel.: +60-3-5514-6345

Abstract: Glioblastoma (GBM) is the most malignant brain tumor and accounts for most adult brain tumors. Current available treatment options for GBM are multimodal, which include surgical resection, radiation, and chemotherapy. Despite the significant advances in diagnostic and therapeutic approaches, GBM remains largely resistant to treatment, with a poor median survival rate between 12 and 18 months. With increasing drug resistance, the introduction of phytochemicals into current GBM treatment has become a potential strategy to combat GBM. Phytochemicals possess multifarious bioactivities with multitarget sites and comparatively marginal toxicity. Among them, curcumin is the most studied compound described as a potential anticancer agent due to its multi-targeted signaling/molecular pathways properties. Curcumin possesses the ability to modulate the core pathways involved in GBM cell proliferation, apoptosis, cell cycle arrest, autophagy, paraptosis, oxidative stress, and tumor cell motility. This review discusses curcumin's anticancer mechanism through modulation of Rb, p53, MAPK, P13K/Akt, JAK/STAT, Shh, and NF-κB pathways, which are commonly involved and dysregulated in preclinical and clinical GBM models. In addition, limitation issues such as bioavailability, pharmacokinetics perspectives strategies, and clinical trials were discussed.

Keywords: curcumin; glioblastoma; anticancer; molecular signaling mechanism

Citation: Wong, S.C.; Kamarudin, M.N.A.; Naidu, R. Anticancer Mechanism of Curcumin on Human Glioblastoma. *Nutrients* **2021**, *13*, 950. https://doi.org/10.3390/nu13030950

Academic Editors: Roberta Masella and Francesca Cirulli

Received: 14 February 2021
Accepted: 10 March 2021
Published: 16 March 2021

Publisher's Note: MDPI stays neutral with regard to jurisdictional claims in published maps and institutional affiliations.

Copyright: © 2021 by the authors. Licensee MDPI, Basel, Switzerland. This article is an open access article distributed under the terms and conditions of the Creative Commons Attribution (CC BY) license (https://creativecommons.org/licenses/by/4.0/).

1. Introduction

Brain tumors can be classified into grade I and II (benign, low-grade), grade III (malignant, high-grade) such as anaplastic astrocytoma, and grade IV (highly aggressive and malignant) such as glioblastoma (GBM) [1]. GBM is the most common and aggressive form of malignant primary adult brain tumor [2]. In the United States alone, the annual age-adjusted incidence of GBM is 3.22 per 100,000 persons based on registry data from 2012 to 2016 [2]. Based on the 2016 WHO classification of the central nervous system tumors, GBM is classified as a grade IV diffuse glioma. GBM is further classified into isocitrate dehydrogenase-wildtype (IDH-wildtype), IDH-mutant, and not otherwise specified (NOS) [3]. IDH-wildtype or primary (de novo) GBM accounts for 90% of the total proportion of GBM cases [3]. The IDH-mutant or secondary GBM, which may arise from a lower grade diffuse glioma, only accounts for about 10% of the total GBM cases [3]. Primary GBM is more common in elderly patients (median age of 62 years), while secondary GBM preferentially arises in younger patients (median age of 44 years) [3].

The standard care for newly diagnosed GBM patients is surgical resection, followed by radiotherapy (60Gy in 30 fractions) with concurrent oral administration of temozolomide (TMZ), followed by six cycles of adjuvant [4]. Additionally, monoclonal antibody bevacizumab and other alkylating agents such as carmustine, lomustine, nimustine, and fotemustine are used in GBM treatment [4]. Unfortunately, these treatments often prove ineffective, given the poor prognosis outcomes of GBM with a five-year survival rate under 10% and a median survival rate of around 12 to 18 months [2,5]. The high infiltration degree of GBM often causes surgical resection incapable of fully resecting the GBM tumor, leaving the residual presence of microscopic foci [6,7]. Moreover, the GBM tumors often develop chemo- and radio-resistance with the formation of glioma stem cells, leading to GBM recurrence [8]. In TMZ-resistant GBM tumors, numerous molecular pathways such as nuclear factor kappa light chain enhancer of activated B cells (NF-κB), p53, and JAK-STAT are found to be commonly dysregulated [8]. In addition, several clinical complications such as pancytopenia, pyrexia, wound healing complications, multi-organ failure, or even death are observed following the chemo-radiation and immunotherapy treatment [8–10].

Thus, in recent years, scientists have been focusing on phytochemicals as potential therapeutic agents in cancer management to minimize drug toxicity and side effects. Flavonoids represent the most common and widely distributed phytochemicals in fruits and vegetables. Various flavonoids such as tannins, quinones, stilbenes, and curcuminoids possess antioxidant, anti-inflammatory, antiviral, antimutagenic, and, most importantly, anticancer properties [11,12]. Among them, curcuminoids (especially curcumin) have been gaining immense attention because of its anticarcinogenic, antitumor, antioxidant, and anti-inflammatory actions [13–15]. Curcuminoids are a family of active compounds found in the turmeric rhizome (*Curcuma longa*), an Indian spice commonly used in cooking. Natural curcuminoids are composed of curcumin, bisdemethoxycurcumin, and desmethoxycurcumin in a proportion of 77:3:17 [16]. Among them, curcumin is the most abundant compound and has been widely studied as a potential therapeutic agent in chronic diseases, such as neurodegenerative, cardiovascular, pulmonary, metabolic, and autoimmune diseases [17]. For instance, curcumin was able to restore oxidative stress and DNA methyltransferase (DNMT) functions against diabetic retinopathy [15]. Curcumin also acts as a wound healing promoting agent by facilitating collagen synthesis and fibroblast migration [18]. Several pre-clinical and clinical studies also reported its anticancer effects in colorectal cancer [19,20], pancreatic cancer [21], lung cancer [22], and GBM [23]. Curcumin can modulate multiple cellular signaling pathways and molecular targets involved in GBM tumor growth, migration, invasion, cell death, and proliferation [24–27]. Retinoblastoma (Rb), p53, MAP kinase (MAPK), P13K/Akt, JAK/STAT, sonic hedgehog (Shh), and NF-κB pathways are the most common targeted dysregulated pathways found in GBM and modulated by curcumin [28–34]. Moreover, curcumin is highly lipophilic and able to cross the blood–brain barrier (BBB) [35,36].

To date, numerous review studies have suggested curcumin as a potential drug for GBM. However, a greater focus on curcumin's anticancer potential in molecular signaling pathways that are commonly dysregulated in GBM is needed to provide a more comprehensive understanding of its therapeutic effects. This review includes the initial until recent pre-clinical and clinical studies of curcumin's mechanisms of action in modulating several molecular pathways such as Rb, p53, MAPK, P13K/Akt, JAK/STAT, Shh, and NF-κB pathways. This review paper also discusses curcumin's related issues such as low bioavailability, pharmacokinetics, and the perspective strategies to overcome these issues.

2. Dysregulated Signaling Pathways Associated with GBM Pathogenesis

Almost all GBMs are found to have dysregulated Rb, p53, JAK/STAT, MAPK, P13K/Ak Shh, and NF-κB pathways. Thus, the following section discusses the mechanisms of action modulated by curcumin via these molecular signaling pathways involved in GBM cell proliferation, apoptosis, cell cycle arrest, autophagy, paraptosis, oxidative stress, and cell motility. The reported observation of in vitro and in vivo studies of curcumin against GBM are summarized in Table 1.

Table 1. Signaling pathways and mechanism of actions targeted by curcumin in vitro and in vivo against glioblastoma (GBM).

Molecular Pathway	GBM Model (In Vitro or In Vivo)	Pathway Mechanism Targeted by Curcumin OR Mechanism of Actions Induced by Curcumin	References
Rb	DBTRG	Induced G1/S phase arrest by upregulating CDKN2A/p16 and downregulating the expression of RB protein	[28]
P53	DBTRG	Induced G2/M phase arrest by upregulating p21 and downregulating cdc2, followed by increasing of p53 protein	[28]
	U251	Induced G2/M phase arrest by upregulating ING4 expression	[37]
	C6	Induced G0/G1 phase arrest and apoptosis by upregulating p53 and p21Waf/Cip1 protein levels	[38]
	U87MG	Enhanced anticancer effect of ETP and TMZ through downregulation of the p53 protein expression	[39]
	A172	Induced paraptosis by regulating the ER-related miRNAs that interact with the p53-BCL2 pathway	[40]
	KSN60 U251MG (KO)	Decreased p21 expression and increased cyclin B1	[41]
	U87MG	Induced apoptosis through increased BAX:BCL2 ratio via mitochondria-mediated pathway	[42]
JAK/STAT	Tu-2449 Tu-9648 Tu-251	Suppressed cell proliferation by inducing G2/M cell cycle arrest and inhibited cell invasion through downregulation of STAT3 target genes *c-Myc*, *MMP-9*, *Snail* and *Twist*, *Ki67*	[29]
	C6B3F1 mice	Reduced growth and midline crossing of intracranially implanted tumors and proliferation of tumor, increased tumor-free long term survival rate by 15% and 38%	
	A-172 MZ-18 MZ-54 MZ-256 MZ-304	Inhibited cell proliferation, migration, invasion by decreasing expression of phosphorylated STAT3 protein and its target genes *c-Myc* and *Ki67*	[43]
	U251 U87	Induce epigenetic modifications through suppression of STAT3 protein activity, followed by *RANK* promoter methylation along with RANK activation	[44]
	Glio3 Glio9	Induction of intracellular ROS production through downregulation of STAT3 activity	[30]
MAPK	U87MG U373MG CRT-MG	Inhibited invasion of GBM cells through downregulation of MAP kinase pathway along with decreased PMA-induced mRNA expression of *MMP-1*, *-3*, *-9*, and *-4*	[45,46]
	Glio3 Glio9	Decreased GSC viability through induction of P38, ERK, and JNK activity	[30]
	U87MG	Inhibited cell proliferation by upregulating Egr-1 expression through ERK and JNK pathway	[47]
	U87MG U373MG	Induced autophagy through induction of ERK1/2 pathway	[48]
	C6	Inhibited neuroinflammatory effect through inhibition of LPS-induced CCL2 expression via JNK pathway	[49]
P13K/Akt	U138MG U87 U373 C6	Induced cell cycle arrest in G2/M phase and apoptosis by inhibiting Akt phosphorylation on Ser473	[33]
	C6 mice	Reduced tumor size and increased number of apoptotic cells in tumor	
	U251	Inhibited cell proliferation, migration, and invasion by decreasing P13K/mTOR protein expression and restoring PTEN expression	[31]
	U87 U87 xenograft	Inhibited tumor growth and increased PTEN expression	

Table 1. *Cont.*

Molecular Pathway	GBM Model (In Vitro or In Vivo)	Pathway Mechanism Targeted by Curcumin OR Mechanism of Actions Induced by Curcumin	References
	U87MG GL261 F98 U373MG	Induced autophagy through suppression of AKT/mTOR/p70S6K pathway	[48,50]
	U87 xenograft	Inhibited tumor growth and induced autophagy	
	U118MG U251MG U87MG	Enhanced anti-proliferation, anti-migration, and proapoptotic activities of ACNU against GBM by suppressing P13K/Akt pathway	[51]
Shh	U87 T98G	Inhibited GBM cell proliferation, migration, and invasion through suppressing core components and GLI1-dependent target genes in Shh/GLI1 pathway	[32]
	U87 xenograft	Inhibited GBM growth and prolonged the survival rate	
	U87 U251	Promoted expression of mi-R326 and enhanced inhibition of SHH/GLI1 pathway	[52]
	U87 xenograft	Inhibited GBM growth and prolonged the survival rate	
NF-κB	U138MG C6	Induced apoptosis through inhibition of NF-κB survival pathways by downregulating the antiapoptotic proteins	[33]
	C6-implanted rats	Decreased brain tumors (growth/size/) without reported tissues, metabolic, oxidative, or hematology toxicity	
	U118MG U251MG U887MG	Blocked GBM tumor growth by inhibiting NF-κB/COX-2 pathway	[51]
	T98G	Induced apoptosis through downregulation of NF-κB, IAPs, and upregulation expression of IκBα	[53]
	NP-2	Induced tumor cell death through downregulation of NF-κB activity and its regulated protein cyclin D1	[54]
	GBM 8401	Induced apoptosis through inhibition of NF-κB activity	[55]
	C6	Induced cytotoxic and antiproliferative activity of PTX through inhibition of NF-κB signaling pathway	[38]
	U87MG	Induced apoptosis through suppression of apoptosis protein inhibitor and downregulation of anti-apoptotic NF-κB dependent genes	[42]

2.1. Retinoblastoma (RB) Pathway

The RB pathway plays a central role in cell proliferation by regulating the cell cycle [56]. This pathway mainly consists of five components, which are CDKN2A/p16Ink4a, cyclin D1, cyclin-dependent protein kinases (cdk4/6), RB-family of pocket proteins (RB, p107, p103), and E2F [57,58]. CDKN2A/p16 is a negative regulator that competes with cyclin D1 to bind to and inhibit the activity of CDK4/6. This, in turn, can induce cell cycle arrest at the G1/S phase by inhibiting phosphorylation of RB protein by the cyclin D1/cdk4/6 complex. The unphosphorylated RB protein binds to E2F protein to inhibit the activation of E2F-regulated gene expression, thereby inhibiting cell cycle progression, DNA replication, and nucleotide biosynthesis [57].

Many of the important components of this pathway are frequently altered in many cancer cells, including GBM [59]. According to The Cancer Genome Atlas (TCGA) pilot project, most of the GBM acquired mutations include homozygous deletion or mutation of *CDKN2A/p16* and *RB1*, and amplification of *CDK4*, *CDK6*, and *cyclin D*, which are associated with the RB signaling pathway [60]. *RB* promoter methylation and gene silencing are found in GBMs and are more frequently reported in secondary GBMs than primary GBMs [61]. The inhibition of the RB pathway via silencing/suppression of its component proteins increases etoposide-induced DNA double strand breaks, p53 activation, and TMZ-induced GBM apoptosis [62–65]. Additionally, the inhibition of cyclin D1 can downregulate P-glycoprotein (pgp) expression, which may help to overcome chemoresistance in GBMs [63].

Thus, the RB signaling pathway is an important drug targeted pathway to improve GBM prognosis and patient outcomes.

As shown in Figure 1a, curcumin inhibits the RB signaling pathway by increasing the negative regulator CDKN2A/p16Ink4a activity, which then suppresses the phosphorylation of RB protein. To date, Chin-Cheng Su and colleagues demonstrated that curcumin significantly inhibited the RB pathway in DBTRG glial cells in a time- and concentration-dependent manner [28]. In this study, curcumin treatment upregulated CDKN2A/p16 and downregulated the phosphorylated RB protein. It has been shown that CDKN2A/p16 protein compete with cyclin D1 to bind to CDK4/6 protein, which then inhibits phosphorylation of RB protein. Unphosphorylated RB protein could not dissociate from its repressor E2F to permit transcription of G1 genes for proceeding from G1 to S phase.

2.2. P53 Pathway

P53 is a tumor suppressor protein that can activate cell cycle arrest or induce cell apoptosis to prevent damaged cells from further dividing and growing [66]. Following DNA damage, p53 is activated to induce transcription of p21Waf/Cip1, a cyclin-dependent kinase inhibitor [58]. This P21Waf/Cip1 protein can induce G1/S and G2/M arrest by binding to and inhibiting the activity of Cdc2, cyclin-CDK2, -CDK1, and CDK4/6 complexes. This allows the damaged cells to undergo DNA repair prior to mitosis. The p53 activity can be inhibited by MDM2, the transcription of which is induced by TP53 through a negative feedback loop [67]. However, P14ARF, which is located in part of the CDKN2A locus, can bind to and inhibit MDM2 from binding to the N-terminal transactivation domain of TP53 [58,68]. On the other hand, activated p53 protein can activate the pro-apoptotic BH3-only members of the Bcl-2 protein family [69]. These pro-apoptotic proteins bind and inhibit the pro-survival Bcl-2 proteins to initiate the pro-apoptotic multi-BH domain members of the Bcl-2 family, such as BAX and BAK, to induce cell apoptosis.

Dysregulation of p53 pathway in GBM is mostly due to *TP53* mutation, amplification of *MDM2*, or loss of expression of CDKN2A-p14ARF [70]. The TCGA project demonstrated that the p53 signaling pathway is altered in most GBM samples with the association of *TP53* mutation or homozygous deletion, *P14ARF* deletions, and amplification of *MDM2* and *MDM4* [60]. According to WHO, the *TP53* mutation is more commonly seen in secondary GBM and is higher in proportion to primary GBM [3]. The clinical study showed that most of the sample cells (from GBM patients with age around 56) were Bcl-2 positive, and most of the Bcl-2 positive cloned cells acquired chemoresistance [71]. Thus, various strategies have been developed to target the p53 pathway, such as inhibition of pro-survival genes or MDM2/p53 interaction, degradation of mutant p53, and restoration of wildtype p53 [72–74].

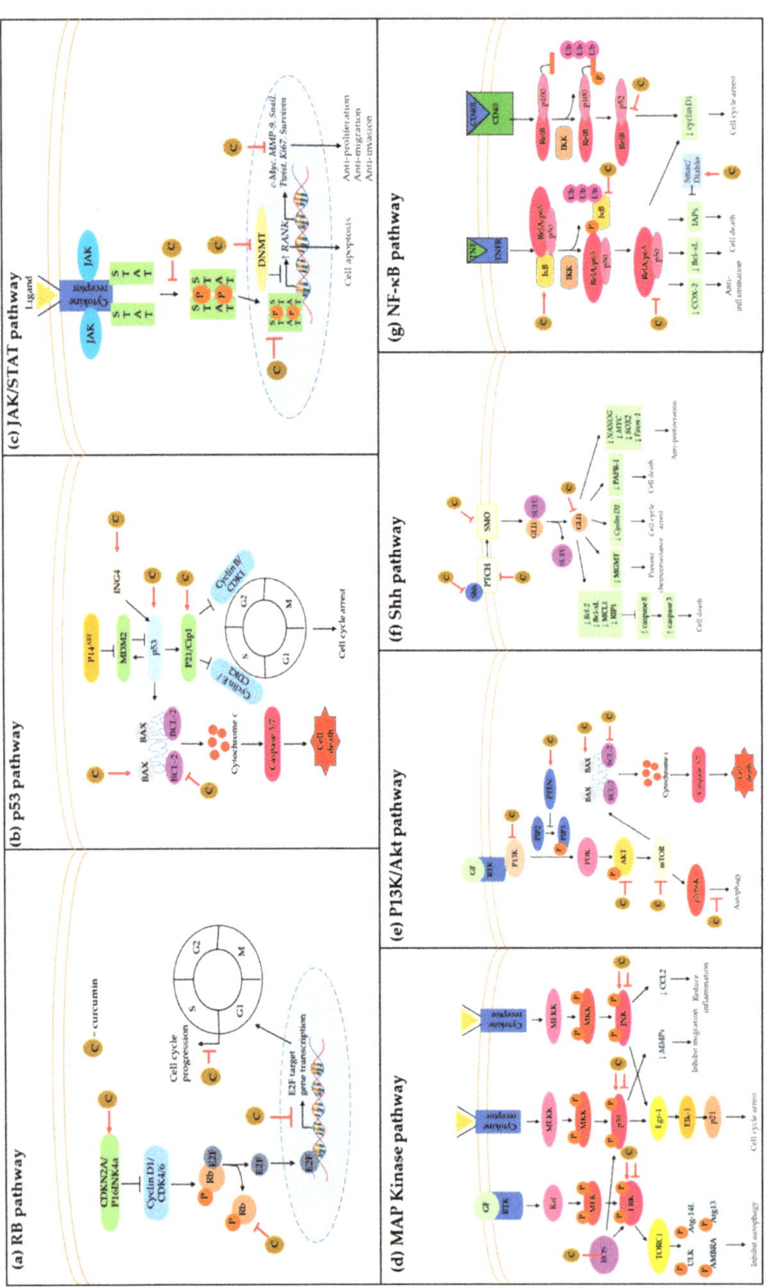

Figure 1. Mode of curcumin actions as anti-cancer agent on the key molecular targets in GBM. Curcumin possesses anti-cancer properties by inhibiting signaling pathways and their downstream molecular targets; **(a)** retinoblastoma (RB) pathway; **(b)** p53 pathway; **(c)** JAK1,2/STAT3 pathway; **(d)** MAP kinase pathway; **(e)** P13K/Akt pathway; **(f)** Shh pathway; **(g)** NF-κB pathway. Molecular targets and signaling pathways that are upregulated and downregulated by curcumin are noted by using → and ⊣, respectively.

As shown in Figure 1b, curcumin upregulates the expression of p53, p21, and ING4 and increases the BAX:BCL2 ratio to induce cell cycle arrest and apoptosis. Curcumin upregulates the expression of p53 and p21 in a time- and concentration-dependent manner, which then induces G2/M arrest in DBTRG cells [28]. Curcumin induces p53 activity by upregulating p21Waf/Cip1 and ING4 protein expression [37]. The upregulation of ING4 expression can increase p53 acetylation at Lys-382 and protein stability [75,76]. The acetylation of p53 inhibits its interaction with MDM2, which eventually induces cell cycle arrest and apoptosis, as observed in U251 cells [37]. In a different study, curcumin enhances paclitaxel (PTX) activity in C6 cells by increasing TP53 and p21 gene expression. In response to the increase of TP53 and p21 gene expression, the cell population in the G0/G1 phase increases while the cell population in the S phase decreases, indicating G0/G1 phase arrest [38]. Moreover, curcumin suppresses A172 cell viability by inducing paraptosis through the regulation of genes associated with the endoplasmic reticulum (ER) stress response [40]. Interestingly, in this study, interaction network analysis (with IPA software) revealed that the altered levels of ER-related miRNAs interact with p53-BCL2 pathways. Thus, it was suggested that the p53-BCL2 pathway might be involved in curcumin anticancer mechanisms. Additionally, curcumin potentiates the cytotoxic and apoptosis-inducing effect of etoposide and TMZ through downregulation of p53 mRNAs and upregulation of BAX-Bcl-2 in T98G and U87MG cells [39,42,55].

However, a contradictory finding showed that curcumin did not induce cell cycle arrest, as it enhanced cyclin B1 and decreased p21 expression in the radioresistant KNS60 and U251MG(KO) cells [41]. These radioresistant cells usually have a high basal p53 level, but the expression of p53 decreased following curcumin treatment. These results showed that mechanism actions of curcumin in radioresistant GBM cells are different.

2.3. *JAK/STAT Pathway*

Cytokines and growth factors can activate the Janus kinase/signal transducers and activators of transcription (JAK/STAT) pathway to regulate cell proliferation, differentiation, migration, and apoptosis. This pathway involves the activation of growth factor receptor kinases, phosphorylation, dimerization, and translocation of STAT proteins into the nucleus to activate the downstream target genes. JAKs (Jak1, JAK2, JAK3, Tyk2) are the cytoplasmic tyrosine kinases that relay intracellular signals originating from extracellular receptors [77–79]. Following JAKs activation and phosphorylation of the tyrosine residues on receptors, STAT is activated through its recruitment and binding to the phosphorylated tyrosine residues. The activation of STATs only lasts from a few minutes to several hours under normal physiological conditions. However, aberrant activation of STAT signaling is found in many GBM tissues compared with normal human astrocytes, white matter, and normal adjacent tissue to the tumor [29,80,81]. Studies showed that the inhibition of either JAK or STAT phosphorylation is associated with reduced levels of anti-apoptotic proteins, resulting in apoptosis in GBM cells [29,43,81–83].

As shown in Figure 1c, curcumin can inhibit STAT activation and its downstream target genes involved in cell proliferation, migration, and invasion. Curcumin inhibits JAK1,2/STAT3 tyrosine-phosphorylation, and STAT3 target genes such as *c-Myc*, *MMP-9*, *Snail*, *Twist*, and *Ki67*, which in turn decrease GBM cell migration, invasion, and proliferation [29]. In the same study, curcumin significantly decreased the tumor cell proliferation and growth of the mid-line crossing in the intracranially implanted tumor-bearing mice compared with the control diet. Moreover, curcumin treatment resulted in 15% and 38% tumor-free long-term survival in Tu-2449-bearing mice and Tu-9648-bearing mice, respectively, where all control mice died. Consistently, another study supported that curcumin is capable of inhibiting cell proliferation through the inhibition of STAT3 protein along with reduction *c-Myc* and *Ki-67* transcription in several glioma cell lines [43]. Moreover, curcumin can inhibit the DNMTanalogue M.Sssl to demethylate the *RANK* CpG sites and transcriptionally upregulate *RANK* gene expression in U251 cells [44]. This study further reported that curcumin induces *RANK* expression through STAT3 suppression. Activation of RANK has been known to be associated with pro-apoptotic and anti-tumorigenesis activities [84,85]. Additionally, inhibition of STAT3 can result in suppression of STAT3-DNMT1 interaction, which then demethylates tumor suppressor gene promoters [86]. Other than that, curcumin decreases the Tyr705 and increases the Ser737 phosphorylated form of STAT3 in human patient-derived GSC lines [30]. This causes the inactivation of STAT3 proteins via suppression of its nucleus translocation, which suppresses the activation of its downstream target genes, such as *survivin*, which inhibits GBM cell proliferation.

2.4. MAP Kinase Pathway

The mitogen-activated protein kinase (MAPK) pathway is a three-layer signaling cascade. This MAPK cascade is comprised of MAPK3, which activates MAPKK2 through serine/threonine phosphorylation, which then activates MAPK through tyrosine/threonine phosphorylation within a conserved Thr–Xxx–Tyr motif in the activation loop of the kinase domain [87]. There are at least 11 members of the MAPK superfamily, which are divided into three main groups: the extracellular signal-regulated protein kinases (ERK), c-Jun N-terminal kinases (JNK), and p38s [88]. Generally, *ERK* genes are activated by growth factors and mitogen, and the signaling cascades include RAF as MAPKKK and MEK as MAPK. The activation of ERK signaling promotes cell growth, apoptosis, differentiation, and development [89]. While p38s and JNK are activated by stress, inflammatory cytokines, and growth factors [90], their signaling cascades include MEKK as MAPKKK and MKK as MAPKK. Activation of either *p38s* or *JNK* genes may support inflammation, cell apoptosis, cell motility, growth, and chromatin remodeling. ERK, p38, and JNK signaling pathways favor both anti-apoptotic and pro-apoptotic proteins, depending on the cell type and condition [91]. Hence, the aberrant activation or deactivation of this MAPK pathway can promote abnormal cell proliferation, contributing to tumorigenesis.

Studies showed that targeting the MEK-ERK1/2 pathway is one of the approaches to block adhesion of GBM cells onto gelatin/collagen component of ECM, therefore decreasing the proliferation and migration of GBM cells [92,93]. The high expression of p38 had a positive correlation with the glioma's malignancy grade, while suppressing p38 expression inhibited proliferation and induced apoptosis in GBM cells [94]. Ken-ichiro Matsuda and colleagues discovered that self-renewing stem-like GBM cells have elevated JNK phosphorylation levels, accompanied by increased c-JUN phosphorylation at the cognate JNK phosphorylation site [95]. Treatment with JNK inhibitor reduced the self-renewing ability of the stem-like GBM cells, suggesting JNK is needed for self-renewal in vitro and in vivo. Several studies showed that the inhibition of ERK, JNK, and p38 MAPK pathways induced GBM cell cycle arrest and inhibited cell proliferation [96–99].

Curcumin can modulate the MAPK signaling pathway to regulate cell proliferation, tumorigenesis, apoptosis, and inflammation, as shown in Figure 1d. So-Young Kim and colleagues discovered that curcumin potently inhibited glioma invasion by inhibiting all the MAPK pathways (JNK, p38, ERK), which then suppressed phorbol myristate acetate (PMA)-induced mRNA expression of *MMP-1, -3, -9,* and *-14* in U87MG and U373MG cells [45,46]. Overexpression of the matrix metalloproteinases (MMPs) facilitates migration and invasion of malignant brain tumor cells to the surrounding brain tissues. These MMPs are upregulated in human malignant gliomas [100]. Among the MMPs, MMP-9 is the most common enzyme that promotes brain tumor invasion and is frequently found in GBM [101]. MMP-1 protein level is increased with the tumor grade and correlated with increased glioma invasiveness [102]. At the same time, the activation of MMP-3 can degrade the brain's hyaluronic acid-rich matrix, which leads to the invasion and migration of tumor cells [103]. Additionally, MMP-14 is a membrane-bound protease that can remodel the ECM to stimulate proMMP-2 activation. Curcumin treatment (10 μM) inhibits cell invasion by more than 90% in U87MG and U373MG cells [45].

In contrast, curcumin increases the phosphorylated ERK, p38, and c-Jun proteins levels, which decreases GBM stem cell (GSCs) proliferation, sphere-forming ability, and colony-forming potential [30]. Curcumin can also promote MAPK pathway activation through the induction of reactive oxygen species (ROS) [30]. The production of intracellular ROS can induce the activation of ERK and p38 MAPK pathways through oxidative modification of intracellular kinases and inactivation of the MAPK phosphatases [104,105]. Moreover, curcumin induced Egr-1 expression through the activation of p38, ERK, and JNK pathways, which mediated the transactivation of Elk-1 in U87MG human GBM cells [47]. Egr-1 binds directly to the p21 promoter to stimulate p21 transcription, inhibiting CDK activity and resulting in cell cycle arrest. Elk-1 is the direct target gene of the p38, ERK, and JNK pathways, which can form a complex with serum response factor on the serum response element of the Egr-1 promoter to activate the p53-independent transcriptional activation of p21Waf/Cip1 protein. Additionally, curcumin can induce autophagy in GBM cells through the inhibition of the ERK1/2 pathway [48]. Another study suggests that the inhibition of the ERK pathway can lead to suppression of TORC1, which plays an important role in inhibiting autophagy initiation through the phosphorylation of Atg13, ULK, AMBRA, and Atg-14L [106]. In a C6 orthotropic xenograft, curcumin suppressed the phosphorylated JNK1 and JNK2 levels, which decreased lipopolysaccharide-induced CCL2 production [49]. The overexpression of CCL2 can contribute to GBM progression by inducing encephalopathy with mild perivascular leukocyte infiltration, impaired BBB function, and increased expression of proinflammatory cytokine expression [107].

2.5. P13K/AKT Pathway

The phosphatidylinositol-3-kinase (P13K)/Akt signaling pathway is critical in regulating cell growth, cell cycle arrest, apoptosis, and mRNA translation to maintain normal physiological conditions. There are three different classes of P13Ks, namely Class I, II, and III, which are categorized accordingly to their different structure and specific substrates [108,109]. Class I P13Ks are heterodimers consisting of a p110 catalytic subunit and p85 adaptor subunit [110,111]. Notably, Class 1 P13Ks is the most common type of P13K, which is incriminated in human cancer. The binding of cytokines or growth factors to the corresponding receptors results in the tyrosine residue autophosphorylation, followed by P13K binding protein recruitment. Upon allosteric activation of the p110 catalytic subunit, P13K catalyzes the phosphorylation of PtdIns(4,5) P2 (PIP2) to PtdIns(3,4,5) P3 (PIP3), which then recruits a subset of signaling proteins with pleckstrin homology (PH), such as AKT and PDK1, to initiate cell proliferation pathways. Protein-phosphatase and tensin homologue (PTEN) act to dephosphorylate PIP3 into PIP2 to prevent activation of the downstream kinases [110,111]. One of Akt's common target proteins is mTOR, which regulates cell growth and proliferation by promoting biosynthesis of multiple proteins such as cyclin D1, HIF, and VEGF [110]. The mTORC1 activates S6K and inactivates 4EBP1, promoting the production and translation of proteins to promote cell growth. While the mechanism of mTORC2 is less well clarified, it has the same responsibility in promoting cell proliferation [110].

Studies have reported that mutations of the core genes involved in P13K/AKT pathways are commonly found in human GBM tissues [60,112]. Alfeu Zanotto-Flho and colleagues reported that the P13K/Akt pathway are highly upregulated (seven- to eight-fold) in C6 and U138MG cell lines compared to the normal astrocytes cells [33]. Consistently, data from the TCGA pilot project showed that most of the GBM samples acquire homozygous deletion or mutation of *PTEN*, *P13K* mutation, and amplification of *AKT* and *FOXO* genes [60]. Another report indicated that genetic alterations such as loss of heterozygosity (LOH), mutation, and methylation have been identified in most GBM patients. LOH or *PTEN* mutation is positively associated with the poor survival of GBM patients [113]. It was reported that the delivery of Akt small-molecule inhibitor to inhibit the P13K/AKT pathway effectively suppressed the growth of both stem and non-stem GBM cell populations [114].

As seen in Figure 1e, the P13K/Akt signaling pathway and its key molecular targets are inhibited by curcumin to prevent GBM progression. Curcumin inhibits 80% of the P13K/Akt pathway's constitutive activation by suppressing the phosphorylation of Akt proteins on Ser473 [33]. Inhibition of the P13K/Akt pathway resulted in the induction of G2/M phase arrest as an early step of the apoptotic mechanism, which could probably explain how curcumin spared the non-transformed and its selectivity towards the tumor cells. In the same study, curcumin decreased GBM tumor size and increased apoptotic tumor cells in C6-implanted Wistar rats. Most importantly, curcumin did not cause any tissue toxicity in the rats' liver, kidney, lungs, or heart. Other than that, curcumin inhibited the P13K/Akt pathway by increasing PTEN expression, which decreased p-Akt and p-mTOR expression, leading to cell apoptosis [31]. In the same study, curcumin also inhibited GBM tumor growth by increasing PTEN protein expression in the U87 xenograft model.

Additionally, curcumin induces autophagy by inhibiting the AKT/mTOR/p70S6K pathway in GBM cell lines and xenograft models [48,50]. In these studies, curcumin significantly decreased the levels of P13Kp85, phosphoP13Kp85, total Akt, p-AKT, mTOR, and p-mTOR. mTOR is not only a major effector of cell growth and proliferation, but it can also inhibit autophagy events in its active form [115]. Thus, inhibiting expression of P13K and AKT, which regulate mTOR expression, is a feasible strategy to induce autophagy–cell death in GBM cells. Notably, curcumin downregulates Bcl-2 and upregulates BAX, leading to the release of cytochrome-c and caspase-3 activation. Curcumin also enhances the anti-cancer effects of nimustine hydrochloride (ACNU) against GBM by inhibiting the phosphorylation of P13K and the AKT (serine/threonine) [51,116].

2.6. Sonic Hedgehog (Shh) Pathway

The hedgehog (Hh) signaling pathway is critical for embryonic development, organogenesis, regeneration, and homeostasis for adult tissue [117]. There are three main types of Hh proteins, which are sonic hedgehog (Shh), Indian hedgehog (Ihh), and desert hedgehog (Dhh). The activation of the Shh pathway can occur either through canonical or non-canonical signaling pathways [117]. The canonical Shh activation occurs by ligand-dependent interaction when Shh binds to the patched transmembrane receptor (PTCH) [117,118]. Following this binding, PTCH is incapable of inhibiting the second transmembrane protein, smoothened (Smo). Smo signals the suppressor of fused (SUFU), which is the negative regulator of glioma-associated oncogene homologue (GLI), to release and activate GLI. The activated GLI translocates into the nucleus and modulates downstream gene expression. On the other hand, the non-canonical Shh activation occurs through either GLI-independent mechanisms or Smo-independent mechanisms [117,118]. In the GLI-independent mechanism, Smo is no longer inhibited by the PTCH, and therefore it can stimulate the release of calcium ions from the ER to control the growth of the actin cytoskeleton [118]. In contrast, the Smo-independent mechanism is involved in cyclin B activation to increase cell proliferation and survival [118].

Under normal physiological conditions, the Shh pathway is minimally active in differentiated adult tissue, as it is a highly conserved development pathway. The Shh pathway is frequently associated with GBM tumorigenesis [119–122]. Studies have reported that most of the GBM patient tissues samples exhibited an aberrant activation of Hh signaling with the presence of GLI1 in both nucleus and cytoplasm [119,123]. Among GLI family members, overexpression of GLI1 is mostly associated with poor prognosis in several cancers, including GBMs [32,124–127]. GLI1 protein can upregulate several target genes such as *PTCH1*, *CycD1*, *MYC*, *Bcl-2*, *NANOG*, and *SOX2* to promote cell proliferation, apoptosis, angiogenesis, and stem cell self-renewal [128–130]. A study showed that inhibiting GLI1 alone significantly decreases the metabolic activity of GBM cells to reduce chemoresistance [119]. This study also revealed that inhibiting the expression of GLI1 proteins can elevate the nuclear p53 level in U87MG cells. Additionally, the overexpression of Smo is significantly associated with poor prognosis in GBM patients [131]. Smo expression inhibition, which suppresses GBM proliferation, migration, invasion, and tumorigenesis further supports this observation [132].

As shown in Figure 1f, curcumin is a potent inhibitor of the SHH/GLI signaling pathway by downregulating the Shh, Smo, PTCH, and GLI protein levels and its downstream target genes such as *cyclin D1, Bcl-2, and Foxm-1* [32,52,133]. Curcumin inhibits GBM cell proliferation, colony formation, migration, and induced apoptosis through downregulation of both mRNA and protein levels of SHH/GLI1 signaling (Shh, Smo, and GLI1) in U87 and T98G cells [32]. Curcumin also inhibits GLI nuclear translocation, which deactivates its downstream target genes including *cyclin D1, Bcl-1,* and *Foxm-1*. The combination treatment of curcumin and miR-326 can further reduce the tumor volume and prolong the survival period of U87-bearing mice by inhibiting GLI1 proteins compared with miR-326 or curcumin treatment alone [52]. In the same study, curcumin-treated GBM cells significantly decreased the expression of GLI1 protein, and this observation was enhanced with combination treatment with miR-326. The curcumin and miR-326 treatment also increased the expression of caspase-3 cleaved anti-poly ADP ribose polymerase 1 (PARP-1) caspase-in GBM cells. Simultaneously, the pro-survival proteins BCL-XL, MCL1, and RIP1 were decreased compared to the control and curcumin only-treated GBM cells.

2.7. NF-κB Pathway

NF-κB is a family of highly conserved transcription factors that regulate the transcription of various genes involved in cellular activities. There are four members under this NF-κB family: NF-κB1(p50/p105), NF-κB2(p52/p100), Rel-like domain-containing protein A (RelA/p65), and c-rel [134]. They form a dimeric complex (either homodimers or heterodimers) and bind to the specific sequences of DNA called response elements (RE) for the transcription of gene involved in cell proliferation, apoptosis, and inflammatory response [133,135]. NF-κB activation can occur through two major signaling pathways: the canonical and the non-canonical NF-κB signaling pathways [136,137]. The canonical pathway is mediated through the nuclear translocation of p50, RelA, and c-Rel into the nucleus and binding to the targeted DNA sequences. In contrast, the non-canonical NF-κB pathway selectively responds to stimulus and activates p100-sequestered NF-κB members, predominantly via translocation of NF-κB p52 and Rel B into the nucleus. NF-κB members normally bind to the DNA sequences of anti-apoptotic, pro-survival, and immune response genes. Several studies have demonstrated that human GBM cells have aberrant NF-κB activity to maintain their tumorigenic activity [138–143].

NF-κB p65 subunit is overexpressed in 81% of 69 samples of GBM and is frequently noted in high-grade compared to low-grade astrocytomas [139]. The constitutive activation of NF-κB p65 is detected in 93% of the GBM cells as compared to normal astrocytes. Studies conducted by Baisakhi et al. showed that inhibition of NF-κB activity resulted in decreased IL-8 transcription, which then inhibited GBM cell invasion and migration [140]. A study conducted by Denise Smith et al. demonstrated that GBM cells that are transfected with short hairpin inhibitory RNAs of RelA and c-Rel for six days displayed reduced tumor growth, signifying the role of RelA and c-Rel in GBM [144]. These studies highlight the importance of inhibiting the overactivation of NF-κB subunits as molecular targets in GBM.

Curcumin modulates the NF-κB pathway to confer the anti-inflammation, anti-proliferation, and apoptotic activity in GBM cells, as shown in Figure 1e. Curcumin can downregulate NF-κB activity by decreasing the expression of anti-apoptotic protein Bcl-xL in GBM cell lines [33]. Reducing the Bcl-xL triggers mitochondrial depolarization, which precedes the losses in mitochondrial membrane integrity. This suggests that curcumin induces mitochondrial-mediated apoptosis in GBM following the inhibition of NF-κB pathways. Other than mitochondrial depolarization, curcumin promotes cell cycle arrest in the G2/M phase prior to cell apoptosis. Most importantly, curcumin acts irrespective of the p53 or PTEN mutational status of the cells. Both PTEN and p53 mutated cells had the same experimental outcomes compared with the wild-type cells after being treated with curcumin. This shows that curcumin exerts p53-independent cell death via inhibition of NF-κB pathways. In the same study, the inhibition of NF-κB by curcumin increases the number of apoptotic cells in tumors, further reducing the tumor size and hemorrhagic areas in C6-implanted Wistar rats. This was in line with other studies showing that curcumin increases the IκB inhibitor proteins and decreases the expression of NF-κB-regulated genes that contribute to GBM chemoresistance [51,53,145]. Curcumin also enhances the anticancer effect of nimustine hydrochloride (ACNU) by suppressing the phosphorylation of IκB, p65, and p50, which then decreases COX-2 expression [51]. Additionally, curcumin's antiproliferative activity might be facilitated through the downregulation of cyclin D1, since the promoter of cyclin D1 is regulated by NF-κB [54]. In the study conducted by Tzuu-Yuan and colleagues, curcumin increased NF-κB transcription factor inhibition in a concentration-dependent manner in GBM 8401 cells [55].

Curcumin improves the cytotoxic effect of PTX by reducing the phosphorylation of IκB and suppressing NF-κB p65 nuclear translocation to inhibit cell growth in C6 rat glioma cells [38]. Furthermore, curcumin upregulates the pro-apoptotic molecular Smac/Diablo to suppress NF-κB and IAPs (cIAP-1 and cIAP-2), which induces apoptosis [42]. Studies suggest a positive feedback system between NF-κB and IAPs, as IAPs can be upregulated by NF-κB and vice versa [146–148]. Hence, downregulation of both NF-κB and IAPs protein might further suppress GBM tumorigenesis.

3. Issues of Curcumin Bioavailability and Methods to Overcome Them

Despite the promising anticancer mechanisms, curcumin efficacy is hindered by its low bioavailability. Various studies have reported that very low curcumin concentration was detected in blood, tumors, or extraintestinal tissues [149–151], which may be due to the poor absorption, rapid metabolism, chemical instability, and rapid systemic elimination characteristics of curcumin [16]. A study reported that orally-administered curcumin at a dose of 500 mg/kg only had 0.06 μg/mL maximum serum concentration, indicating only 1% oral bioavailability [152]. Due to its chemical structure, curcumin has low solubility in neutral or acidic pH. It is fully protonated, unlike in alkaline conditions where it can be hydrolyzed, especially in the intestinal (pH 6.8). Additionally, rapid metabolism and systemic elimination happen through the formation of glucuronides and sulphates by conjugation in the intestine. Studies reported that free curcumin was undetectable, but curcumin glucuronides and sulphates were highly detected in most of the subjects' serum samples who had been administered with curcumin, and this indicates rapid metabolism of curcumin [153,154].

The first step of pharmaceutical strategies is to improve curcumin solubility and its absorption to overcome this problem. The incorporation of curcumin in solid dispersion, nanoparticles, micelles, conjugates, liposomes, and phytosomal formulations have increased curcumin's solubility and absorption rate in GBM cells [155–161]. Studies showed that curcumin-loaded noisome nanoparticles (CM-NP) can more effectively suppress the viability, proliferation, and migration of GSCs by inducing cell cycle arrest and apoptosis [155]. The CM-NP also efficiently increases ROS-suppression of tumor growth and inhibits monocyte chemoattractant protein 1 (MCP1) to reduce the invasiveness of GSCs compared to curcumin alone. Additionally, rats injected with curcumin-loaded PLGA nanoparticles have significantly smaller tumor size after five days of injection, while the group injected with curcumin alone displayed no significant change [156]. Another in vivo study reported that the combination of antisense-oligonucleotide against miR-21 with curcumin-loaded DP micelle complex reduced the tumor volume more effectively than single therapy curcumin or miR21ASO alone [157].

Additionally, relative to natural curcumin, solid lipid curcumin particles can promote cell death and DNA fragmentation by increasing the levels of caspase-3, Bax, and p53 with downregulation of Bcl-2, c-Myc, and Akt proteins in GBM cell lines [159]. The antibody-conjugated biodegradable polymeric nanoparticles (Mab-PLGA NPs) could enhance the photodynamic efficiency of curcumin on DKMG/EGFRvIII GBM cells compared to curcumin loaded biodegradable polymeric nanoparticles alone (56% vs. 24%) [158]. Furthermore, curcumin analogue induces FBXL2-mediated AR ubiquitination, ROS, lipid peroxidation, and suppression of glutathione peroxidase 4 to inhibit growth of TMZ-sensitive and -resistant GBM in vitro and in vivo [160].

Moreover, curcumin liposomes can significantly improve the anti-tumor effects of curcumin by enhancing the uptake effects, apoptosis effects, and endocytic effects of C6 glioma cells and C6 glioma stem cells. Curcumin liposomes were also shown to inhibit tumor growth and increase the survival period of brain glioma-bearing mice [161]. Other than that, a study showed that curcumin-loaded targeted liposomes cross the BBB two-fold higher than the non-targeted liposomes loaded with curcumin [162]. Curcumin-loaded targeted liposomes can more effectively inhibit GBM tumor growth and increase the survival rate of U87 GBM tumor-bearing mice compared to free curcumin as well as the non-targeted liposome-loaded curcumin [162]. Additionally, curcumin phytosome meriva (CCP) has been shown to improve curcumin bioavailability [163], which then help to activate natural killer cells and mediate elimination of GBM and GBM stem cells [164]. Based on the results of the preclinical studies, the use of the conjugate, nanoparticles, micelles, solid lipid, analogues, liposomes, or phytosomal formulation could certainly be clinically developed to benefit GBM patients.

4. Clinical Trials

Currently, there is only one clinical study investigating the curcumin effects on 13 newly diagnosed pre-operative GBM patients [23]. In 2014, this clinical study reported the highest serum and intratumoral concentrations of curcumin detected using the micellar curcumin formulation. It was reported that the intratumoral concentration of curcumin detected might not be sufficient to cause short-term antitumor effects. Still, it might help to control tumor growth in a long-term way. Moreover, intratumoral inorganic phosphate was significantly increased by curcumin. This might indicate increased demand for high-energy phosphates or mitochondrial dysfunction, since inorganic phosphate is used for ATP generation [23]. In addition, the side effects of taking curcumin are significantly less severe than the current chemotherapeutic drugs [23]. Thus, the oral administration of micellar curcumin is relatively safer and well-tolerated. However, this clinical trial only involved a small number of patients with a small dose of curcumin. Further clinical trials should be carried out to strengthen the statistical validity with a larger sample size. In the future, phase I/II clinical trials should be carried out to determine the safety and ideal dosage for GBM treatment, and phase III to IV to examine curcumin's efficacy and potential side effects with a larger sample population. Other than that, randomized controlled trials can be carried out to avoid bias and provide higher accuracy results.

5. Conclusions and Future Perspectives

Taken together, curcumin possesses the ability to modulate various core signaling pathways that are commonly dysregulated in GBM. However, among these signaling pathways, a greater emphasis on Rb and Shh pathways could be of focus for future pre-clinical studies, since the current data are still limited. Additionally, the contradictory findings on curcumin modulation of the p53 pathway warrant future investigation and suggest that curcumin use may be selective against radioresistant GBM tumor. Nevertheless, the combination of curcumin with standard chemotherapeutic drugs mainly results in the modulation of multiple signaling pathways that promote their anti-cancer effects. Curcumin's ability to modulate the major signaling pathways while promoting the efficacy of standard chemotherapeutic drugs warrants its use as a potential nutraceutical-based adjuvant drug for GBM treatment.

Since the clinical studies of curcumin in GBM patients are lacking, it is worthwhile for future clinical studies to incorporate curcumin as a potential neo-adjuvant in GBM. Although issues such as bioavailability, poor absorption, and rapid systemic elimination may hinder its efficacy, the pre-clinical use of nanodelivery has shown great promise while increasing the efficacy of curcumin and chemotherapeutic drugs. Considering this, a greater emphasis should also be given towards the nanoformulations of curcumin in future clinical studies, in combination with the standard chemotherapeutic drugs. Therefore, the multimodal modulation of signaling pathways via nanoformulation of targeted curcumin delivery that can synergize chemotherapeutic drugs efficacy may provide a clinical perspective in GBM therapy.

Author Contributions: S.C.W. designed the outline of the manuscript and wrote the manuscript. R.N. designed the outline of the manuscript and edited and revised the manuscript. M.N.A.K. designed the outline of the manuscript and edited and revised the manuscript. All authors have read and agreed to the published version of the manuscript.

Funding: This research was funded by a JCSMHS—BRIMS collaboration grant, Monash University Malaysia.

Acknowledgments: The authors are thankful to Monash University Malaysia, for providing financial support to conduct this study.

Conflicts of Interest: The authors declare no conflict of interest.

References

1. Hanif, F.; Muzaffar, K.; Perveen, K.; Malhi, S.M.; Simjee, S.U. Glioblastoma Multiforme: A Review of its Epidemiology and Pathogenesis through Clinical Presentation and Treatment. *Asian Pac. J. Cancer Prev.* **2017**, *18*, 3–9.
2. Ostrom, Q.T.; Cioffi, G.; Gittleman, H.; Patil, N.; Waite, K.; Kruchko, C. CBTRUS Statistical Report: Primary Brain and Other Central Nervous System Tumors Diagnosed in the United States in 2012–2016. *Neuro Oncol.* **2019**, *21*, v1–v100. [CrossRef]
3. Louis, D.N.; Perry, A.; Reifenberger, G.; Von Deimling, A.; Figarella-Branger, D.; Cavenee, W.K. The 2016 World Health Organization Classification of Tumors of the Central Nervous System: A summary. *Acta Neuropathol.* **2016**, *131*, 803–820. [CrossRef] [PubMed]
4. Fernandes, C.; Costa, A.; Osório, L. Current Standards of Care in Glioblastoma Therapy. In *Glioblastoma [Internet]*; De Vleeschouwer, S., Ed.; Codon Publications: Brisbane, AU, USA, 27 September 2017; Chapter 11. Available online: https://www.ncbi.nlm.nih.gov/books/NBK469987/ (accessed on 10 February 2021). [CrossRef]
5. Roy, S.; Lahiri, D.; Maji, T.; Biswas, J. Recurrent Glioblastoma: Where we stand. *South Asian J. Cancer* **2015**, *4*, 163–173. [CrossRef] [PubMed]
6. Wilson, T.A.; Karajannis, M.A.; Harter, D.H. Glioblastoma multiforme: State of the art and future therapeutics. *Surg. Neurol. Int.* **2014**, *5*, 64.
7. Davis, M.E. Glioblastoma: Overview of Disease and Treatment. *Clin. J. Oncol. Nurs.* **2016**, *20*, S2–S8. [CrossRef] [PubMed]
8. Lee, S.Y. Temozolomide resistance in glioblastoma multiforme. *Genes Dis.* **2016**, 3, 198–210. [CrossRef] [PubMed]
9. Gilbert, M.R.; Dignam, J.J.; Armstrong, T.S.; Wefel, J.S.; Blumenthal, D.T.; Vogelbaum, M.A. A Randomized Trial of Bevacizumab for Newly Diagnosed Glioblastoma. *N. Engl. J. Med.* **2014**, *370*, 699–708. [CrossRef] [PubMed]
10. Poulsen, H.S.; Urup, T.; Michaelsen, S.R.; Staberg, M.; Villingshøj, M.; Lassen, U. The impact of bevacizumab treatment on survival and quality of life in newly diagnosed glioblastoma patients. *Cancer Manag. Res.* **2014**, *6*, 373–387. [CrossRef] [PubMed]
11. Chahar, M.K.; Sharma, N.; Dobhal, M.P.; Joshi, Y.C. Flavonoids: A versatile source of anticancer drugs. *Pharmacogn. Rev.* **2011**, *5*, 1–12.
12. Panche, A.N.; Diwan, A.D.; Chandra, S.R. Flavonoids: An overview. *J. Nutr. Sci.* **2016**, *5*, e47. [CrossRef] [PubMed]
13. Hewlings, S.J.; Kalman, D.S. Curcumin: A Review of Its Effects on Human Health. *Foods* **2017**, *6*, 92. [CrossRef]
14. Gupta, S.C.; Patchva, S.; Aggarwal, B.B. Therapeutic roles of curcumin: Lessons learned from clinical trials. *AAPS J.* **2013**, *15*, 195–218. [CrossRef]
15. Maugeri, A.; Mazzone, M.G.; Giuliano, F.; Vinciguerra, M.; Basile, G.; Barchitta, M. Curcumin Modulates DNA Methyltransferase Functions in a Cellular Model of Diabetic Retinopathy. *Oxid. Med. Cell. Longev.* **2018**, *2018*, 5407482. [CrossRef]
16. Ma, Z.; Wang, N.; He, H.; Tang, X. Pharmaceutical strategies of improving oral systemic bioavailability of curcumin for clinical application. *J. Control. Release* **2019**, *316*, 359–380. [CrossRef] [PubMed]
17. Aggarwal, B.B.; Harikumar, K.B. Potential therapeutic effects of curcumin, the anti-inflammatory agent, against neurodegenerative, cardiovascular, pulmonary, metabolic, autoimmune and neoplastic diseases. *Int. J. Biochem. Cell Biol.* **2009**, *41*, 40–59. [CrossRef] [PubMed]
18. Barchitta, M.; Maugeri, A.; Favara, G.; Magnano San Lio, R.; Evola, G.; Agodi, A. Nutrition and Wound Healing: An Overview Focusing on the Beneficial Effects of Curcumin. *Int. J. Mol. Sci.* **2019**, *20*, 1119. [CrossRef] [PubMed]
19. Sharma, R.A.; McLelland, H.R.; Hill, K.A.; Ireson, C.R.; Euden, S.A.; Manson, M.M. Pharmacodynamic and Pharmacokinetic Study of Oral Curcuma Extract in Patients with Colorectal Cancer. *Clin. Cancer Res.* **2001**, *7*, 1894. [PubMed]
20. He, Z.-Y.; Shi, C.-B.; Wen, H.; Li, F.-L.; Wang, B.-L.; Wang, J. Upregulation of p53 Expression in Patients with Colorectal Cancer by Administration of Curcumin. *Cancer Investig.* **2011**, *29*, 208–213. [CrossRef] [PubMed]
21. Kanai, M.; Yoshimura, K.; Asada, M.; Imaizumi, A.; Suzuki, C.; Matsumoto, S. A phase I/II study of gemcitabine-based chemotherapy plus curcumin for patients with gemcitabine-resistant pancreatic cancer. *Cancer Chemother. Pharmacol.* **2011**, *68*, 157–164. [CrossRef] [PubMed]

22. Polasa, K.; Raghuram, T.C.; Krishna, T.P.; Krishnaswamy, K. Effect of turmeric on urinary mutagens in smokers. *Mutagenesis* **1992**, *7*, 107–109. [CrossRef]
23. Dützmann, S.; Schiborr, C.; Kocher, A.; Pilatus, U.; Hattingen, E.; Weissenberger, J. Intratumoral Concentrations and Effects of Orally Administered Micellar Curcuminoids in Glioblastoma Patients. *Nutr. Cancer* **2016**, *68*, 943–948. [CrossRef] [PubMed]
24. Giordano, A.; Tommonaro, G. Curcumin and Cancer. *Nutrients* **2019**, *11*, 2376. [CrossRef] [PubMed]
25. Barati, N.; Momtazi-Borojeni, A.A.; Majeed, M.; Sahebkar, A. Potential therapeutic effects of curcumin in gastric cancer. *J. Cell. Physiol.* **2019**, *234*, 2317–2328. [CrossRef]
26. Hesari, A.; Azizian, M.; Sheikhi, A.; Nesaei, A.; Sanaei, S.; Mahinparvar, N. Chemopreventive and therapeutic potential of curcumin in esophageal cancer: Current and future status. *Int. J. Cancer* **2019**, *144*, 1215–1226. [CrossRef] [PubMed]
27. Jalili-Nik, M.; Soltani, A.; Moussavi, S.; Ghayour-Mobarhan, M.; Ferns, G.A.; Hassanian, S.M. Current status and future prospective of Curcumin as a potential therapeutic agent in the treatment of colorectal cancer. *J. Cell. Physiol.* **2018**, *233*, 6337–6345. [CrossRef] [PubMed]
28. Su, C.-C.; Wang, M.-J.; Chiu, T.-L. The anti-cancer efficacy of curcumin scrutinized through core signaling pathways in glioblastoma. *Int. J. Mol. Med.* **2010**, *26*, 217–224.
29. Weissenberger, J.; Priester, M.; Bernreuther, C.; Rakel, S.; Glatzel, M.; Seifert, V. Dietary Curcumin Attenuates Glioma Growth in a Syngeneic Mouse Model by Inhibition of the JAK1,2/STAT3 Signaling Pathway. *Clin. Cancer Res.* **2010**, *16*, 5781. [CrossRef] [PubMed]
30. Gersey, Z.C.; Rodriguez, G.A.; Barbarite, E.; Sanchez, A.; Walters, W.M.; Ohaeto, K.C. Curcumin decreases malignant characteristics of glioblastoma stem cells via induction of reactive oxygen species. *BMC Cancer* **2017**, *17*, 99. [CrossRef] [PubMed]
31. Wang, Z.; Liu, F.; Liao, W.; Yu, L.; Hu, Z.; Li, M. Curcumin suppresses glioblastoma cell proliferation by p-AKT/mTOR pathway and increases the PTEN expression. *Arch. Biochem. Biophys.* **2020**, *689*, 108412. [CrossRef] [PubMed]
32. Du, W.-Z.; Feng, Y.; Wang, X.-F.; Piao, X.-Y.; Cui, Y.-Q.; Chen, L.-C. Curcumin suppresses malignant glioma cells growth and induces apoptosis by inhibition of SHH/GLI1 signaling pathway in vitro and vivo. *CNS Neurosci. Ther.* **2013**, *19*, 926–936. [CrossRef] [PubMed]
33. Zanotto-Filho, A.; Braganhol, E.; Edelweiss, M.I.; Behr, G.A.; Zanin, R.; Schröder, R.; Simões-Pires, A.; Battastini, A.M.O.; Moreira, J.C.F. The curry spice curcumin selectively inhibits cancer cells growth in vitro and in preclinical model of glioblastoma. *J. Nutr. Biochem.* **2012**, *23*, 591–601. [CrossRef]
34. Mao, H.; Lebrun, D.G.; Yang, J.; Zhu, V.F.; Li, M. Deregulated signaling pathways in glioblastoma multiforme: Molecular mechanisms and therapeutic targets. *Cancer Investig.* **2012**, *30*, 48–56. [CrossRef] [PubMed]
35. Chiu, S.S.; Lui, E.; Majeed, M.; Vishwanatha, J.K.; Ranjan, A.P.; Maitra, A. Differential Distribution of Intravenous Curcumin Formulations in the Rat Brain. *Anticancer Res.* **2011**, *31*, 907. [PubMed]
36. Priyadarsini, K.I. The chemistry of curcumin: From extraction to therapeutic agent. *Molecules* **2014**, *19*, 20091–20112. [CrossRef]
37. Liu, E.; Wu, J.; Cao, W.; Zhang, J.; Liu, W.; Jiang, X. Curcumin induces G2/M cell cycle arrest in a p53-dependent manner and upregulates ING4 expression in human glioma. *J. Neuro-Oncol.* **2007**, *85*, 263–270. [CrossRef]
38. Fratantonio, D.; Molonia, M.S.; Bashllari, R.; Muscarà, C.; Ferlazzo, G.; Costa, G. Curcumin potentiates the antitumor activity of Paclitaxel in rat glioma C6 cells. *Phytomedicine* **2019**, *55*, 23–30. [CrossRef] [PubMed]
39. Ramachandran, C.; Nair, S.M.; Escalon, E.; Melnick, S.J. Potentiation of Etoposide and Temozolomide Cytotoxicity by Curcumin and Turmeric Force in Brain Tumor Cell Lines. *J. Complementary Integr. Med.* **2012**, *9*. [CrossRef]
40. Garrido-Armas, M.; Corona, J.C.; Escobar, M.L.; Torres, L.; Ordóñez-Romero, F.; Hernández-Hernández, A. Paraptosis in human glioblastoma cell line induced by curcumin. *Toxicol. In Vitro* **2018**, *51*, 63–73. [CrossRef] [PubMed]
41. Khaw, A.K.; Hande, M.P.; Kalthur, G.; Hande, M.P. Curcumin inhibits telomerase and induces telomere shortening and apoptosis in brain tumour cells. *J. Cell. Biochem.* **2013**, *114*, 1257–1270. [CrossRef]
42. Karmakar, S.; Banik, N.L.; Ray, S.K. Curcumin Suppressed Anti-apoptotic Signals and Activated Cysteine Proteases for Apoptosis in Human Malignant Glioblastoma U87MG Cells. *Neurochem. Res.* **2007**, *32*, 2103–2113. [CrossRef] [PubMed]
43. Senft, C.; Polacin, M.; Priester, M.; Seifert, V.; Kögel, D.; Weissenberger, J. The nontoxic natural compound Curcumin exerts anti-proliferative, anti-migratory, and anti-invasive properties against malignant gliomas. *BMC Cancer* **2010**, *10*, 491. [CrossRef] [PubMed]
44. Wu, B.; Yao, X.; Nie, X.; Xu, R. Epigenetic reactivation of RANK in glioblastoma cells by curcumin: Involvement of STAT3 inhibition. *DNA Cell Biol.* **2013**, *32*, 292–297. [CrossRef]
45. Kim, S.-Y.; Jung, S.-H.; Kim, H.-S. Curcumin is a potent broad spectrum inhibitor of matrix metalloproteinase gene expression in human astroglioma cells. *Biochem. Biophys. Res. Commun.* **2005**, *337*, 510–516. [CrossRef]
46. Woo, M.-S.; Jung, S.-H.; Kim, S.-Y.; Hyun, J.-W.; Ko, K.-H.; Kim, W.-K. Curcumin suppresses phorbol ester-induced matrix metalloproteinase-9 expression by inhibiting the PKC to MAPK signaling pathways in human astroglioma cells. *Biochem. Biophys. Res. Commun.* **2005**, *335*, 1017–1025. [CrossRef] [PubMed]
47. Choi, B.H.; Kim, C.G.; Bae, Y.-S.; Lim, Y.; Lee, Y.H.; Shin, S.Y. p21Waf1/Cip1 Expression by Curcumin in U-87MG Human Glioma Cells: Role of Early Growth Response-1 Expression. *Cancer Res.* **2008**, *68*, 1369. [CrossRef] [PubMed]
48. Aoki, H.; Takada, Y.; Kondo, S.; Sawaya, R.; Aggarwal, B.B.; Kondo, Y. Evidence That Curcumin Suppresses the Growth of Malignant Gliomas in Vitro and in Vivo through Induction of Autophagy: Role of Akt and Extracellular Signal-Regulated Kinase Signaling Pathways. *Mol. Pharmacol.* **2007**, *72*, 29. [CrossRef]

49. Zhang, Z.-J.; Zhao, L.-X.; Cao, D.-L.; Zhang, X.; Gao, Y.-J.; Xia, C. Curcumin Inhibits LPS-Induced CCL2 Expression via JNK Pathway in C6 Rat Astrocytoma Cells. *Cell. Mol. Neurobiol.* **2012**, *32*, 1003–1010. [CrossRef] [PubMed]
50. Maiti, P.; Scott, J.; Sengupta, D.; Al-Gharaibeh, A.; Dunbar, G.L. Curcumin and Solid Lipid Curcumin Particles Induce Autophagy, but Inhibit Mitophagy and the PI3K-Akt/mTOR Pathway in Cultured Glioblastoma Cells. *Int. J. Mol. Sci.* **2019**, *20*, 399. [CrossRef]
51. Zhao, J.; Zhu, J.; Lv, X.; Xing, J.; Liu, S.; Chen, C. Curcumin potentiates the potent antitumor activity of ACNU against glioblastoma by suppressing the PI3K/AKT and NF-κB/COX-2 signaling pathways. *OncoTargets Ther.* **2017**, *10*, 5471–5482. [CrossRef] [PubMed]
52. Yin, S.; Du, W.; Wang, F.; Han, B.; Cui, Y.; Yang, D. MicroRNA-326 sensitizes human glioblastoma cells to curcumin via the SHH/GLI1 signaling pathway. *Cancer Biol. Ther.* **2018**, *19*, 260–270. [CrossRef]
53. Karmakar, S.; Banik, N.L.; Patel, S.J.; Ray, S.K. Curcumin activated both receptor-mediated and mitochondria-mediated proteolytic pathways for apoptosis in human glioblastoma T98G cells. *Neurosci. Lett.* **2006**, *407*, 53–58. [CrossRef] [PubMed]
54. Nagai, S.; Kurimoto, M.; Washiyama, K.; Hirashima, Y.; Kumanishi, T.; Endo, S. Inhibition of Cellular Proliferation and Induction of Apoptosis by Curcumin in Human Malignant Astrocytoma Cell Lines. *J. Neuro-Oncol.* **2005**, *74*, 105–111. [CrossRef] [PubMed]
55. Huang, T.-Y.; Tsai, T.-H.; Hsu, C.-W.; Hsu, Y.-C. Curcuminoids Suppress the Growth and Induce Apoptosis through Caspase-3-Dependent Pathways in Glioblastoma Multiforme (GBM) 8401 Cells. *J. Agric. Food Chem.* **2010**, *58*, 10639–10645. [CrossRef] [PubMed]
56. Giacinti, C.; Giordano, A. RB and cell cycle progression. *Oncogene* **2006**, *25*, 5220–5227. [CrossRef]
57. Knudsen, E.S.; Wang, J.Y.J. Targeting the RB-pathway in cancer therapy. *Clin. Cancer Res.* **2010**, *16*, 1094–1099. [CrossRef] [PubMed]
58. Nakada, M.; Kita, D.; Watanabe, T.; Hayashi, Y.; Teng, L.; Pyko, I.V. Aberrant Signaling Pathways in Glioma. *Cancers* **2011**, *3*, 3242–3278. [CrossRef]
59. Biernat, W.; Tohma, Y.; Yonekawa, Y.; Kleihues, P.; Ohgaki, H. Alterations of cell cycle regulatory genes in primary (de novo) and secondary glioblastomas. *Acta Neuropathol.* **1997**, *94*, 303–309. [CrossRef]
60. Cancer Genome Atlas Research Network. Comprehensive genomic characterization defines human glioblastoma genes and core pathways. *Nature* **2008**, *455*, 1061–1068. [CrossRef]
61. Grzmil, M.; Hemmings, B.A. Deregulated signalling networks in human brain tumours. *Biochim. Biophys. Acta BBA Proteins Proteom.* **2010**, *1804*, 476–483. [CrossRef]
62. Biasoli, D.; Kahn, S.A.; Cornélio, T.A.; Furtado, M.; Campanati, L.; Chneiweiss, H. Retinoblastoma protein regulates the crosstalk between autophagy and apoptosis, and favors glioblastoma resistance to etoposide. *Cell Death Dis.* **2013**, *4*, e767. [CrossRef] [PubMed]
63. Zhang, D.; Dai, D.; Zhou, M.; Li, Z.; Wang, C.; Lu, Y. Inhibition of Cyclin D1 Expression in Human Glioblastoma Cells is Associated with Increased Temozolomide Chemosensitivity. *Cell. Physiol. Biochem.* **2018**, *51*, 2496–2508. [CrossRef]
64. Fry, D.W.; Harvey, P.J.; Keller, P.R.; Elliott, W.L.; Meade, M.; Trachet, E. Specific inhibition of cyclin-dependent kinase 4/6 by PD 0332991 and associated antitumor activity in human tumor xenografts. *Mol. Cancer Ther.* **2004**, *3*, 1427.
65. Michaud, K.; Solomon, D.A.; Oermann, E.; Kim, J.-S.; Zhong, W.-Z.; Prados, M.D. Pharmacologic inhibition of cyclin-dependent kinases 4 and 6 arrests the growth of glioblastoma multiforme intracranial xenografts. *Cancer Res.* **2010**, *70*, 3228–3238. [CrossRef]
66. Chen, J. The Cell-Cycle Arrest and Apoptotic Functions of p53 in Tumor Initiation and Progression. *Cold Spring Harb. Perspect. Med.* **2016**, *6*, a026104. [CrossRef]
67. Shangary, S.; Wang, S. Small-Molecule Inhibitors of the MDM2-p53 Protein-Protein Interaction to Reactivate p53 Function: A Novel Approach for Cancer Therapy. *Annu. Rev. Pharmacol. Toxicol.* **2009**, *49*, 223–241. [CrossRef] [PubMed]
68. Kubbutat, M.H.G.; Jones, S.N.; Vousden, K.H. Regulation of p53 stability by Mdm2. *Nature* **1997**, *387*, 299–303. [CrossRef]
69. Hemann, M.T.; Lowe, S.W. The p53-Bcl-2 connection. *Cell Death Differ.* **2006**, *13*, 1256–1259. [CrossRef] [PubMed]
70. Zawlik, I.; Kita, D.; Vaccarella, S.; Mittelbronn, M.; Franceschi, S.; Ohgaki, H. Common Polymorphisms in the MDM2 and TP53 Genes and the Relationship between TP53 Mutations and Patient Outcomes in Glioblastomas. *Brain Pathol.* **2009**, *19*, 188–194. [CrossRef]
71. Fels, C.; Schäfer, C.; Hüppe, B.; Bahn, H.; Heidecke, V.; Kramm, C.M. Bcl-2 Expression in Higher-grade Human Glioma: A Clinical and Experimental Study. *J. Neuro-Oncol.* **2000**, *48*, 207–216. [CrossRef] [PubMed]
72. Zhang, Y.; Dube, C.; Gibert, M., Jr.; Cruickshanks, N.; Wang, B.; Coughlan, M. The p53 Pathway in Glioblastoma. *Cancers* **2018**, *10*, 297. [CrossRef] [PubMed]
73. Ghaemi, S.; Arefian, E.; Rezazadeh Valojerdi, R.; Soleimani, M.; Moradimotlagh, A.; Jamshidi Adegani, F. Inhibiting the expression of anti-apoptotic genes BCL2L1 and MCL1, and apoptosis induction in glioblastoma cells by microRNA-342. *Biomed. Pharmacother.* **2020**, *121*, 109641. [CrossRef] [PubMed]
74. Pareja, F.; Macleod, D.; Shu, C.; Crary, J.F.; Canoll, P.D.; Ross, A.H. PI3K and Bcl-2 Inhibition Primes Glioblastoma Cells to Apoptosis through Downregulation of Mcl-1 and Phospho-BAD. *Mol. Cancer Res.* **2014**, *12*, 987. [CrossRef]
75. Doyon, Y.; Cayrou, C.; Ullah, M.; Landry, A.-J.; Côté, V.; Selleck, W. ING Tumor Suppressor Proteins Are Critical Regulators of Chromatin Acetylation Required for Genome Expression and Perpetuation. *Mol. Cell* **2006**, *21*, 51–64. [CrossRef] [PubMed]
76. Kim, S. HuntING4 New Tumor Suppressors. *Cell Cycle* **2005**, *4*, 516–517. [CrossRef] [PubMed]
77. Yamaoka, K.; Saharinen, P.; Pesu, M.; Holt, V.E.T., 3rd; Silvennoinen, O.; O'Shea, J.J. The Janus kinases (Jaks). *Genome Biol.* **2004**, *5*, 253. [CrossRef]

78. Zhou, Y.-J.; Chen, M.; Cusack, N.A.; Kimmel, L.H.; Magnuson, K.S.; Boyd, J.G. Unexpected Effects of FERM Domain Mutations on Catalytic Activity of Jak3: Structural Implication for Janus Kinases. *Mol. Cell* **2001**, *8*, 959–969. [CrossRef]
79. Rawlings, J.S.; Rosler, K.M.; Harrison, D.A. The JAK/STAT signaling pathway. *J. Cell Sci.* **2004**, *117*, 1281. [CrossRef]
80. Schaefer, L.K.; Ren, Z.; Fuller, G.N.; Schaefer, T.S. Constitutive activation of Stat3α in brain tumors: Localization to tumor endothelial cells and activation by the endothelial tyrosine kinase receptor (VEGFR-2). *Oncogene* **2002**, *21*, 2058–2065. [CrossRef]
81. Rahaman, S.O.; Harbor, P.C.; Chernova, O.; Barnett, G.H.; Vogelbaum, M.A.; Haque, S.J. Inhibition of constitutively active Stat3 suppresses proliferation and induces apoptosis in glioblastoma multiforme cells. *Oncogene* **2002**, *21*, 8404–8413. [CrossRef]
82. Zhang, L.; Alizadeh, D.; Van Handel, M.; Kortylewski, M.; Yu, H.; Badie, B. Stat3 inhibition activates tumor macrophages and abrogates glioma growth in mice. *Glia* **2009**, *57*, 1458–1467. [CrossRef]
83. Kim, H.Y.; Park, E.J.; Joe, E.-h.; Jou, I. Curcumin Suppresses Janus Kinase-STAT Inflammatory Signaling through Activation of Src Homology 2 Domain-Containing Tyrosine Phosphatase 2 in Brain Microglia. *J. Immunol.* **2003**, *171*, 6072. [CrossRef]
84. Papanastasiou, A.D.; Sirinian, C.; Kalofonos, H.P. Identification of novel human receptor activator of nuclear factor-kB isoforms generated through alternative splicing: Implications in breast cancer cell survival and migration. *Breast Cancer Res.* **2012**, *14*, R112. [CrossRef]
85. Von dem Knesebeck, A.; Felsberg, J.; Waha, A.; Hartmann, W.; Scheffler, B.; Glas, M. RANK (TNFRSF11A) is epigenetically inactivated and induces apoptosis in gliomas. *Neoplasia* **2012**, *14*, 526–534. [CrossRef] [PubMed]
86. Lee, H.; Zhang, P.; Herrmann, A.; Yang, C.; Xin, H.; Wang, Z. Acetylated STAT3 is crucial for methylation of tumor-suppressor gene promoters and inhibition by resveratrol results in demethylation. *Proc. Natl. Acad. Sci. USA* **2012**, *109*, 7765–7769. [CrossRef] [PubMed]
87. Soares-Silva, M.; Diniz, F.F.; Gomes, G.N.; Bahia, D. The Mitogen-Activated Protein Kinase (MAPK) Pathway: Role in Immune Evasion by Trypanosomatids. *Front. Microbiol.* **2016**, *7*. [CrossRef] [PubMed]
88. Teramoto, H.; Gutkind, J.S. Mitogen-Activated Protein Kinase Family. In *Encyclopedia of Biological Chemistry*, 2nd ed.; Lennarz, W.J., Lane, M.D., Eds.; Academic Press: Cambridge, MA, USA, 2013; pp. 176–180.
89. Mebratu, Y.; Tesfaigzi, Y. How ERK1/2 activation controls cell proliferation and cell death: Is subcellular localization the answer? *Cell Cycle* **2009**, *8*, 1168–1175. [CrossRef] [PubMed]
90. Huang, G.; Shi, L.Z.; Chi, H. Regulation of JNK and p38 MAPK in the immune system: Signal integration, propagation and termination. *Cytokine* **2009**, *48*, 161–169. [CrossRef] [PubMed]
91. Vo, V.A.; Lee, J.-W.; Lee, H.J.; Chun, W.; Lim, S.Y.; Kim, S.-S. Inhibition of JNK Potentiates Temozolomide-induced Cytotoxicity in U87MG Glioblastoma Cells via Suppression of Akt Phosphorylation. *Anticancer Res.* **2014**, *34*, 5509.
92. Ramaswamy, P.; Nanjaiah, N.D.; Borkotokey, M. Role of MEK-ERK signaling mediated adhesion of glioma cells to extra-cellular matrix: Possible implication on migration and proliferation. *Ann. Neurosci.* **2019**, *26*, 52–56. [CrossRef] [PubMed]
93. Lo, H.-W. Targeting Ras-RAF-ERK and its interactive pathways as a novel therapy for malignant gliomas. *Curr. Cancer Drug Targets* **2010**, *10*, 840–848. [CrossRef] [PubMed]
94. Yang, K.; Liu, Y.; Liu, Z.; Liu, J.; Liu, X.; Chen, X. p38γ overexpression in gliomas and its role in proliferation and apoptosis. *Sci. Rep.* **2013**, *3*, 2089. [CrossRef] [PubMed]
95. Matsuda, K.-I.; Sato, A.; Okada, M.; Shibuya, K.; Seino, S.; Suzuki, K. Targeting JNK for therapeutic depletion of stem-like glioblastoma cells. *Sci. Rep.* **2012**, *2*, 516. [CrossRef] [PubMed]
96. Ke, X.-X.; Pang, Y.; Chen, K.; Zhang, D.; Wang, F.; Zhu, S. Knockdown of arsenic resistance protein 2 inhibits human glioblastoma cell proliferation through the MAPK/ERK pathway. *Oncol. Rep.* **2018**, *40*, 3313–3322. [CrossRef]
97. Ouyang, Z.; Xu, G. Antitumor effects of Sweroside in human glioblastoma: Its effects on mitochondrial mediated apoptosis, activation of different caspases, G0/G1 cell cycle arrest and targeting JNK/p38 MAPK signal pathways. *J. BUON* **2019**, *24*, 2141–2146. [PubMed]
98. Heiland, D.H.; Haaker, G.; Delev, D.; Mercas, B.; Masalha, W.; Heynckes, S. Comprehensive analysis of PD-L1 expression in glioblastoma multiforme. *Oncotarget* **2017**, *8*, 42214–42225. [CrossRef] [PubMed]
99. Wurm, J.; Behringer, S.P.; Ravi, V.M.; Joseph, K.; Neidert, N.; Maier, J.P. Astrogliosis Releases Pro-Oncogenic Chitinase 3-Like 1 Causing MAPK Signaling in Glioblastoma. *Cancers* **2019**, *11*, 1437. [CrossRef] [PubMed]
100. Munaut, C.; Noël, A.; Hougrand, O.; Foidart, J.-M.; Boniver, J.; Deprez, M. Vascular endothelial growth factor expression correlates with matrix metalloproteinases MT1-MMP, MMP-2 and MMP-9 in human glioblastomas. *Int. J. Cancer* **2003**, *106*, 848–855. [CrossRef] [PubMed]
101. Sawaya, R.; Go, Y.; Kyritisis, A.P.; Uhm, J.; Venkaiah, B.; Mohanam, S. Elevated Levels of Mr92,000 Type IV Collagenase during Tumor Growthin Vivo. *Biochem. Biophys. Res. Commun.* **1998**, *251*, 632–636. [CrossRef]
102. Lorenzl, S.; Albers, D.S.; Chirichigno, J.W.; Augood, S.J.; Beal, M.F. Elevated levels of matrix metalloproteinases-9 and -1 and of tissue inhibitors of MMPs, TIMP-1 and TIMP-2 in postmortem brain tissue of progressive supranuclear palsy. *J. Neurol. Sci.* **2004**, *218*, 39–45. [CrossRef]
103. Bignami, A.; Hosley, M.; Dahl, D. Hyaluronic acid and hyaluronic acid-binding proteins in brain extracellular matrix. *Anat. Embryol.* **1993**, *188*, 419–433. [CrossRef]
104. Keshari, R.S.; Verma, A.; Barthwal, M.K.; Dikshit, M. Reactive oxygen species-induced activation of ERK and p38 MAPK mediates PMA-induced NETs release from human neutrophils. *J. Cell. Biochem.* **2013**, *114*, 532–540. [CrossRef]

105. McCubrey, J.A.; LaHair, M.M.; Franklin, R.A. Reactive Oxygen Species-Induced Activation of the MAP Kinase Signaling Pathways. *Antioxid. Redox Signal.* **2006**, *8*, 1775–1789. [CrossRef] [PubMed]
106. Escamilla-Ramírez, A.; Castillo-Rodríguez, R.A.; Zavala-Vega, S.; Jimenez-Farfan, D.; Anaya-Rubio, I.; Briseño, E. Autophagy as a Potential Therapy for Malignant Glioma. *Pharmaceuticals* **2020**, *13*, 156. [CrossRef] [PubMed]
107. Chang, A.L.; Miska, J.; Wainwright, D.A.; Dey, M.; Rivetta, C.V.; Yu, D. CCL2 Produced by the Glioma Microenvironment Is Essential for the Recruitment of Regulatory T Cells and Myeloid-Derived Suppressor Cells. *Cancer Res.* **2016**, *76*, 5671–5682. [CrossRef]
108. Engelman, J.A.; Luo, J.; Cantley, L.C. The evolution of phosphatidylinositol 3-kinases as regulators of growth and metabolism. *Nat. Rev. Genet.* **2006**, *7*, 606–619. [CrossRef] [PubMed]
109. Katso, R.; Okkenhaug, K.; Ahmadi, K.; White, S.; Timms, J.; Waterfield, M.D. Cellular Function of Phosphoinositide 3-Kinases: Implications for Development, Immunity, Homeostasis, and Cancer. *Annu. Rev. Cell Dev. Biol.* **2001**, *17*, 615–675. [CrossRef] [PubMed]
110. Porta, C.; Paglino, C.; Mosca, A. Targeting PI3K/Akt/mTOR Signaling in Cancer. *Front. Oncol.* **2014**, *4*, 64. [CrossRef]
111. Yang, J.; Nie, J.; Ma, X.; Wei, Y.; Peng, Y.; Wei, X. Targeting PI3K in cancer: Mechanisms and advances in clinical trials. *Mol. Cancer* **2019**, *18*, 26. [CrossRef] [PubMed]
112. Brennan, C.W.; Verhaak, R.G.W.; McKenna, A.; Campos, B.; Noushmehr, H.; Salama, S.R. The somatic genomic landscape of glioblastoma. *Cell* **2013**, *155*, 462–477. [CrossRef] [PubMed]
113. Koul, D. PTEN Signaling pathways in glioblastoma. *Cancer Biol. Ther.* **2008**, *7*, 1321–1325. [CrossRef]
114. Gallia, G.L.; Tyler, B.M.; Hann, C.L.; Siu, I.M.; Giranda, V.L.; Vescovi, A.L. Inhibition of Akt inhibits growth of glioblastoma and glioblastoma stem-like cells. *Mol. Cancer Ther.* **2009**, *8*, 386–393. [CrossRef] [PubMed]
115. Kim, Y.C.; Guan, K.-L. mTOR: A pharmacologic target for autophagy regulation. *J. Clin. Investig.* **2015**, *125*, 25–32. [CrossRef] [PubMed]
116. Bava, S.V.; Puliyappadamba, V.T.; Deepti, A.; Nair, A.; Karunagaran, D.; Anto, R.J. Sensitization of taxol-induced apoptosis by curcumin involves down-regulation of nuclear factor-κB and the serine/threonine kinase Akt and is independent of tubulin polymerization. *J. Biol. Chem.* **2018**, *293*, 12283. [CrossRef]
117. Carballo, G.B.; Honorato, J.R.; De Lopes, G.P.F. A highlight on Sonic hedgehog pathway. *Cell Commun. Signal.* **2018**, *16*, 11. [CrossRef]
118. Robbins, D.J.; Fei, D.L.; Riobo, N.A. The Hedgehog signal transduction network. *Sci. Signal.* **2012**, *5*, re6. [CrossRef] [PubMed]
119. Melamed, J.R.; Morgan, J.T.; Ioele, S.A.; Gleghorn, J.P.; Sims-Mourtada, J.; Day, E.S. Investigating the role of Hedgehog/GLI1 signaling in glioblastoma cell response to temozolomide. *Oncotarget* **2018**, *9*, 27000–27015. [CrossRef] [PubMed]
120. Ulasov, I.V.; Nandi, S.; Dey, M.; Sonabend, A.M.; Lesniak, M.S. Inhibition of Sonic hedgehog and Notch pathways enhances sensitivity of CD133(+) glioma stem cells to temozolomide therapy. *Mol. Med.* **2011**, *17*, 103–112. [CrossRef]
121. Takezaki, T.; Hide, T.; Takanaga, H.; Nakamura, H.; Kuratsu, J.-I.; Kondo, T. Essential role of the Hedgehog signaling pathway in human glioma-initiating cells. *Cancer Sci.* **2011**, *102*, 1306–1312. [CrossRef] [PubMed]
122. Takebe, N.; Miele, L.; Harris, P.J.; Jeong, W.; Bando, H.; Kahn, M. Targeting Notch, Hedgehog, and Wnt pathways in cancer stem cells: Clinical update. *Nat. Rev. Clin. Oncol.* **2015**, *12*, 445–464. [CrossRef]
123. Rossi, M.; Magnoni, L.; Miracco, C.; Mori, E.; Tosi, P.; Pirtoli, L. β-catenin and Gli1 are prognostic markers in glioblastoma. *Cancer Biol. Ther.* **2011**, *11*, 753–761. [CrossRef]
124. Honorato, J.R.; Hauser-Davis, R.A.; Saggioro, E.M.; Correia, F.V.; Sales-Junior, S.F.; Soares, L.O.S. Role of Sonic hedgehog signaling in cell cycle, oxidative stress, and autophagy of temozolomide resistant glioblastoma. *J. Cell. Physiol.* **2020**, *235*, 3798–3814 [CrossRef] [PubMed]
125. Zhou, A.; Lin, K.; Zhang, S.; Ma, L.; Xue, J.; Morris, S.-A. Gli1-induced deubiquitinase USP48 aids glioblastoma tumorigenesis by stabilizing Gli1. *EMBO Rep.* **2017**, *18*, 1318–1330. [CrossRef]
126. Ruiz i Altaba, A.; Stecca, B.; Sánchez, P. Hedgehog–Gli signaling in brain tumors: Stem cells and paradevelopmental programs in cancer. *Cancer Lett.* **2004**, *204*, 145–157. [CrossRef]
127. Cheng, J.; Gao, J.; Tao, K. Prognostic role of Gli1 expression in solid malignancies: A meta-analysis. *Sci. Rep.* **2016**, *6*, 22184. [CrossRef] [PubMed]
128. Hui, C.-C.; Angers, S. Gli Proteins in Development and Disease. *Annu. Rev. Cell Dev. Biol.* **2011**, *27*, 513–537. [CrossRef] [PubMed]
129. Scales, S.J.; De Sauvage, F.J. Mechanisms of Hedgehog pathway activation in cancer and implications for therapy. *Trends Pharmacol Sci.* **2009**, *30*, 303–312. [CrossRef] [PubMed]
130. Stecca, B.; Ruiz i Altaba, A. Context-dependent regulation of the GLI code in cancer by HEDGEHOG and non-HEDGEHOG signals. *J. Mol. Cell Biol.* **2010**, *2*, 84–95. [CrossRef]
131. Tu, Y.; Niu, M.; Xie, P.; Yue, C.; Liu, N.; Qi, Z. Smoothened is a poor prognosis factor and a potential therapeutic target in glioma *Sci. Rep.* **2017**, *7*, 42630. [CrossRef] [PubMed]
132. Jeng, K.-S.; Sheen, I.S.; Leu, C.-M.; Tseng, P.-H.; Chang, C.-F. The Role of Smoothened in Cancer. *Int. J. Mol. Sci.* **2020**, *21* 6863. [CrossRef]
133. Elamin, M.H.; Shinwari, Z.; Hendrayani, S.-F.; Al-Hindi, H.; Al-Shail, E.; Khafaga, Y. Curcumin inhibits the Sonic Hedgehog signaling pathway and triggers apoptosis in medulloblastoma cells. *Mol. Carcinog.* **2010**, *49*, 302–314. [CrossRef]
134. Puliyappadamba, V.T.; Hatanpaa, K.J.; Chakraborty, S.; Habib, A.A. The role of NF-κB in the pathogenesis of glioma. *Mol. Cell Oncol.* **2014**, *1*, e963478. [CrossRef] [PubMed]

135. Mitchell, S.; Vargas, J.; Hoffmann, A. Signaling via the NFκB system. *Wiley Interdiscip. Rev. Syst. Biol. Med.* **2016**, *8*, 227–241. [CrossRef] [PubMed]
136. Xia, Y.; Shen, S.; Verma, I.M. NF-κB, an active player in human cancers. *Cancer Immunol. Res.* **2014**, *2*, 823–830. [CrossRef] [PubMed]
137. Sun, S.-C. The non-canonical NF-κB pathway in immunity and inflammation. *Nat. Rev. Immunol.* **2017**, *17*, 545–558. [CrossRef]
138. Shoichi, N.; Kazuo, W.; Masanori, K.; Akira, T.; Shunro, E.; Toshiro, K. Aberrant nuclear factor-κB activity and its participation in the growth of human malignant astrocytoma. *J. Neurosurg.* **2002**, *96*, 909–917.
139. Wang, H.; Wang, H.; Zhang, W.; Huang, H.J.; Liao, W.S.L.; Fuller, G.N. Analysis of the activation status of Akt, NFκB, and Stat3 in human diffuse gliomas. *Lab. Investig.* **2004**, *84*, 941–951. [CrossRef] [PubMed]
140. Raychaudhuri, B.; Han, Y.; Lu, T.; Vogelbaum, M.A. Aberrant constitutive activation of nuclear factor κB in glioblastoma multiforme drives invasive phenotype. *J. Neuro-Oncol.* **2007**, *85*, 39–47. [CrossRef] [PubMed]
141. Atkinson, G.P.; Nozell, S.E.; Harrison, D.K.; Stonecypher, M.S.; Chen, D.; Benveniste, E.N. The prolyl isomerase Pin1 regulates the NF-kappaB signaling pathway and interleukin-8 expression in glioblastoma. *Oncogene* **2009**, *28*, 3735–3745. [CrossRef]
142. Kim, S.-H.; Ezhilarasan, R.; Phillips, E.; Gallego-Perez, D.; Sparks, A.; Taylor, D. Serine/Threonine Kinase MLK4 Determines Mesenchymal Identity in Glioma Stem Cells in an NF-κB-dependent Manner. *Cancer Cell* **2016**, *29*, 201–213. [CrossRef]
143. Xu, R.X.; Liu, R.Y.; Wu, C.M.; Zhao, Y.S.; Li, Y.; Yao, Y.Q. DNA Damage-Induced NF-κB Activation in Human Glioblastoma Cells Promotes miR-181b Expression and Cell Proliferation. *Cell. Physiol. Biochem.* **2015**, *35*, 913–925. [CrossRef]
144. Smith, D.; Shimamura, T.; Barbera, S.; Bejcek, B.E. NF-κB controls growth of glioblastomas/astrocytomas. *Mol. Cell. Biochem.* **2008**, *307*, 141–147. [CrossRef] [PubMed]
145. Jiang, Z.; Zheng, X.; Rich, K.M. Down-regulation of Bcl-2 and Bcl-xL expression with bispecific antisense treatment in glioblastoma cell lines induce cell death. *J. Neurochem.* **2003**, *84*, 273–281. [CrossRef] [PubMed]
146. Mitsiades, N.; Mitsiades, C.S.; Poulaki, V.; Chauhan, D.; Richardson, P.G.; Hideshima, T. Biologic sequelae of nuclear factor–κB blockade in multiple myeloma: Therapeutic applications. *Blood* **2002**, *99*, 4079–4086. [CrossRef]
147. Zou, T.; Rao, J.N.; Guo, X.; Liu, L.; Zhang, H.M.; Strauch, E.D. NF-κB-mediated IAP expression induces resistance of intestinal epithelial cells to apoptosis after polyamine depletion. *Am. J. Physiol. Cell Physiol.* **2004**, *286*, C1009–C1018. [CrossRef] [PubMed]
148. Jeremias, I.; Kupatt, C.; Baumann, B.; Herr, I.; Wirth, T.; Debatin, K.M. Inhibition of Nuclear Factor κB Activation Attenuates Apoptosis Resistance in Lymphoid Cells. *Blood* **1998**, *91*, 4624–4631. [CrossRef]
149. Kakran, M.; Sahoo, N.G.; Tan, I.-L.; Li, L. Preparation of nanoparticles of poorly water-soluble antioxidant curcumin by antisolvent precipitation methods. *J. Nanopart. Res.* **2012**, *14*, 757. [CrossRef]
150. Dubey, S.K.; Sharma, A.K.; Narain, U.; Misra, K.; Pati, U. Design, synthesis and characterization of some bioactive conjugates of curcumin with glycine, glutamic acid, valine and demethylenated piperic acid and study of their antimicrobial and antiproliferative properties. *Eur. J. Med. Chem.* **2008**, *43*, 1837–1846. [CrossRef] [PubMed]
151. Lao, C.D.; Ruffin, M.T., IV; Normolle, D.; Heath, D.D.; Murray, S.I.; Bailey, J.M.; Boggs, M.E.; Crowell, J.; Rock, C.L.; Brenner, D.E. Dose escalation of a curcuminoid formulation. *BMC Complementary Altern. Med.* **2006**, *6*, 10. [CrossRef] [PubMed]
152. Tapal, A.; Tiku, P.K. Complexation of curcumin with soy protein isolate and its implications on solubility and stability of curcumin. *Food Chem.* **2012**, *130*, 960–965. [CrossRef]
153. Noguchi-Shinohara, M.; Hamaguchi, T.; Yamada, M. Chapter 5—The Potential Role of Curcumin in Treatment and Prevention for Neurological Disorders. In *Curcumin for Neurological and Psychiatric Disorders*; Farooqui, T., Farooqui, A.A., Eds.; Academic Press: Cambridge, MA, USA, 2019; pp. 85–103.
154. Dhillon, N.; Aggarwal, B.B.; Newman, R.A.; Wolff, R.A.; Kunnumakkara, A.B.; Abbruzzese, J.L. Phase II Trial of Curcumin in Patients with Advanced Pancreatic Cancer. *Clin. Cancer Res.* **2008**, *14*, 4491. [CrossRef] [PubMed]
155. Sahab-Negah, S.; Ariakia, F.; Jalili-Nik, M.; Afshari, A.R.; Salehi, S.; Samini, F. Curcumin Loaded in Niosomal Nanoparticles Improved the Anti-tumor Effects of Free Curcumin on Glioblastoma Stem-like Cells: An In Vitro Study. *Mol. Neurobiol.* **2020**, *57*, 3391–3411. [CrossRef]
156. Orunoğlu, M.; Kaffashi, A.; Pehlivan, S.B.; Şahin, S.; Söylemezoğlu, F.; Oğuz, K.K. Effects of curcumin-loaded PLGA nanoparticles on the RG2 rat glioma model. *Mater. Sci. Eng. C* **2017**, *78*, 32–38. [CrossRef]
157. Tan, X.; Kim, G.; Lee, D.; Oh, J.; Kim, M.; Piao, C. A curcumin-loaded polymeric micelle as a carrier of a microRNA-21 antisense-oligonucleotide for enhanced anti-tumor effects in a glioblastoma animal model. *Biomater. Sci.* **2018**, *6*, 407–417. [CrossRef]
158. Jamali, Z.; Khoobi, M.; Hejazi, S.M.; Eivazi, N.; Abdolahpour, S.; Imanparast, F. Evaluation of targeted curcumin (CUR) loaded PLGA nanoparticles for in vitro photodynamic therapy on human glioblastoma cell line. *Photodiagn. Photodyn. Ther.* **2018**, *23*, 190–201. [CrossRef] [PubMed]
159. Maiti, P.; Al-Gharaibeh, A.; Kolli, N.; Dunbar, G.L. Solid Lipid Curcumin Particles Induce More DNA Fragmentation and Cell Death in Cultured Human Glioblastoma Cells than Does Natural Curcumin. *Oxid. Med. Cell. Longev.* **2017**, *2017*, 9656719. [CrossRef] [PubMed]
160. Chen, T.-C.; Chuang, J.-Y.; Ko, C.-Y.; Kao, T.-J.; Yang, P.-Y.; Yu, C.-H. AR ubiquitination induced by the curcumin analog suppresses growth of temozolomide-resistant glioblastoma through disrupting GPX4-Mediated redox homeostasis. *Redox Biol.* **2020**, *30*, 101413. [CrossRef] [PubMed]
161. Wang, Y.; Ying, X.; Xu, H.; Yan, H.; Li, X.; Tang, H. The functional curcumin liposomes induce apoptosis in C6 glioblastoma cells and C6 glioblastoma stem cells in vitro and in animals. *Int. J. Nanomed.* **2017**, *12*, 1369–1384. [CrossRef] [PubMed]

162. Gabay, M.; Weizman, A.; Zeineh, N.; Kahana, M.; Obeid, F.; Allon, N. Liposomal Carrier Conjugated to APP-Derived Peptide for Brain Cancer Treatment. *Cell. Mol. Neurobiol.* **2020**. [CrossRef]
163. Mirzaei, H.; Shakeri, A.; Rashidi, B.; Jalili, A.; Banikazemi, Z.; Sahebkar, A. Phytosomal curcumin: A review of pharmacokinetic, experimental and clinical studies. *Biomed. Pharmacother.* **2017**, *85*, 102–112. [CrossRef]
164. Mukherjee, S.; Fried, A.; Hussaini, R.; White, R.; Baidoo, J.; Yalamanchi, S. Phytosomal curcumin causes natural killer cell-dependent repolarization of glioblastoma (GBM) tumor-associated microglia/macrophages and elimination of GBM and GBM stem cells. *J. Exp. Clin. Cancer Res.* **2018**, *37*, 168. [CrossRef] [PubMed]

Review

Potential Therapeutic Effects of Curcumin on Glycemic and Lipid Profile in Uncomplicated Type 2 Diabetes—A Meta-Analysis of Randomized Controlled Trial

Emma Altobelli [1,2,*], Paolo Matteo Angeletti [1,3], Ciro Marziliano [1], Marianna Mastrodomenico [1], Anna Rita Giuliani [1] and Reimondo Petrocelli [4]

1 Department of Life, Public Health and Environmental Sciences, University of L'Aquila, 67100 L'Aquila, Italy; paolomatteoangeletti@gmail.com (P.M.A.); c.marziliano@univaq.it (C.M.); mariannamastrodomenico@gmail.com (M.M.); annarita.giuliani@univaq.it (A.R.G.)
2 Epidemiology and Biostatistics Unit, Local Health Unit, 64100 Teramo, Italy
3 Rianimazione e TIPO Cardiochirurgica, Ospedale G. Mazzini, Local Health Unit, 64100 Teramo, Italy
4 Public Health Unit, ASREM, 86100 Campobasso, Italy; reimondo.petrocelli@asrem.org
* Correspondence: emma.altobelli@univaq.it; Tel.: +39-0862-434666; Fax: +39-0862-432903

Abstract: Diabetes mellitus is an important issue for public health, and it is growing in the world. In recent years, there has been a growing research interest on efficacy evidence of the curcumin use in the regulation of glycemia and lipidaemia. The molecular structure of curcumins allows to intercept reactive oxygen species (ROI) that are particularly harmful in chronic inflammation and tumorigenesis models. The aim of our study performed a systematic review and meta-analysis to evaluate the effect of curcumin on glycemic and lipid profile in subjects with uncomplicated type 2 diabetes. The papers included in the meta-analysis were sought in the MEDLINE, EMBASE, Scopus, Clinicaltrials.gov, Web of Science, and Cochrane Library databases as of October 2020. The sizes were pooled across studies in order to obtain an overall effect size. A random effects model was used to account for different sources of variation among studies. Cohen's *d*, with 95% confidence interval (CI) was used as a measure of the effect size. Heterogeneity was assessed while using Q statistics. The ANOVA-Q test was used to value the differences among groups. Publication bias was analyzed and represented by a funnel plot. Curcumin treatment does not show a statistically significant reduction between treated and untreated patients. On the other hand, glycosylated hemoglobin, homeostasis model assessment (HOMA), and low-density lipoprotein (LDL) showed a statistically significant reduction in subjects that were treated with curcumin, respectively ($p = 0.008$, $p < 0.001$, $p = 0.021$). When considering HBA1c, the meta-regressions only showed statistical significance for gender ($p = 0.034$). Our meta-analysis seems to confirm the benefits on glucose metabolism, with results that appear to be more solid than those of lipid metabolism. However, further studies are needed in order to test the efficacy and safety of curcumin in uncomplicated type 2 diabetes.

Keywords: curcuma; turmeric; type 2 diabetes; dyslipidemia; meta-analysis; randomized control trial

Citation: Altobelli, E.; Angeletti, P.M.; Marziliano, C.; Mastrodomenico, M.; Giuliani, A.R.; Petrocelli, R. Potential Therapeutic Effects of Curcumin on Glycemic and Lipid Profile in Uncomplicated Type 2 Diabetes—A Meta-Analysis of Randomized Controlled Trial. *Nutrients* **2021**, *13*, 404. https://doi.org/10.3390/nu13020404

Academic Editor: Roberta Masella
Received: 3 January 2021
Accepted: 25 January 2021
Published: 27 January 2021

Publisher's Note: MDPI stays neutral with regard to jurisdictional claims in published maps and institutional affiliations.

Copyright: © 2021 by the authors. Licensee MDPI, Basel, Switzerland. This article is an open access article distributed under the terms and conditions of the Creative Commons Attribution (CC BY) license (https://creativecommons.org/licenses/by/4.0/).

1. Introduction

Type 2 diabetes (T2DM) is an important issue for public health, and it is growing in the world; in fact, according to the World Health Organization (WHO) report in 2016, 422-million people are diagnosed with diabetes [1]. In Europe, 50% of Countries show T2DM prevalence rates in the range of 8–9% [2]. Bommer et al. have demonstrated that the global costs of T2DM and its consequences are large, and they will substantially increase by 2030 [3].

Lifestyle risk factors that are related to diabetes, as obesity and overweight are two major risk factors. The treatment of T2DM includes the use of anti-diabetic drugs and prevention based on lifestyle habits. In recent years, there has been growing research interest on the efficacy evidence of curcumin use in the regulation of glycemia and lipidaemia [4].

Curcumin is the main bioactive component that is extracted from the rhizome of Curcuma Longa. It is a product used since ancient times, in cuisine, as in traditional medicine [5]. The properties that are attributed to curcumin are remarkable: in fact, it has an antioxidant and anti-inflammatory effect [6].

Its molecular structure makes it possible to intercept reactive oxygen species (ROI) that are particularly harmful in chronic inflammation and tumorigenesis models [5].

Curcumin can have a therapeutic effect on some chronic diseases, such as rheumatoid arthritis, coronary artery disease, atherosclerosis, T2DM, and obesity [7].

Ramirez-Bosca et al. showed that daily treatment with curcumin can decrease the low-density lipoprotein (LDL) in healthy subjects [8]. In addition, Mohammadi et al. [9] demonstrated that one-month oral administration of curcumin (1 gram/day) could reduce the triglycerides concentrations in obese subjects. Rahimi et al., in a randomized trial, highlighted that curcumin reduces HbA1c during three months of therapy [10].

In order to evaluate its effectiveness, some studies have been conducted on the effects of curcumin on glycemic and lipid control in subjects with uncomplicated T2DM [10–16].

In this systematic review and meta-analysis, we aim to evaluate the effect of curcumin on glycemic and lipid profile in subjects with uncomplicated T2DM.

2. Materials and Methods

The papers that were included in the meta-analysis were sought in the MEDLINE, EMBASE, Scopus, Clinicaltrials.gov, Web of Science, and Cochrane Library databases as of October 2020. The search terms used were: curcuma or curcumins or turmeric AND type 2 diabetes or diabetes; ("Diabetes Mellitus, Type 2" [Mesh]) AND ("Curcuma" [Mesh]) and applied the following filters: humans, published articles from 2000 to 2020, and clinical trials.

The papers were selected while using the Preferred Reporting Items for Systematic Reviews and Meta-Analyses (PRISMA) flowchart (Figure 1) and the PRISMA checklist (Table S1) [17]. A manual search of possible references of interest was also performed. Only studies that were published in English were considered. The papers were selected by two independent reviewers (P.M.A. and C.M); a methodologist (E.A.) resolved any disagreements. Bias was assessed using the Cochrane Collaboration tool for assessing the risk of bias (Table S2) [18].

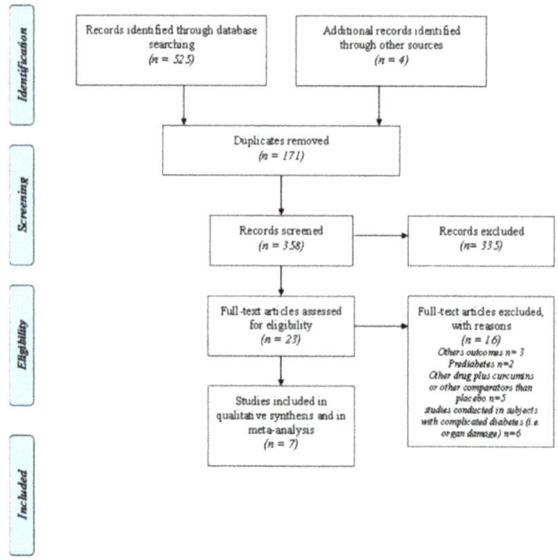

Figure 1. Prisma Flow-chart.

3. Statistical Analysis

The sizes were pooled across studies in order to obtain an overall effect size. A random effects model was used to account for the different sources of variation among studies [19]. Cohen's d, with 95% confidence interval (CI) and *p*-value, was used as a measure of effect size [20]. Heterogeneity was assessed using Q statistics, I^2, Tau, and Tau2. The stability of study findings was checked with moderator analysis. A subgroups analysis was also performed while considering the country of primary studies, because four of seven studies were conducted in Iran. The ANOVA-Q test was used to value the differences among groups. Publication bias was analyzed and represented by a funnel plot; funnel plot symmetry was assessed with Egger's test [21]. Finally, publication bias was checked using the trim and fill procedure; we used Rosenthal's estimator and the fail-safe number to analyze publication bias [22]. Finally, meta-regression analyses were utilized for the following variables: article publication year, gender, age, and dose. Regression models were applied for continuous variables. Meta-regressions were performed when the number of studies containing the variables to be analyzed was ≥4. PROMETA 3 software (IDo Statistics-Internovi, Cesena, Italy) was used. The considered outcomes were body mass index (BMI), homeostasis model assessment-insulin resistance index (HOMA-IR), glycosylated hemoglobin (Hb1Ac), Triglycerides (TG), Total Cholesterol (TC), High-density lipoprotein (HDL), and LDL. All of the values were reported in mg/mL while using a conversion formula [23].

4. Results

The literature search highlighted the presence of 529 references (Figure 1). After removing the duplicates, 358 papers were screened. Twenty-three full texts were verified. 16 were excluded and seven were included in the meta-analysis.

Table 1 reports the characteristics of the primary studies and results of outcomes. Table 2 shows the results of meta-analysis. We highlight that the papers included in this meta-analysis showed a low risk of bias: Supplementary Table S2 reports the results of risk bias assessment. The results of meta-regressions are showed in Table S3. Figures S1–S7 show results of sensitivity analysis.

Table 1. The characteristics of included studies, according to intervention group (curcumins) and control groups (placebo) and results for each selected outcome.

Author, Year, Country	Mean Age	Male %	Diabetes Duration (Years)	Groups	BMI Variation Mean (SD)	HOMA-IR	Hb1Ac Mean (SD)	TG Mean (SD)	TC Mean (SD)	HDL Mean (SD)	LDL Mean (SD)
Hodaei 2019 Iran [10]	59	50	1–10	Intervention Group N = 21 (1500 mg)	29.2 (3.76) 28.9 (3.73)	62 (63) 62.4 (42)	11.3 (1.6) 11 (2.0)	-	-	-	-
				Control Group N = 23	28.2 (2.5) 28.1 (2.5)	53 (40) 65 (44)	11.2 (1.3) 11.1 (1.8)	-	-	-	-
Adibian 2019 Iran [11]	59	50	1–10	Intervention Group N = 21 (1500 mg)	-	-	-	124 (36) 109 (36)	167 (34) 163 (39)	30 (2) 30 (2)	112 (31) 108 (36)
				Control Group N = 23	-	-	-	126 (52) 121 (44)	180 (47) 175 (47)	30 (2) 30 (2)	125 (44) 118 (47)

Table 1. *Cont.*

Author, Year, Country	Mean Age	Male %	Diabetes Duration (Years)	Groups	BMI Variation Mean (SD)	HOMA-IR	Hb1Ac Mean (SD)	TG Mean (SD)	TC Mean (SD)	HDL Mean (SD)	LDL Mean (SD)
Adab 2018 Iran [12]	55	51	5–10	Intervention Group N = 39 (2100 mg)	28.98 (3.68) 28.26. (3.45)	2.42 (1.73) 2.21 (1.43)	7.06 (1.01) 7.04 (0.98)	181.56 (79.9) 141.74 (52.02)	148.85 (36.11) 149.82 (35.67)	38.79 (10.30) 37.07 (9.12)	82.56 (20.99) 75.23 (18.84)
				Control Group N= 36	28.82 (4.96) 28.68 (4.86)	2.24 (1.48) 2.69 (2.02)	6.79 (1.08) 7.28 (1.59)	164.05 (81.19) 197.05 (96.98)	155.36 (36.27) 176.88 (37.58)	44.63 (10.66) 42.11 (9.39)	86.61 (21.99) 89.05 (21.46)
Rahimi 2015 Iran [13]	58.64	45	NR	Intervention Group (80 mg) N = 35	26.92 (2.71) 25.57 (2.71)	-	7.59 (1.74) 7.31 (1.54)	109 (94.75) 131 (60.27)	163.4 (33.94) 158.62 (44.06)	54.30 (14.02) 60.95 (15.68)	96.57 (33.94) 91.04 (28.72)
				Control Group N = 35	27.27 (3.59) 27.50 (3.38)	-	7.49 (1.75) 9.05 (2.33)	142 (97.5) 113 (58)	85.5 (15.3) 80.5 (9.1)	60.35 (15.96) 55.00 (11.09)	98.78 (30.33) 99.78 (30.33) 84.00 (12.59)
Chuengsamarn Thailand 2014 [14]	59	48	12	Intervention Group N = 107	-	9.58 (4.3) 4,32 (1.8)	-	219.12 (97.52) 141.99 (67.79)	-	-	-
				Control Groups N = 106	-	6.89 (2.67) 6.78 (2.5)	-	252.72 (114.12) 252 (114.12)	-	-	-
Na China 2013 [15]	55.07	49	7.6	Intervention Group (300 Mg) N = 50	-	5.80 (3.35) 4.14 (1.81)	7.77 (1.82) 7.02 (2.04)	223.8 (46.8) 157.5 (49.5)	540 (100) 493 (41.7)	52.9 (10.04) 54.8 (11.2)	166.0 (46.3) 146.7 (39.7)
				Control Group N= 50	-	5.82 (3.90) 5.49 (2.15)	7.72 (2.12) 7.99 (2.86)	193.8 (92.04) 186.7 (66.3)	538 (109) 522 (105.3)	51.35 (10.81) 51.74 (8.8)	166.8 (44.4) 160.2 (45.17)
Usharani India 2008 [16]		52	8	Intervention Group N = 23 (300 mg)	-	-	8.04 (0.85) 8.03 (0.76)	176.39 (27.61) 165.26 (25.78)	195.0 (41.16) 185.34 (34.35)	38.78 (7.69) 39.91 (0.68)	120.35 (42.13) 111.34 (37.65)
				Control Group N= 21	-	-	7.82 (0.57) 7.80 (0.62)	170.14 (47.54) 168.14 (47.10)	195.95 (35.72) 198.76 (35.09)	36.38 (7.67) 37.04 (5.92)	124.59 (34.94) 122.18 (35.56)

%: Males enrolled in each study.TG: triglycerides, TC: total cholesterol.

Table 2. Meta-Analysis results and moderator analysis (country: Iran versus outside Iran).

Outcome	K	Total Sample Size	Effect Size		Heterogeneity				Publication Bias			
			(95% CI)	p	I^2	p	T^2	T	Egger (p)	BEGGS (p)	Fail Safe (n)	Rosenthal (n)
BMI	3	168	−0.30 (−0.62, 0.02)	0.067	0.00	0.514	0.00	0.00	0.842	0.602	0	25
Hb1Ac	5	333	−0.42 (−0.77, −0.11)	0.008	42.42	0.107	0.06	0.24	0.501	0.327	12	35
Iran	3	189	−0.52 (−1.00, −0.04)	0.032	61.35	0.075	0.11	0.33				
Outside Iran	2	144	−0.28 (−0.67, 0.10)	0.153	22.22	0.257	0.02	0.14				
ANOVA Q TEST $p = 0.443$												
HOMA	4	432	−0.41 (−0.66, −0.22)	<0.001	0.00	0.916	0.00	0.00	0.073	0.042	12	30
Iran	2	119	−0.33 (−0.69, 0.04)	0.078	0.00	0.667	0.00	0.00				
Outside Iran	2	313	−0.45 (−0.67, −0.22)	<0.001	0.00	0.885	0.00	0.00				

Table 2. *Cont.*

Outcome	K	Total Sample Size	Effect Size		Heterogeneity				Publication Bias			
			(95% CI)	p	I^2	p	T^2	T	Egger (p)	BEGGS (p)	Fail Safe (n)	Rosenthal (n)
						ANOVA Q TEST $p = 0.580$						
HDL	5	333	0.22 (−0.08, 0.52)	0.143	45.91	0.116	0.05	0.27	0.856	0.327	1	35
Iran	3	189	0.31 (−0.21, 0.839	0.241	68.28	0.043	0.22	0.41				
Outside Iran	2	145	0.11 (−0.22, 0.43)	0.527	0.00	0.713	0.00	0.00				
						ANOVA Q TEST: $p = 0.512$						
LDL	5	300	−0.28 (−0.52, −0.04)	0.021	0.00	0.083	0.00	0.00	0.646	0.624	1	35
Iran	3	156	−0.32 (−0.67, 0.03)	0.077	0.00	0.582	0.00	0.00				
Outside Iran	2	144	−0.25 (−0.77, 0.42)	0.130	0.00	0.754	0.00	0.00				
						ANOVA Q TEST: $p = 0.793$						
TG	5	476	−0.57 (−0.83, −0.31)	<0.001	41.56	0.144	0.04	0.19	0.943	0.322	37	35
Iran	2	119	−0.59 (−1.22, 0.03)	0.063	63.77	0.099	0.13	0.36				
Outside Iran	3	357	−0.55 (−0.88, −0.22)	<0.001	49.19	0.140	0.04	0.20				
						ANOVA Q TEST $p = 0.904$						
TC	5	312	−0.30 (−0.53, −0.07)	0.01	0.00	0.573	0.00	0.00	0.975	1.00	3	30
Iran	3	168	−0.27 (−0.69, 0.15)	0.211	30.47	0.237	0.10	0.31				
Outside Iran	2	144	−0.32 (−0.65, 0.01)	0.056	0.00	0.979	0.00	0.00				
						ANOVA Q TEST $p = 0.847$						

4.1. BMI

BMI was investigated in three studies [10,12,13] involving a total of 168 patients. Overall, curcumin treatment does not show a statistically significant reduction between the treated and untreated patients: this results in the absence of statistical heterogeneity (Table 2, Figure 2).

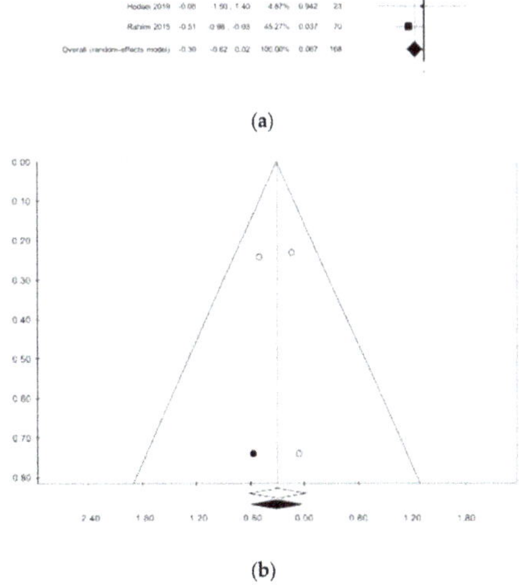

(a)

(b)

Figure 2. Meta-analysis results according to body mass index (BMI): (**a**) forest plot (**b**) funnel plot.

4.2. *Hb1Ac*

Glycosylated hemoglobin was evaluated in five studies [10,12,13,15,16]. A statistically significant reduction was found in subjects that were treated with curcumin: −0.42 (−0.77; −0.11) $p = 0.008$, with moderate heterogeneity ($I^2 = 42.42$), but not statistically significant ($p = 0.107$). Publication bias analysis did not highlight any differences between the observed and estimated values. It should be emphasized that there is a difference between the studies conducted in Iran and those conducted in other countries (Table 2, Figure 3); however, this difference is not statistically significant (ANOVA Q test $p = 0.443$). The meta-regressions only showed statistical significance for gender ($p = 0.034$).

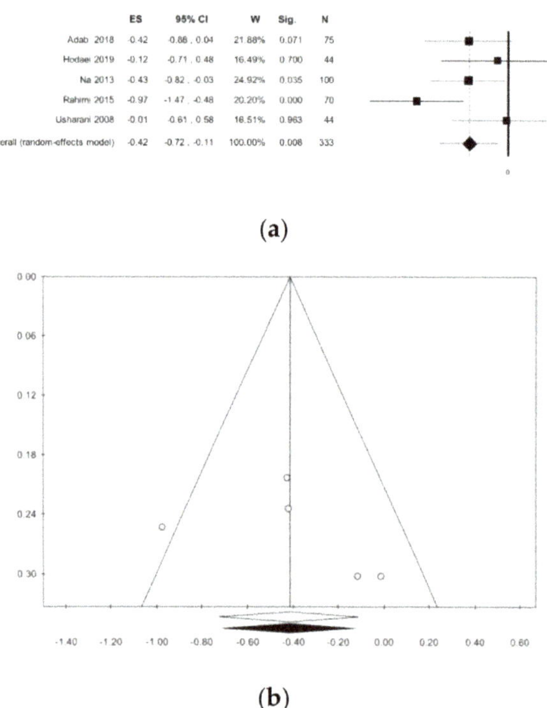

Figure 3. Meta-analysis results according to glycosylated hemoglobin (Hb1Ac): (**a**) forest plot (**b**) funnel plot.

4.3. *HOMA*

HOMA was detected in four studies [10,12,14,15] for a total of 432 patients. There is a statistically significant reduction of this index, without statistical heterogeneity: −0.42 (−0.77; −0.11) $p < 0.001$, (Table 2, Figure 4). Publication bias analysis highlighted a difference between the observed and estimated values: 0.45 (−0.61; −0.28) $p < 0.001$, with two trimmed studies. The subgroup analysis showed a difference between the studies that were conducted in Iran and those conducted in other countries, however this difference is not statistically significant (Table 2). The meta-regressions, concerning the selected moderators, did not show any statistical significance (Table S3).

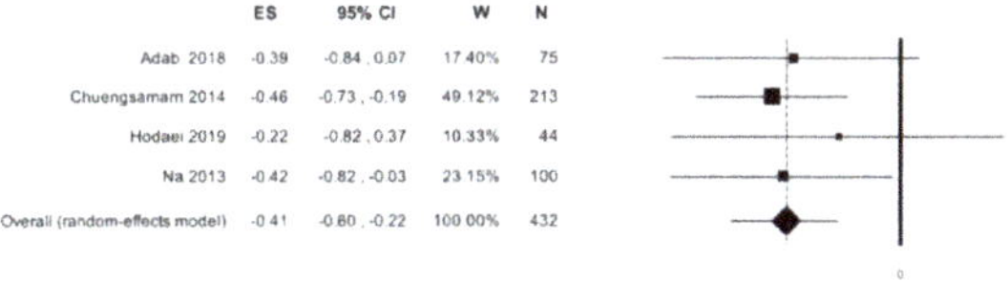

(**a**)

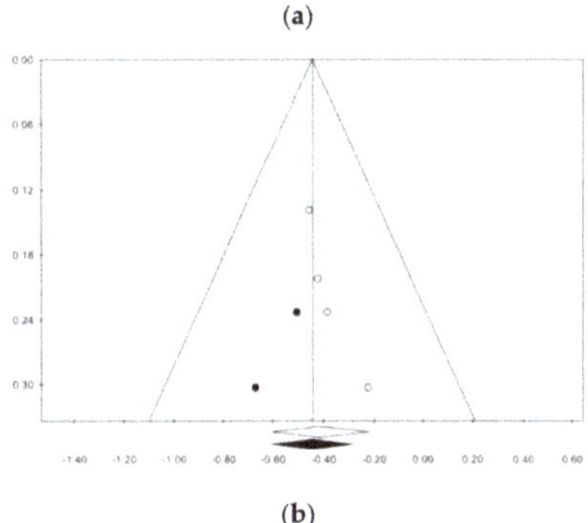

(**b**)

Figure 4. Meta-analysis results according to homeostasis model assessment (HOMA): (**a**) forest plot (**b**) funnel plot.

4.4. HDL

HDL was evaluated in five studies for a total of 333 patients [11–13,15,16]. The analysis did not show statistically significant differences (Table 2, Figure 5). The subgroup analysis did not show a difference between the studies that were conducted in Iran and those conducted in other countries. Publication bias analysis did not highlight any differences between the observed and estimated values. Meta-regressions, regarding the selected moderators, did not show any statistical significance, with the exception of gender ($p = 0.002$) (Table S3).

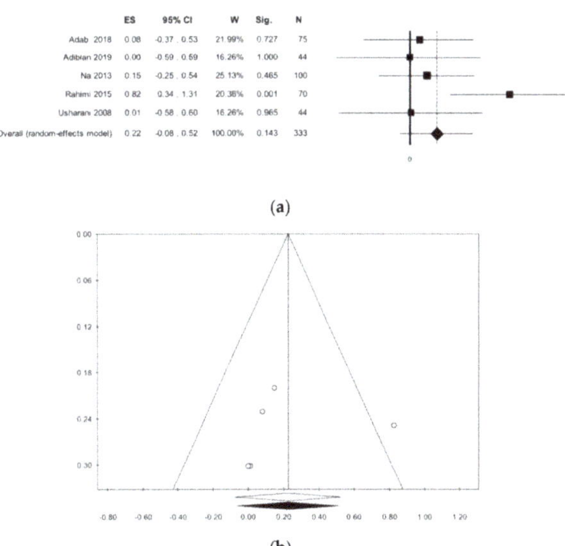

(**a**)

(**b**)

Figure 5. Meta-analysis results according to high-density lipoprotein (HDL): (**a**) forest plot (**b**) funnel plot.

4.5. LDL

The LDL dosage was evaluated in 300 patients for a total of five studies [11–13,15,16]. The meta-analysis showed a statistically significant reduction in curcumin-treated patients when compared to the placebo, with no statistical heterogeneity: −0.28 (−0.52; −0.04) $p = 0.021$, I^2 0.00 (Table 2, Figure 6). The subgroup analysis, as shown in Table 2, highlights a difference between studies that were conducted in Iran and those conducted outside Iran, but there is not a statistically significant difference. Publication bias analysis did not highlight any differences between the observed and estimated values. The meta-regressions, for the selected moderators, did not show any statistical significance (Table S3).

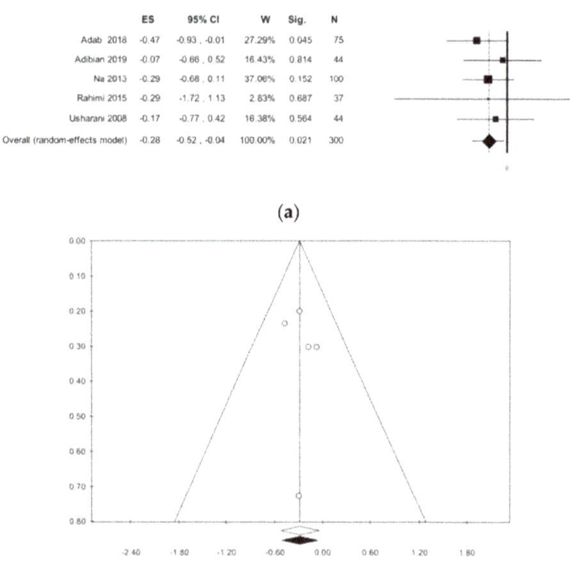

Figure 6. Meta-analysis results according to low-density lipoprotein (LDL): (**a**) forest plot (**b**) funnel plot.

4.6. Triglycerides

Triglycerides were evaluated in five primary studies, involving a total of 476 patients [11–16]. In patients treated with curcumin, a non-significant reduction in plasma triglyceride concentrations, without statistical heterogeneity, was identified (Table 2, Figure 7). The publication bias analysis highlighted a difference between observed and estimated values, respectively: −0.62 (−0.87; −0.37) ($p < 0.001$), −0.57 (−0.83; −0.31), with 1 trimmed study. The subgroup analysis revealed a non-statistically significant difference.

4.7. Total Cholesterol

The total cholesterol was investigated in five studies for a total of 312 statistical units [11–13,15,16]. There is a reduction in cholesterol in curcumin-treated patients as compared to the placebo-treated patients, with no statistical heterogeneity (Table 2, Figure 8). Publication bias analysis highlighted a difference between the observed and estimated values, respectively: −0.30 (−0.53; −0.07) $p < 0.001$, −0.40 (−0.62; −0.28) ($p < 0.001$) with two trimmed studies. No statistically significant associations were found in meta-regressions (Table S3).

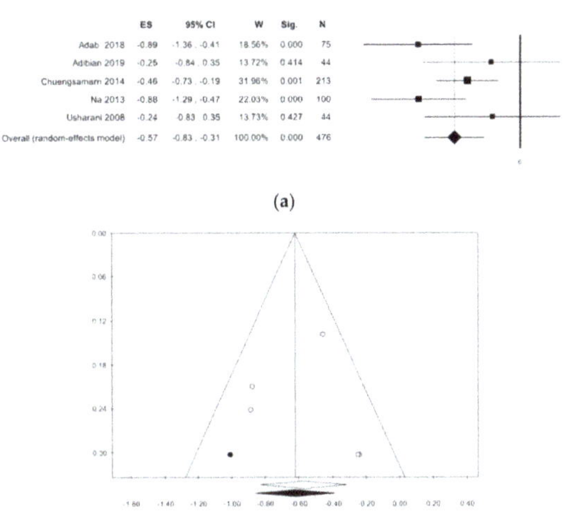

(a)

(b)

Figure 7. Meta-analysis results according to Triglycerides: (**a**) forest plot (**b**) funnel plot.

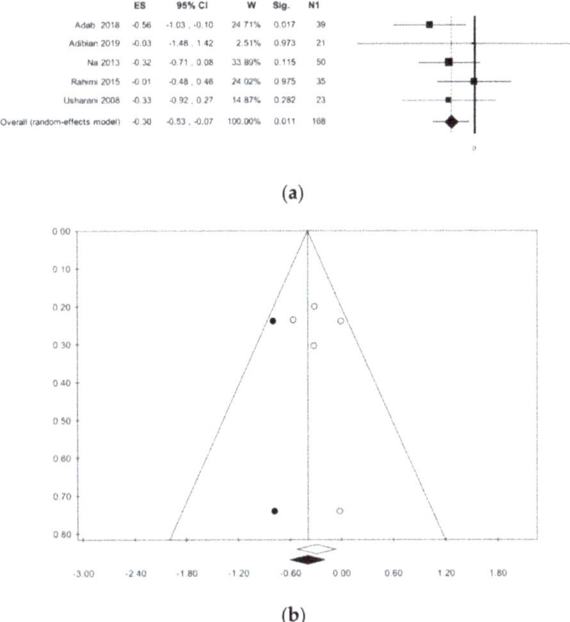

(a)

(b)

Figure 8. Meta-analysis results according to Total Cholesterol: (**a**) forest plot (**b**) funnel plot.

5. Discussion

In recent decades, there has been considerable interest among researchers in nutraceuticals and in particular, in naturally derived products, also known as natural health products (NHP), for the prevention, cure, and treatment of cardiovascular and metabolic diseases [24–26]. Some research indicates that chronically ill people tend to consume more NHP, and some surveys confirm that patients with T2DM are not excluded from this [27–31]. The pandemic spread of non-communicable diseases (NCD), and T2DM in particular, makes this population a target market of considerable interest for producers. The use of NHP in diabetics is linked to the co-treatment of hyperglycemia, dyslipidemia, and complications of diabetes. Among the various products, there is also curcumin. Currently, the FDA and EFSA recommend doses of curcumin of maximum 3 mg/kg/day, including the onset of toxicity for higher quantities (in particular, teratogenic effects, astrocytic cell abnormalities are reported, and it is not recommended in gallbladder stones) [5]. Overall, curcumin has a low bioavailability due to its hydrophobicity, so the pharmacologically active form is administered with a lipid vehicle or in association with piperine [5,32]. Specifically, for diabetes, curcumin also has an effect on hepatic lipogenesis, blocking the activity of the sterol regulatory element-binding protein gene (SREBP1) [6,33] and simultaneously activating the enzymes carnitine palmitoyltransferase 1 (CPT1) and acyl-CoA cholesterol acyltransferase (ACAT) that are involved in lipid mobilization [6,33].

The results of our meta-analysis seem to confirm this modulating capacity on lipid metabolism. The trials considered highlighted an overall reduction in LDL, TG, and TC in patients with uncomplicated T2DM. This result does not seem to be affected by statistical heterogeneity. There is a moderate publication bias. The low fail-safe for LDL, HDL, and TC indicates caution in the interpretation of the overall result, even if a high Rosenthal value allows for the observation found to be considered valid [22]. In order to confirm this, the subgroup analysis shows some differences that could be explained by the following considerations: method of conducting the study, quantity of curcumin administered, execution techniques, analysis, and collection of the blood chemistry method. The lack of influence of curcumin on HDL can be motivated by the fact that, notoriously, the increase in HDL is due globally to a more active lifestyle [34]. Similarly, the non-influence on BMI could be interpreted [35].

Because hypoglycemic properties of curcumin have been known since 1972 [36,37], the action is probably mediated by the inhibition of Phosporilase Kinase, which is to say, avoiding the mobilization of glucose from glycogen reserves [32,37–39]. Furthermore, curcumin would have a role in reducing the accumulation of advanced glycation end products [40] and the accumulation of these same metabolites at the level of the pancreatic insulae [41]. The inhibition of this process would mediate the Peroxisome Proliferator Activated Receptor Gamma (PPAR-Y), which, by increasing the amount of glutathione, would prevent oxidative damage that is caused by the state of hyperglycemia [42].

Hb1Ac and HOMA show significant results without statistical heterogeneity and publication bias. Regarding Hb1Ac, the subgroup analysis shows a difference between the studies that were conducted in Iran and those conducted outside Iran; on the contrary, this is not the case for HOMA, which shows concordant results. The hypoglycemic capacity of curcumin has also been tested in prediabetic populations with some success [43].

Concerning Hb1Ac and HDL, the results of the meta-regressions only show statistical significance with respect to gender.

6. Conclusions

Our meta-analysis seems to confirm the benefits on glucose metabolism, with results that appear to be more solid than those of lipid metabolism. In conclusion, the daily supplement of curcumin could improve some metabolic aspects of uncomplicated T2DM patients.

However, further studies are needed in order to test the efficacy and safety of curcumin in uncomplicated T2DM. The limitations of the present work can be attributed to some biases present in the primary studies: the small numbers of enrolled patients and the possible impact of grey literature. These aspects cannot be totally corrected through the meta-analytic technique.

Supplementary Materials: The following are available online at https://www.mdpi.com/2072-6643/13//404/s1, Table S1: Prisma Checklist, Table S2: Risk of bias assessment, Table S3: Meta regressions Results, Figures S1–S7: Sensitivity Analysis.

Author Contributions: E.A.: Guarantor of the article, study concept and design, literature search, data analysis, and manuscript writing. P.M.A.: literature search, data abstraction, participant manuscript writing. M.M., A.R.G. and C.M. participant systematic review (PRISMA). R.P. participant study concept and manuscript writing. All authors have read and agreed to the published version of the manuscript.

Funding: This research received no external funding.

Founding: The founding sponsors had no role in the design of the study; in the collection, analyses, or interpretation of data; in the writing of the manuscript, and in the decision to publish the results.

Data Availability Statement: Not applicable.

Conflicts of Interest: The authors declare no conflict of interest.

References

1. World Health Organization (WHO). *World Health Statistics 2016: Monitoring Health for the SDGs Sustainable Development Goals*; World Health Organization: Geneva, Switzerland, 2016.
2. Altobelli, E.; Angeletti, P.M.; Profeta, V.F.; Petrocelli, R. Lifestyle risk factors for type 2 diabetes mellitus and national diabetes care systems in European countries. *Nutrients* **2020**, *12*, 2806. [CrossRef] [PubMed]
3. Bommer, C.; Sagalova, V.; Heesemann, E.; Manne-Goehler, J.; Atun, R.; Bärnighausen, T.; Davies, J.; Vollmer, S. Global economic burden of diabetes in adults: Projections from 2015 to 2030. *Diabetes Care* **2018**, *41*, 963–970. [CrossRef] [PubMed]
4. Kim, H.S.; Hwang, Y.C.; Koo, S.H.; Park, K.S.; Lee, M.S.; Kim, K.W.; Lee, M.K. PPAR-gamma activation increases insulin secretion through the up-regulation of the free fatty acid receptor GPR40 in pancreatic beta-cells. *PLoS ONE* **2013**, *8*, e50128.
5. Shaikh, J.; Ankola, D.D.; Beniwal, V.; Singh, D.; Kumar, M.N. Nanoparticle encapsulation improves oral bioavailability of curcumin by at least 9-fold when compared to curcumin administered with piperine as absorption enhancer. *Eur. J. Pharm. Sci.* **2009**, *37*, 223–230. [CrossRef] [PubMed]
6. Pivari, F.; Mingione, A.; Brasacchio, C.; Soldati, L. Curcumin and type 2 diabetes mellitus: Prevention and treatment. *Nutrients* **2019**, *11*, 1837. [CrossRef]
7. Aggarwal, B.B.; Harikumar, K.B. Potential therapeutic effects of curcumin, the anti-inflammatory agent, against neurodegenerative, cardiovascular, pulmonary, metabolic, autoimmune and neoplastic diseases. *Int. J. Biochem. Cell Biol.* **2009**, *41*, 40–59. [CrossRef]
8. Ramírez-Boscá, A.; Soler, A.; Carrión, M.A.; Díaz-Alperi, J.; Bernd, A.; Quintanilla, C.; Quintanilla Almagro, E.; Miquel, J. An hydroalcoholic extract of curcuma longa lowers the apo B/apo A ratio. Implications for atherogenesis prevention. *Mech. Ageing Dev.* **2000**, *20*, 41–47.
9. Mohammadi, A.; Sahebkar, A.; Iranshahi, M.; Amini, M.; Khojasteh, R.; Ghayour-Mobarhan, M.; Ferns, G.A. Effects of supplementation with curcuminoids on dyslipidemia in obese patients: A randomized crossover trial. *Phytother. Res.* **2013**, *27*, 374–379. [CrossRef]
10. Hodaei, H.; Adibian, M.; Nikpayam, O.; Hedayati, M.; Sohrab, G. The effect of curcumin supplementation on anthropometric indices, insulin resistance and oxidative stress in patients with type 2 diabetes: A randomized, double-blind clinical trial. *Diabetol. Metab. Syndr.* **2019**, *27*, 41. [CrossRef]
11. Adibian, M.; Hodaei, H.; Nikpayam, O.; Sohrab, G.; Hekmatdoost, A.; Hedayati, M. The effects of curcumin supplementation on high-sensitivity C-reactive protein, serum adiponectin, and lipid profile in patients with type 2 diabetes: A randomized, double-blind, placebo-controlled trial. *Phytother. Res.* **2019**, *33*, 1374–1383. [CrossRef]
12. Adab, Z.; Eghtesadi, S.; Vafa, M.R.; Heydari, I.; Shojaii, A.; Haqqani, H.; Arablou, T.; Eghtesadi, M. Effect of turmeric on glycemic status, lipid profile, Hs-CRP, and total antioxidant capacity in hyperlipidemic type 2 diabetes mellitus patients. *Phytother. Res.* **2019**, *33*, 1173–1181. [CrossRef] [PubMed]
13. Rahimi, H.R.; Mohammadpour, A.H.; Dastani, M.; Jaafari, M.R.; Abnous, K.; Ghayour Mobarhan, M.; Kazemi Oskuee, R. The effect of nano-curcumin on HbA1c, fasting blood glucose, and lipid profile in diabetic subjects: A randomized clinical trial. *Avicenna J. Phytomed.* **2016**, *6*, 567–577. [PubMed]
14. Chuengsamarn, S.; Rattanamongkolgul, S.; Phonrat, B.; Tungtrongchitr, R.; Jirawatnotai, S. Reduction of atherogenic risk in patients with type 2 diabetes by curcuminoid extract: A randomized controlled trial. *J. Nutr. Biochem.* **2014**, *25*, 144–150. [CrossRef] [PubMed]
15. Na, L.X.; Li, Y.; Pan, H.Z.; Zhou, X.L.; Sun, D.J.; Meng, M.; Li, X.X.; Sun, C.H. Curcuminoids exert glucose-lowering effect in type 2 diabetes by decreasing serum free fatty acids: A double-blind, placebo-controlled trial. *Mol. Nutr. Food Res.* **2013**, *57*, 1569–1577. [CrossRef] [PubMed]
16. Usharani, P.; Mateen, A.A.; Naidu, M.U.; Raju, Y.S.; Chandra, N. Effect of NCB-02, atorvastatin and placebo on endothelial function, oxidative stress and inflammatory markers in patients with type 2 diabetes mellitus: A randomized, parallel-group, placebo-controlled, 8-week study. *Drugs Res. Dev.* **2008**, *9*, 243–250. [CrossRef]
17. Moher, D.; Liberati, A.; Tetzlaff, J.; Altman, D.G. PRISMA Group. *Preferred reporting items for systematic reviews and meta-analyses: The PRISMA statement. PLoS Med.* **2009**, *21*, e1000097.

18. Higgins, J.P.T.; Green, S.; Altman, D.G. *Cochrane Handbook for Systematic Reviews of Interventions: Cochrane Book Series Chapter 8*; Assessing Risk of Bias in Included Studies Published Online; Wiley and Sons: Hoboken, NJ, USA, 2008.
19. Higgins, J.P.T.; Thompson, S.G.; Spiegelhalter, D.J. A re-evaluation of random-effects meta-analysis. *J. R. Stat. Soc.* **2009**, *172*, 137–159. [CrossRef]
20. Rhea, M.R. Determining the magnitude of treatment effects in strength training research through the use of the effect size. *J. Strength Cond. Res.* **2004**, *18*, 918–920.
21. Rothstein, H.R.; Sutton, A.J.; Borenstein, M. *Publication Bias in Meta-Analysis*; Wiley and Sons: Chichester, UK, 2005.
22. Fragkos, K.C.; Tsagris, M.; Frangos, C.C. Publication bias in meta-analysis: Confidence intervals for Rosenthal's fail-safe number. *Int. Sch. Res. Notices.* **2014**, *2014*, 825383. [CrossRef]
23. Rugge, B.; Balshem, H.; Sehgal, R.; Relevo, R.; Gorman, P.; Helfand, M. *Screening and Treatment of Subclinical Hypothyroidism or Hyperthyroidism*; Comparative Effectiveness Reviews, No. 24; Agency for Healthcare Research and Quality: Rockville, MD, USA, 2011. Available online: https://www.ncbi.nlm.nih.gov/books/NBK83496/ (accessed on 25 January 2021).
24. Waltenberger, B.; Mocan, A.; Šmejkal, K.; Heiss, E.H.; Atanasov, A.G. Natural products to counteract the epidemic of cardiovascular and metabolic disorders. *Molecules* **2016**, *22*, 807. [CrossRef]
25. Dwyer, J.T.; Coates, P.M.; Smith, M.J. Dietary supplements: Regulatory challenges and research resources. *Nutrients* **2018**, *10*, 41. [CrossRef] [PubMed]
26. Kantor, E.D.; Rehm, C.D.; Du, M.; White, E.; Giovannucci, E.L. Trends in dietary supplement use among US adults from 1999–2012. *JAMA Intern. Med.* **2016**, *316*, 1464–1474. [CrossRef] [PubMed]
27. Geil, P.; Shane-McWhorter, L. Dietary supplements in the management of diabetes: Potential risks and benefits. *J. Am. Diet. Assoc.* **2008**, *8*, S59–S65. [CrossRef] [PubMed]
28. Alherbish, A.; Charrois, T.L.; Ackman, M.L.; Tsuyuki, R.T.; Ezekowitz, J.A. The prevalence of natural health product use in patients with acute cardiovascular disease. *PLoS ONE* **2011**, *6*, e19623. [CrossRef]
29. Necyk, C.; Zubach-Cassano, L. Natural health products and diabetes: A practical review. *Can. J. Diabetes* **2017**, *41*, 642–647. [CrossRef]
30. Damnjanovic, I.; Kitic, D.; Stefanovic, N.; Zlatkovic-Guberinic, S.; Catic-Djordjevic, A.; Velickovic-Radovanovic, R. Herbal self-medication use in patients with diabetes mellitus type 2. *Turk. J. Med. Sci.* **2015**, *45*, 964–971. [CrossRef]
31. Heller, T.; Müller, N.; Kloos, C.; Wolf, G.; Müller, U.A. Self-medication and use of dietary supplements in adult patients with endocrine and metabolic disorders. *Exp. Clin. Endocrinol. Diabetes* **2012**, *120*, 540–546. [CrossRef]
32. Kunnumakkara, A.B.; Bordoloi, D.; Padmavathi, G.; Monisha, J.; Roy, N.K.; Prasad, S.; Aggarwal, B.B. Curcumin, the golden nutraceutical: Multitargeting for multiple chronic diseases. *Br. J. Pharmacol.* **2017**, *174*, 1325–1348. [CrossRef]
33. Seo, K.I.; Choi, M.S.; Jung, U.J.; Kim, H.J.; Yeo, J.; Jeon, S.M.; Lee, M.K. Effect of curcumin supplementation on blood glucose, plasma insulin, and glucose homeostasis related enzyme activities in diabetic Db/Db mice. *Mol. Nut. Food Res.* **2008**, *52*, 995–1004. [CrossRef]
34. Escolà-Gil, J.C.; Julve, J.; Griffin, B.A.; Freeman, D.; Blanco-Vaca, F. HDL and lifestyle interventions. *Handb. Exp. Pharmacol.* **2015**, *224*, 569–592.
35. Lv, N.; Azar, K.M.J.; Rosas, L.G.; Wulfovich, S.; Xiao, L.; Ma, J. Behavioral lifestyle interventions for moderate and severe obesity: A systematic review. *Prev. Med.* **2017**, *100*, 180–193. [CrossRef] [PubMed]
36. Gupta, S.C.; Patchva, S.; Aggarwal, B.B. Therapeutic roles of curcumin: Lessons learned from clinical trials. *AAPS J.* **2013**, *15*, 195–218. [CrossRef] [PubMed]
37. Srinivasan, M. Effect of curcumin on blood sugar as seen in a diabetic subject. *Indian J. Med. Sci.* **1972**, *26*, 269–270. [PubMed]
38. Aggarwal, B.B.; Deb, L.; Prasad, S. Curcumin differs from tetrahydrocurcumin for molecular targets, signaling pathways and cellular responses. *Molecules* **2014**, *20*, 185–205. [CrossRef] [PubMed]
39. Reddy, S.; Aggarwal, B.B. Curcumin is a non-competitive and selective inhibitor of phosphorylase kinase. *FEBS Lett.* **1994**, *341*, 19–22. [CrossRef]
40. Sajithlal, G.B.; Chithra, P.; Chandrakasan, G. Advanced glycation end products induce crosslinking of collagen in vitro. *Biochim. Biophys. Acta* **1998**, *1407*, 215–224. [CrossRef]
41. Meghana, K.; Sanjeev, G.; Ramesh, B. Curcumin prevents streptozotocin-induced islet damage by scavenging free radicals: A prophylactic and protective role. *Eur. J. Pharmacol.* **2007**, *577*, 183–191. [CrossRef]
42. Stefanska, B. Curcumin ameliorates hepatic fibrosis in type 2 diabetes mellitus—Insights into its mechanisms of action. *Br. J. Pharmacol.* **2012**, *166*, 2209–2211. [CrossRef]
43. Lee, M.S.; Wahlqvist, M.L.; Chou, Y.C.; Fang, W.H.; Lee, J.T.; Kuan, J.C.; Liu, H.Y.; Lu, T.M.; Xiu, L.; Hsu, C.C.; et al. Turmeric improves post-prandial working memory in pre-diabetes independent of insulin. *Asia Pac. J. Clin. Nutr.* **2014**, *23*, 581–591.

 nutrients

Review

Curcumin from Turmeric Rhizome: A Potential Modulator of DNA Methylation Machinery in Breast Cancer Inhibition

Krystyna Fabianowska-Majewska [1,†], Agnieszka Kaufman-Szymczyk [2,†], Aldona Szymanska-Kolba [2], Jagoda Jakubik [2], Grzegorz Majewski [1] and Katarzyna Lubecka [2,*]

1 Faculty of Medicine, Lazarski University, 02-662 Warsaw, Poland; krystyna.fabianowska-majewska@lazarski.pl (K.F.-M.); grzegorz.majewski@lazarski.pl (G.M.)
2 Department of Biomedical Chemistry, Faculty of Health Sciences, Medical University of Lodz, 92-215 Lodz, Poland; agnieszka.kaufman-szymczyk@umed.lodz.pl (A.K.-S.); szymanska.aldona@gmail.com.pl (A.S.-K.); jagoda.jakubik@stud.umed.lodz.pl (J.J.)
* Correspondence: katarzyna.lubecka@umed.lodz.pl
† These authors contributed equally to this work.

Abstract: One of the most systematically studied bioactive nutraceuticals for its benefits in the management of various diseases is the turmeric-derived compounds: curcumin. Turmeric obtained from the rhizome of a perennial herb *Curcuma longa* L. is a condiment commonly used in our diet. Curcumin is well known for its potential role in inhibiting cancer by targeting epigenetic machinery, with DNA methylation at the forefront. The dynamic DNA methylation processes serve as an adaptive mechanism to a wide variety of environmental factors, including diet. Every healthy tissue has a precise DNA methylation pattern that changes during cancer development, forming a cancer-specific design. Hypermethylation of tumor suppressor genes, global DNA demethylation, and promoter hypomethylation of oncogenes and prometastatic genes are hallmarks of nearly all types of cancer, including breast cancer. Curcumin has been shown to modulate epigenetic events that are dysregulated in cancer cells and possess the potential to prevent cancer or enhance the effects of conventional anti-cancer therapy. Although mechanisms underlying curcumin-mediated changes in the epigenome remain to be fully elucidated, the mode of action targeting both hypermethylated and hypomethylated genes in cancer is promising for cancer chemoprevention. This review provides a comprehensive discussion of potential epigenetic mechanisms of curcumin in reversing altered patterns of DNA methylation in breast cancer that is the most commonly diagnosed cancer and the leading cause of cancer death among females worldwide. Insight into the other bioactive components of turmeric rhizome as potential epigenetic modifiers has been indicated as well.

Keywords: curcumin; turmeric; bioactive nutraceutical; nutriepigenomics; DNA methylation; breast cancer; chemoprevention

Citation: Fabianowska-Majewska, K.; Kaufman-Szymczyk, A.; Szymanska-Kolba, A.; Jakubik, J.; Majewski, G.; Lubecka, K. Curcumin from Turmeric Rhizome: A Potential Modulator of DNA Methylation Machinery in Breast Cancer Inhibition. *Nutrients* **2021**, *13*, 332. https://doi.org/10.3390/nu13020332

Academic Editors: Roberta Masella and Francesca Cirulli

Received: 3 December 2020
Accepted: 19 January 2021
Published: 23 January 2021

Publisher's Note: MDPI stays neutral with regard to jurisdictional claims in published maps and institutional affiliations.

Copyright: © 2021 by the authors. Licensee MDPI, Basel, Switzerland. This article is an open access article distributed under the terms and conditions of the Creative Commons Attribution (CC BY) license (https://creativecommons.org/licenses/by/4.0/).

1. Introduction

The rates of incidence and mortality from female breast cancer are swiftly increasing worldwide. The variety of causes of mammary cancer reflects the heterogeneity, extended life expectancy, and rapid expansion of the global population. Thus, the differences in the prevalence of the risk factors for breast cancer are often associated with socioeconomic development, followed by deleterious lifestyle changes and environmental exposures, including diet, that are major determinants of any type of cancer. Mammary cancer is the most commonly diagnosed cancer and the leading cause of cancer death among females [1,2]. According to Global Cancer Statistics (GLOBOCAN) 2018, the estimates of incidence and mortality worldwide for breast cancer show about 2.1 million newly diagnosed female breast cancer cases, accounting for almost 25% of all cancer cases, as well as over 626,000 deaths, accounting for 15% of all cancer deaths among women [1,2].

Among various anti-cancer strategies, the early detection (screening and surveillance) remains the best approach to enhance and manage the mammary cancer outcomes, although adherence to the breast cancer screening guidelines is still low. Thus, the application of different therapeutics is still an effective therapy against breast cancer. As over 70% of breast cancer cases are estrogen receptor (ER) positive type, hormonal therapy or aromatase inhibitors are often used as the main treatment. However, there is another group of triple-negative breast cancer (TNBC) patients that lack the expression of ER, PR (PRGR; Progesterone receptor), and HER-2 (ERBB2; Receptor tyrosine-protein kinase erbB-2) receptors. Five subtypes of breast cancer (Luminal A, Luminal B, basal, EERB2-overexpressing, and normal breast-like subtypes) have been described based on the gene expression profile [3,4]. Since breast cancer comprises a heterogeneous population of cells (breast cancer heterogeneity) [3,4], it makes it more difficult to treat with the current standard of therapy, including surgery, radiation, and chemotherapeutic drugs [5]. In cancer cells, including mammary cancer cells, diverse genetic and epigenetic alterations, simultaneous activation of numerous cell-surface receptors and interconnected, multiple, complex signaling pathways have been observed [6]. Hence, using anti-cancer drugs targeting a single gene product or cell signaling pathway from the whole intricate cancer-related signaling network may lead to activation of alternative pathways followed by the development of drug resistance and tumor recurrence that are common in all breast cancer subtypes [3–6]. All this, together with the high toxicity of the current conventional single-target chemotherapeutics [5] entails the necessity of finding novel anti-cancer agents with low toxicity and enhanced efficacy, targeting multiple cancer-related genes and signaling pathways.

The personal or family history of mammary malignancies and inherited genetic mutations in breast cancer susceptibility genes, mostly BRCA1 (BRCA1 DNA Repair Associated) and BRCA2 (BRCA2 DNA Repair Associated), account for 5% to 10% of breast tumor cases. The studies of people migration have revealed that nonhereditary factors (i.e., demographic, social, economic, environmental, and lifestyle factors) are the main drivers of the observed international and interethnic differences in cancer incidence, including breast cancer incidence. Increased incidence rates of breast cancer in successive generations of the developed countries with higher HDI (The Human Development Index) and transitioned countries are attributed to a raised prevalence of known risk factors [1,2]. Those risk factors are related to menstruation (early menarche, late menopause), reproduction (nulliparity, the postponement of childbearing, having fewer children), exogenous hormone intake (prolonged oral contraception and hormone replacement therapy), nutrition (poor diet with excessive consumption of processed meat and red meat, alcohol abuse), anthropometry (overweight in adulthood, greater levels of obesity, mass and distribution of body fat), cigarette smoking, and physical inactivity. Thus, breastfeeding (with longer duration), healthy diet, and physical activity are known preventive factors with potential beneficial health effects [7].

Therefore, it is necessary to ask why certain types of cancer, such as breast cancer, are more prevalent in some countries than in others. The incidence rate of breast neoplastic diseases is predominantly higher in women of Western nations when compared to women in Asian countries. The exact reason for this disparity is not clear, but dietary factors have been conceived to account for approximately 30% of cancer cases in Western nations [8,9]. It has been hypothesized that different dietary patterns related to specific culture and ethnicity including consumption of a variety of vegetables, herbs and spices abundant in natural bioactive compounds (i.e., polyphenols and vitamins) would be a pivotal reason [8,9]. Of all the spices, turmeric (*Curcuma longa* L.) has been gaining more and more attention due to its beneficial health effects. The turmeric-derived polyphenol, curcumin, is one of the most systematically studied bioactive nutraceuticals for its utility in the management of various diseases, with cancer at the forefront [10]. Importantly, recent reports have shown that some phytochemicals, such as curcumin have the ability to target many breast cancer-related signaling pathways, including Wnt/β-Catenin, Notch, Hedgehog, JAK-STAT, and PI3K/Akt/mTOR [11]. The hitherto studies on curcumin chemopreventive activities suggests that curcumin is one of the most relevant compounds to manage the challenges of breast cancer treatment [10–12].

Moreover, curcumin is well known for its potential role in inhibiting cancer by targeting epigenetic machinery, especially DNA methylation machinery. The dynamic DNA methylation processes serve as an adaptive mechanism to a wide variety of environmental factors, including diet. Every healthy tissue has a precise DNA methylation pattern that changes during cancer development forming a cancer-specific design. Hypermethylation of tumor suppressor genes (TSGs), global DNA demethylation, and hypomethylation of oncogenes and prometastatic genes are hallmarks of nearly all types of cancer, including breast cancer. Curcumin has been shown to modulate epigenetic events that are dysregulated in cancer cells and possess the potential to prevent cancer or enhance the effects of conventional anti-cancer therapy [12,13].

Although mechanisms underlying curcumin-mediated changes in the epigenome remain to be fully elucidated, the mode of action targeting both hypermethylated and hypomethylated genes in cancer is promising for cancer chemoprevention.

This review provides a comprehensive discussion of potential epigenetic mechanisms of curcumin in reversing altered patterns of DNA methylation in breast cancer. Insight into the other bioactive components of turmeric rhizome as potential epigenetic modifiers has been indicated as well. The turmeric rhizome contains not only curcumin but also other antioxidative agents, such as C and E vitamins, several minerals, as well as B-group vitamins (i.e., B2, B6 and B9 vitamins) participating in one-carbon metabolism via regulation of S-adenosyl-L-methionine (SAM, a ubiquitous methyl group donor) pool and DNA methylation reaction.

2. Curcumin: Chemical Structure and Physical Properties

Curcumin (synonym: diferuloylmethane; molecular formula: $C_{21}H_{20}O_6$; molecular weight: 368.38 g/mol) is a well-known dietary polyphenol (IUPAC name: 1,7-bis(4-hydroxy3-methoxyphenyl) hepta-1,6-diene-3,5-dione) derived from the rhizome of turmeric, *Curcuma longa* L. [14] and other *Curcuma species* of the ginger family, Zingiberaceae. Both turmeric and curcumin have a history of human application in foods, supplements, and cosmetics, as well as for therapeutic goals in Asia, Europe and the United States of America. Products containing curcumin are available on the market all over the world. The bright orange-yellow powder known as turmeric is prepared from boiled and dried rhizomes of the Asian, perennial, herbaceous plant *Curcuma longa* L. It includes a mixture of three diarylheptanoids, together called curcuminoids, i.e., curcumin, demethoxycurcumin, and bis-demethoxycurcumin [12,15]. The literature data indicates the level of curcumin and other curcuminoids in turmeric powder at approximately 3–5% [16]. Commercially, in turmeric extract, mostly used in preclinical studies and clinical trials, the curcuminoid content is often increased to as high as 95%, including approximately 75% (a/a) curcumin, 20% (a/a) demethoxycurcumin, and 5% (a/a) bisdemethoxycurcumin (HPLC, area%). These phytochemicals are practically insoluble in water at acidic and neutral pH, but soluble in methanol, ethanol, acetone and dimethylsulfoxide (DMSO).

The curcumin itself is a colored, yellow to orange, crystalline compound that is commonly used as a coloring and flavoring agent, and food additive. In 2004, the Joint FAO (The Food and Agriculture Organization)/WHO (The World Health Organization) Expert Committee on Food Additives (JECFA) established an acceptable daily intake (ADI) for curcumin of 0–3 mg/kg body weight. The ADI estimation for curcumin was based on the NOAEL (no-observed-adverse-effect level) of 250–320 mg/kg body weight/day from the reproductive toxicity study for a decreased body weight gain in the F2 rat generation observed at the maximum dose, and an uncertainty factor equal to 100 [17]. JECFA and the European Food Safety Authority (EFSA) Panel on Food Additives and Nutrient Sources added to Food (ANS)) agreed that curcumin (symbol E-100, EFSA) is not carcinogenic and genotoxic. The EFSA Panel perceived that the normal diet provides the curcumin amount of less than 7% of the aforementioned ADI [18].

Curcumin is a hydrophobic molecule, practically insoluble in the aqueous phase of the digestive fluids. Curcumin is rapidly eliminated from the digestive tract, having poor oral bioavailability, due to low absorption from the intestine and rapid degradation in the liver [19] reported in human and animal studies [15]. Under physiological pH conditions, such as 0.1 M phosphate buffer (pH 7.2) at 37 °C over 90% of curcumin is degraded within 30 min [20]. The in vivo studies indicate that following curcumin reduction to dihydrocurcumin and tetrahydrocurcumin, quick conversion to mono-glucuronidated conjugates occurs [21]. Thus, the main curcumin metabolites reported in vivo are the curcumin-, dihydrocurcumin-, and tetrahydrocurcumin-glucuronides, as well as tetrahydrocurcumin [22].

Moreover, the studies in humans revealed that it is unlikely that substantial concentrations of curcumin occur in the body after its ingestion at high doses up to 12 g/person, equivalent to 200 mg/kg body weight for a 60 kg individual. Even upon the oral exposure of 10–12 g of curcumin, the detected plasma concentration of this polyphenol was low, in the nanomolar range, with the highest level of less than 160 nmol/L [23]. The numerous studies supported the safety of high doses of curcumin, depicting only gastric disturbance [15].

According to reports of the U.S. Department of Agriculture (USDA), 100 g of turmeric rhizome contain from 2% to 9% of curcumin and other curcuminoids, as well as several vitamins. Those vitamins include some lipid-soluble vitamins such as E and K vitamins, as well as some water-soluble vitamins such as vitamin C and the B-group vitamins, i.e., folic acid (B9), riboflavin (B2), and pyridoxine (B6). Moreover, the *Curcuma longa* rhizome comprises the minerals (Fe, Mg, Zn, K, Na, and Ca) and macromolecules (proteins, lipids, carbohydrates, and dietary fiber) essential for human health (Table 1) [19]. The co-occurrence of several bioactive compounds such as curcumin and vitamins C and E in natural turmeric enhances its antioxidative properties and encourage greater consumption of natural Curcuma-derived products rather than dietary supplements with pure turmeric-extracted curcumin.

Due to the poor bioavailability and low absorption of pure bioactive curcumin, many researchers have focused on studies to ameliorate its bioavailability, pharmacological properties, chemopreventive activity and therapeutic utility [24–26]. The novel curcumin derivatives have been developed. Thus, curcumin and its derivatives have been still extensively investigated as potential anti-cancer, antioxidant, anti-bacterial, anti-inflammatory, analgesic, accelerating wound healing and improving digestion processes agents. The recent studies have revealed that bioavailability of pure curcumin may be enhanced by various natural or synthetic adjuvants, i.e., piperine from black pepper [24] or folic acid [25]. Moreover, taking into account curcumin hydrophobic properties, its bioavailability and retention time can be improved by applying different forms of conjugates, including liposomes, polymeric micelles, phospholipid complexes, microemulsions and nanoparticles [26]. Further studies are needed to establish their clinical application and effectiveness.

Table 1. Nutrients in 100 g of turmeric, *Curcuma longa* L. (FoodData Central: Spices, turmeric, ground; Data Type: SR Legacy; Food Category: Spices and Herbs; FDC ID: 172231; NDB Number: 2043; FDC Published: 4/1/2019; U.S. Department of Agriculture (USDA), Agricultural Research Service [19].)

Name	Amount (Min-Max)	Unit
Water	12.850	g
Energy	312.000	kcal
Protein	9.680	g
Total lipid (fat)	3.250	g
Carbohydrate	67.140	g
Fiber, total dietary	22.700	g
Calcium, Ca	168.000	mg
Iron, Fe	55.000	mg
Magnesium, Mg	208.000	mg
Phosphorus, P	299.000	mg
Potassium, K	2080.000	mg
Sodium, Na	27.000	mg
Zinc, Zn	4.500	mg
Copper, Cu	1.300	mg
Manganese, Mn	19.800	mg
Selenium, Se	6.200	µg
Vitamin C, total ascorbic acid	0.700	mg
Vitamin B_1 (thiamin)	0.058	mg
Vitamin B_2 (riboflavin)	0.150	mg
Vitamin B_3 (niacin)	1.350	mg
Vitamin B_5 (pantothenic acid)	0.542	mg
Vitamin B_6 (pyridoxine)	0.107 (0.034–0.180)	mg
Folate, total	20.000	µg
Choline, total	49.200	mg
Betaine, total	9.700	mg
Vitamin E (alpha-tocopherol)	4.430	mg
Vitamin K (phylloquinone)	13.400	µg
Fatty acids, total saturated	1.838	g
Fatty acids, total monounsaturated	0.449	g
Fatty acids, total polyunsaturated	0.756	g
Curcuminoids	2.000–9.000	g

3. DNA Methylation and Demethylation Processes

DNA methylation is one of the most important, epigenetic modifications without any alterations in the primary DNA sequence and is frequently associated with silencing of gene expression [27]. This process is tightly connected with replication during the normal cell growth and plays an important role in the regulation of crucial cell functions such as DNA repair, cell cycle, cell differentiation, intracellular signal transduction, and cell apoptosis. Moreover, the aberrations of DNA methylation patterns can be implicated in neoplastic processes of both normal cells and cells with pathological changes. The aberrant DNA methylation patterns are observed at very early stages of pre-cancerous transformation of cells that are still not exhibiting any cancerous phenotype [28]. The tumor-specific alterations of DNA methylation pattern can include global DNA hypomethylation and loci-specific hypomethylation of oncogenes and pro-metastatic genes, as well as loci-specific hypermethylation within TSG regulatory regions such as proximal promoter regions frequently containing CpG islands and/or enhancers [28–31].

The ageing or cancer-related global DNA hypomethylation may result in microsatellite instability, transposon activation and stimulation of oncogene expression. In cancer, the hypermethylation of TSG promoters is often associated with the overexpression of DNA methyltransferases (DNMTs). The DNMT enzymes catalyze a reaction in which methyl group is transferred from S-adenosyl-L-methionine (SAM) to cytosine located in CpG dinucleotide sequences (CpGs) giving 5-methylcytosine (5mC). In normal cells, CpG-rich regions, called CpG islands, are mostly unmethylated and are located within regulatory regions of house-keeping genes, tissue-specific genes, and TSGs [32]. The elevated promoter methylation is a predominant mechanism of chromatin inactivation. Epigenetic modifications, apart from DNA methylation, include covalent post-translational modifications of histone tails (mainly methylation/demethylation and acetylation/deacetylation). DNA hypermethylation and histone deacetylation are concerted to determine the transcriptional activity of certain genes, leading frequently to condensation of chromatin structure making the DNA inaccessible to complexes of transcription proteins [33].

Neoplastic development might be associated with both hypermethylation of gene promoters and increase in the number of silent genes, mainly TSGs encoding proteins that control the normal cell functions and the balance between cell proliferation and apoptosis Additionally, DNA hypermethylation of CpG island within the gene promoter can cause genetic alterations. The spontaneous deamination of 5mC and its transition to thymine within CpG sequences of *TP53* (*Tumor Protein P53*) gene have been observed in cells of various types of cancer [34].

In mammalian cells, the reaction of DNA methylation is the post-replicative DNA modification taking place at the replication fork. It is catalyzed by the enzymes of the DNMT family, including DNMT1, DNMT2, DNMT3A, DNMT3B, and DNMT3L (DNA Methyltransferase 3 Like) [32]. DNMT1 is the main DNA methylating enzyme responsible for the maintenance DNA methylation in normal cells, as well as for the maintenance and de novo DNA methylation in cancer cells. DNMT3A and DNMT3B catalyze de novo DNA methylation. The differences between the DNMT1 and other DNMTs are in the length of the N-terminal regulatory domain. The C-terminal catalytic domain of DNMT1 protein contains the following regions: one implicated in the binding of SAM (methyl donor), another one responsible for binding to DNA, and an active center containing proline and cysteine. Whereas, the N-terminal regulatory domain of DNMT1 can interact with numerous proteins like DMAP1 (DNA methyltransferase 1-associated protein 1), PCNA (Proliferating cell nuclear antigen), and RB (Retinoblastoma-associated protein) It is multifunctional and contains a DNA binding region, a cysteine-rich region, several Zn-binding domains, and two regions responsible for the localization to replication foci There are also regions implicated in DNMT1 interactions with histone deacetylases HDAC1 and HDAC2, as well as the other DNA methyltransferases, DNMT3A and DNMT3B [13].

In mammalian cells, two additional DNMTs have been described, i.e., DNMT2 (TRDMT1) and DNMT3L. They do not possess catalytic activity towards DNA. The DNMT3L has been shown to interact with de novo DNMTs, DNMT3A and likely DNMT3B, what supports their stability and stimulates DNMT3A-mediated DNA methylation [35] The DNMT2 activity does not involve DNA methylation but RNA methylation, specifically cytosine 38 in the anticodon loop of aspartic acid tRNA [36]. The role of DNMT3L in breast tumorigenesis is not well understood. Girault et al. reported that DNMT3L mRNA levels were very low (only detectable but not quantifiable) in a subgroup of 46 breast tumors [37].

Upon DNA replication, within the transcriptionally active DNA (euchromatic DNA) the DNMT1 cooperates with the PCNA protein to form the DNA-DNMT1-PCNA complex responsible for the maintenance of the DNA methylation patterns [38]. In normal somatic cells, the DNA methylation can be inhibited by CDN1A (Cyclin-dependent kinase inhibitor 1, encoded by *CDKN1A* (*P21*) gene) protein, which is the inhibitor of cyclin-dependent kinases. In this case, the CDN1A (P21) protein disrupts the PCNA-DNMT1 complex and forms a new P21-PCNA complex. It leads to inhibition of polymerase δ activity followed by inhibition of both DNA methylation and replication [39]. All this allows the repair processes of DNA double-strand breaks. It is noteworthy to indicate that the P21 and DNMT1 proteins compete for binding to the same motif of the PCNA protein. Several studies have revealed an inverse relation between P21 and DNMT1 concentrations in normal and cancer cells. In normal cells, the protein level of P21 is much higher than DNMT1, whereas in cancer cells the relation is opposite [40].

In normal cells, the DNMT1 activity can be also inhibited by interaction of RB protein with the N-terminal regulatory domain of the DNMT1 protein. The RB binding to DNMT1 blocks the formation of the DNA-DNMT1-PCNA complex and consequently DNA methylation reaction. It might be one of the mechanisms preventing some DNA fragments (e.g., promoters of TSGs) from methylation during normal cell development [41].

The DMAP1 (transcriptional co-repressor) protein affects DNMT1 activity as well. It interacts with DNMT1 at replication foci when the DNMT1 enzyme is bound to PCNA. During this interaction, DNMT1 binds to the co-repressor DMAP1 and simultaneously to HDAC2. The repressive activity of this complex is probably responsible for the appearance of the condensed chromatin state (heterochromatin) and the maintenance of transcriptional silencing of methylated genes upon DNA replication [42].

Among important proteins interacting with the regulatory domain of DNMT1 are the nuclear proteins that are characterized by the presence of a methyl-CpG binding domain (MBD). The MBD2 (Methyl-CpG-binding domain protein 2), MBD3 (Methyl-CpG-binding domain protein 3) and MECP2 (Methyl-CpG-binding protein 2) proteins exert a two-way effect on the epigenome. On the one hand, MECP2 can bind fully methylated DNA mediating transcriptional repression through interaction with HDAC and the corepressor SIN3A (Paired amphipathic helix protein Sin3a) [43]. On the other hand, MBD proteins can bind to hemimethylated DNA and then form the complex with DNMT1 that is forced to catalyze methylation of a newly synthesized DNA strand and cover recognition elements [44].

To achieve transcriptional silencing of the selected genes, DNMT1 cooperates directly with de novo DNMTs, DNMT3A and DNMT3B. These two enzymes are responsible for the de novo DNA methylation in normal cell development. However, in cancer cells, DNMT3A and DNMT3B enzymes synergistically enhance the DNMT1 activity and shift its activity towards de novo DNA methylation [40].

Moreover, the studies revealed that DNMT1 directly interacts with histone-modifying enzymes such as histone deacetylases HDAC1 and HDAC2, and histone methyltransferase SUV91 (Histone-lysine N-methyltransferase SUV39H1) [45]. The studies revealed that methylation of heterochromatic DNA might be triggered by histone modifications, i.e., methylation of histone H3, and deacetylation of histone H3 and H4 [46].

Thus, the interactions of DNMT1 with the aforementioned proteins, i.e., MBPs, DNMT3A and DNMT3B, HDAC1 and HDAC2, and SUV39B may result in transcriptional repression of many genes in cancer cells, mainly TSGs, thereby stabilizing the heterochromatic state [32]. The aberrant DNMTs activity followed by the altered DNA methylation patterns plays a pivotal role during carcinogenesis, including breast cancer development. The literature data indicated that more than 100 genes have been hypermethylated in primary breast tumors and breast cancer cell lines, i.e., *PTEN* (Phosphatase And Tensin Homolog), *RARB* (Retinoic Acid Receptor Beta), *APC* (APC Regulator Of WNT Signaling Pathway), *CDKN2A* (*P16*; Cyclin Dependent Kinase Inhibitor 2A), *CDH1* (Cadherin 1), *DAPK1* (Death Associated Protein Kinase 1), *GSTP1* (Glutathione S-Transferase Pi 1), *RASSF1* (Ras Association Domain Family Member 1), *TIMP3* (TIMP Metallopeptidase Inhibitor 3), and *MGMT* (O-6-Methylguanine-DNA Methyltransferase) [13,47].

The role of altered expression of TET (Methylcytosine dioxygenase TET) demethylating enzymes (i.e., TET1, TET2 and TET3) in cancer is less well understood. The downregulation of TET expression and reduced 5hmC (5-hydroxymethylcytosine) levels have been shown to be associated with breast, gastric, liver, and lung tumors [48]. Moreover, high levels of TET expression estimated in over 160 samples of breast cancer tissue correlated with increased patient survival, probably resulting from potential DNA demethylation–mediated upregulation of TSGs [49]. Yang et al. reported that the TETs expression, mostly TET1, was significantly decreased in human breast tumors, and the 5hmC levels were broadly diminished in breast cancer tissues [50].

The mechanism of DNA methylation and demethylation processes have been depicted in Figure 1.

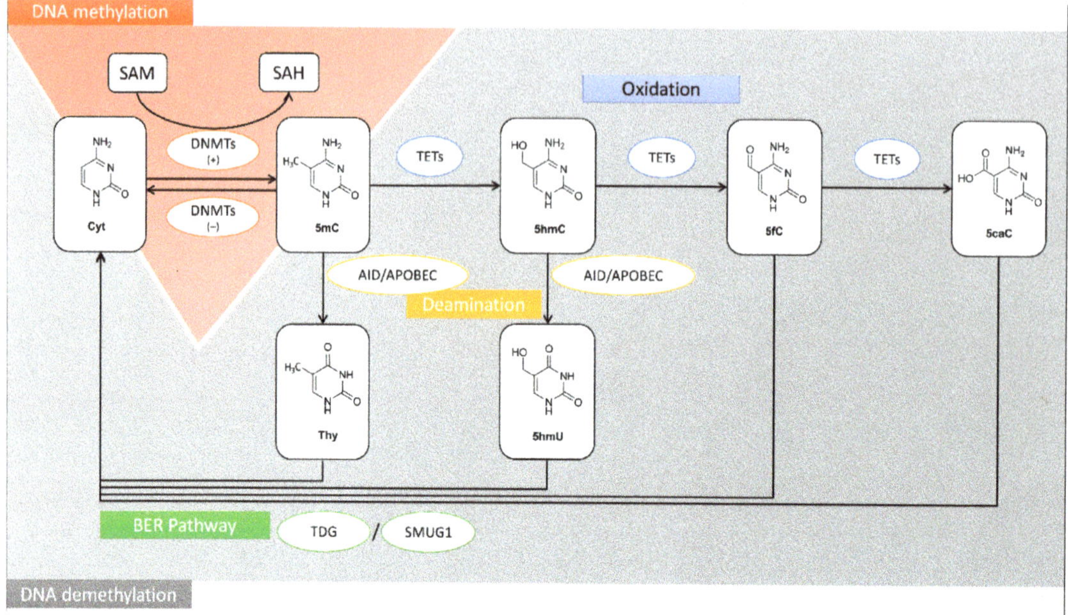

Figure 1. The mechanisms of DNA methylation and demethylation processes. SAM is a methyl donor in the DNMT-catalyzed methylation reactions of cytosine (Cyt) to 5mC. TET proteins catalyze the multi-stage process of 5-methylcytosine oxidation to 5hmC, 5fC, and 5caC. Furthermore, 5mC and 5hmC may be transformed into thymine (Thy) and 5hmU in deamination process, respectively. These misincorporations are recognized and replaced with cytosine by BER pathway and glycosylases, i.e., TDG and SMUG1. The other products of 5mC oxidation process, 5fC and 5caC undergo BER as well. 5mC, 5-methylcytosine; 5hmC, 5-hydroxymethylcytosine; 5hmU, 5-hydroxymethyluracil; 5fC, 5-formylcytosine; 5caC, 5-carboxylcytosine; AID, activation-induced deaminase; APOBEC, DNA dC->dU-editing enzyme APOBEC; BER, base excision repair pathway; Cyt, cytosine; DNMT, DNA methyltransferase; SAH, S-adenosyl-L-homocysteine; SAM, S-adenosyl-L-methionine; SMUG1, Single-strand selective monofunctional uracil DNA glycosylase; Thy, thymine; TET, Methylcytosine dioxygenase TET; TDG, G/T mismatch-specific thymine DNA glycosylase.

4. Curcumin as an Epigenetic Inhibitor of Mammary Cancer

The numerous studies have shown that natural spices and their bioactive components, such as turmeric and curcumin, may induce epigenetic remodeling in breast cancer cells leading to TSG reactivation and oncogene downregulation. Those changes in expression of genes encoding proteins involved in the regulation of intracellular oncogenic signaling pathways may cause inhibition of breast cancer cell proliferation via cell cycle arrest and simultaneously induce cell apoptosis [9,12,15,51].

Several in vitro and in vivo studies on breast cancer indicated that the main bioactive component of turmeric rhizome, curcumin is a potent breast cancer inhibitor possessing anti-proliferative and proapoptotic properties [52–54].

The studies in two tumor cell lines, human breast cancer MCF-7 and T cell lymphoma EL4 of murine origin, as well as in two types of normal cells, including mouse spleen lymphocytes and NIH3T3 mouse fibroblast cells, have revealed the potent cytotoxic cancer-specific activity of curcumin. Total curcumin uptake was significantly higher in both tumor cell lines comparing to normal cells. Moreover, localization of curcumin in the human breast cancer MCF-7 cells has been determined using laser confocal microscopy [55]. The different subcellular distribution of curcumin was observed in MCF-7 cells. It has been documented that curcumin accumulates mainly in the cell membrane and the uptake of curcumin by other cell components is in the following order: cytoplasm > nucleus

> mitochondria. Most likely, the lipophilic curcumin interacts with cellular membrane lipids, what explains why the polyphenol is mainly located in the cell membrane. These findings are consistent with the results of the other studies showing that curcumin uptake is significantly higher in tumor cells compared to normal cells [55–57]. It could be caused by different composition of the lipid droplets in non-malignant and malignant human breast epithelial cell lines. Abramczyk et al. analyzed the chemical composition of lipids and proteins in non-malignant (MCF10A), mildly malignant (MCF7) and malignant (MDA-MB-231) breast cancer cells by Raman imaging. Results of the studies demonstrate increased lipid contents in malignant breast cancer cells compared to non-malignant cells. The number of cytoplasmic lipid droplets correlates with increased aggressiveness of cancer. In malignant breast cells MCF7 and MDA-MB-231 it is respectively 2 and 4 times higher, than in non-malignant MCF10A cells. The higher content of lipid droplets in breast cancer cells may explain better solubility of curcumin and its higher intracellular concentration [58]. Thus, increased uptake of curcumin is consistent with its increased toxicity [55].

4.1. *Curcumin and DNMTs*

Alterations in methylation of gene promoters play an important role in gene transcriptional activity. Hypermethylation-mediated silencing of TSGs and hypomethylation-mediated activation of oncogenes and pro-metastatic genes are probably the most consistent epigenetic hallmarks of human cancers, including breast cancer. Significance of epigenetic changes in cancer development, especially aberrant DNA methylation patterns, is comparable to the relevance of genetic mutations.

The DNA methylation is mediated by specific DNMTs. Increased level of DNMTs was observed in cancer patients. In the study of Mirza et al., the levels of DNMT1, DNMT3A, and DNMT3B mRNA were observed to be 1.2- to 4.4-folds, 1.1- to 3.77-folds, and 1.06- to 4.01-folds elevated in the most of the analyzed breast cancer tissues, respectively, as compared to the adjacent normal breast tissues [59].

In another study, the expression of DNMT1, DNMT3A, and DNMT3B in 256 breast cancer and 36 breast fibroadenoma cases were investigated. The DNMT1 and DNMT3A expression levels were significantly higher in breast cancer than in fibroadenoma samples. The DNMT1 and DNMT3A overexpression was associated with promoter hypermethylation and downregulation of ERα and BRCA1 [60].

The DNMT1 up-regulation was also revealed in most cancer-associated fibroblasts in relation to their corresponding adjacent normal fibroblasts. The ectopic expression of DNMT1 activated primary normal breast fibroblasts and promoted their pro-carcinogenic effects, both in vitro and in orthotopic tumor xenografts whereas DNMT1 knockdown normalized breast myofibroblasts. DNMT1 seems to be critical for the activation of breast stromal fibroblasts as well as the persistence of their active status [61].

Therefore, in recent years, DNA methylation has emerged as an attractive target for the anti-cancer therapeutics. Moreover, natural bioactive compounds, including curcumin, have received increasing attention as potential modulators of epigenetic machinery in cancer cells. The numerous studies have shown that curcumin exerts robust epigenetic anti-cancer effects against breast cancer (Table 2).

In 2009, Liu et al. based on the results of molecular docking (interaction of curcumin and DNMT1) suggested that curcumin covalently blocks the catalytic thiolate of C1226 of DNMT1 to exert its inhibitory effect [62]. Moreover, they observed in in vitro tests that curcumin and one of its major metabolites—tetrahydrocurcumin, can inhibit the activity of CpG Methyltransferase M.SssI (a DNMT1 analog with a structurally similar catalytic domain) [62].

Table 2. Curcumin impact on epigenetic machinery in breast cancer inhibition.

DNMTs Expression	Change	Model	Curcumin Treatment	Reference
mRNA level				
	decrease in all DNMTs (DNMT1, DNMT3A, DNMT3B)	MCF-7 MDA-MB-231	IC50—10 μM/96 h	Mirza S. et al., J Breast Cancer, 2013 [59]
	decrease in all DNMTs (DNMT1, DNMT3A, DNMT3B)	MCF-7	2 and 20 μM/12 and 24 h	Chatterjee B. et al., J Cell Biochem, 2019 [63]
	decrease in DNMT1 (without changes in DNMT3A, DNMT3B)	MDA-MB-361 MDA-MB-231 MCF-7	40 μM/48 h	Liu Y. et al., Mol Cell Biochem, 2017 [65]
	decrease in DNMT1	MCF-7	10 and 20 μM/72 h	Du L. et al., Nutr Cancer, 2012 [66]
protein level				
	2-fold decrease in DNMT1	MCF-7 MDA-MB-231	IC50—10 μM/96 h	Mirza S. et al., J Breast Cancer, 2013 [59]
	decrease in all DNMTs (DNMT1, DNMT3A, DNMT3B)	MCF-7	2 and 20 μM/12 and 24 h	Chatterjee B. et al., J Cell Biochem, 2019 [63]
	reduction in DNMT1 protein level increase in DNMT3A and DNMT3B protein level	HCC-38 UACC-3199 T47D	5 and 10 μM/6 days	Al-Yousef N. et al., Oncol Rep, 2020 [64]
	decrease in DNMT1 (without changes in DNMT3A, DNMT3B)	MDA-MB-361 MDA-MB-231	40 μM/48 h	Liu Y. et al., Mol Cell Biochem, 2017 [65]
	decrease in DNMT1	MCF-7	10 and 20 μM/72 h	Du L. et al., Nutr Cancer, 2012 [66]
Other proteins				
	increase in *TET1* mRNA and TET1 protein level	HCC-38	5 and 10 μM/6 days	Al-Yousef N. et al., Oncol Rep, 2020 [64]
	3-fold decrease in HDAC1 protein level	MCF-7 MDA-MB-231	IC50—10 μM/96 h	Mirza S. et al., J Breast Cancer, 2013 [59]
	decrease in HDAC1 and HDAC2 protein level	MCF-7 MDA-MB-231	50 μM/24 h	Mukherjeea S. et al., Int. J. Green Nanotechnol, 2012 [67]
oncogene	decrease in *SNCG* mRNA (down to 2-fold) and SNCG protein level	T47D HCC-38	5 and 10 μM/6 days	Al-Yousef N. et al., Oncol Rep, 2020 [64]
tumor suppressor	induction of *DLC1* expression on mRNA and protein level	MDA-MB-361	20 and 40 μM/48 h	Liu Y. et al., Mol Cell Biochem, 2017 [65]
tumor suppressor	increase in *BRCA1* mRNA level up to 2-fold with consequent high increase in BRCA1 protein level	HCC-38 UACC-3199	5 and 10 μM/6 days	Al-Yousef N. et al., Oncol Rep, 2020 [64]

Table 2. *Cont.*

DNMTs Expression	Change	Model	Curcumin Treatment	Reference
tumor suppressor	increased level of *CDKN1A* (*p21*, 2-fold in MDA-MB-231 and 4-fold in MCF-7)	MCF-7 MDA-MB-231	IC50—10 μM/96 h	Mirza S. et al., J Breast Cancer, 2013 [59]
tumor suppressor	increased level of *CDKN1A (p21)*	MCF-7 MDA-MB-231	50 μM/24 h	Mukherjeea S. et al., Int. J. Green Nanotechnol, 2012 [67]
tumor suppressor	increased expression of *TP53* and *KLF4* on mRNA and protein levels	MCF-7	2 and 20 μM/12 and 24 h	Chatterjee B. et al., J Cell Biochem, 2019 [63]
tumor suppressor	enhanced mRNA and the protein levels of *RASSF1A*	MCF-7 MDA-MB-231	10 and 20 μM/72 h	Du L. et al., Nutr Cancer, 2012 [66]
transcription factor	reduction in *SP1* expression	MDA-MB-361	40 μM/48 h	Liu Y. et al., Mol Cell Biochem, 2017 [65]
DNMTs activity				
	methylation activity of DNMT1 in nuclear extract decreased by about 70% (compared to the control)	MCF-7	10 and 20 μM/72 h	Du L. et al., Nutr Cancer, 2012 [66]
Promoter methylation				
	demethylation of the proximal promoter of *CDKN1A* (*p21*)	MCF-7	2 and 20 μM/12 and 24 h	Chatterjee B. et al., J Cell Biochem, 2019 [63]
	hypermethylation of the *SNCG* promoter	T47D	5 and 10 μM/6 days	Al-Yousef N. et al., Oncol Rep, 2020 [64]
	partial hypomethylation of the *BRCA1* promoter	HCC-38 UACC-3199	5 and 10 μM/6 days	Al-Yousef N. et al., Oncol Rep, 2020 [64]
	demethylation of *DLC1* promoter	MDA-MB-361	20 and 40 μM/48 h	Liu Y. et al., Mol Cell Biochem, 2017 [65]
	decrease in *RASSF1A* promoter methylation	MCF-7	10 μM/72 h	Du L. et al., Nutr Cancer, 2012 [66]
Global DNA methylation				
	hypomethylation	MCF-7	2 and 20 μM/12 and 24 h	Chatterjee B. et al., J Cell Biochem, 2019 [63]
	the global DNA methylation (GDM) decreased by about 30–35%	MCF-7	10 μM/72 h	Du L. et al., Nutr Cancer, 2012 [66]

Table 2. *Cont.*

DNMTs Expression	Change	Model	Curcumin Treatment	Reference
miRNA				
	downregulation of oncogenic miR-19 (modulates downstream proteins: PTEN, AKT1, MDM2, TP53)	MCF-7	1 μM/4 days	Li X. et al., Phytother Res, 2014 [68]
	upregulation of miR-29b	T47D	5 and 10 μM/6 days	Al-Yousef N. et al., Oncol Rep, 2020 [64]
	upregulation of miR-34a (reduction in *BCL2* and *BMI1* expression)	MDA-MB-231 MDA-MB-435	30 or 34 μM/24 h	Guo J. et al., Mol Cell Biochem, 2013 [69]
	upregulation of miR181b (reduction in *CXCL1, CXCL2, MMPs* expression)	MDA-MB-231	25 μM/24 h	Kronski E. et al., Mol Oncol, 2014 [70]
	upregulation of miR-15a and miR-16 (reduction in *BCL2* expression)	MCF-7	10–60 μM/24 h	Yang J. et al., Med Oncol, 2010 [71]

AKT1 (AKT Serine/Threonine Kinase 1); BCL2 (BCL2 Apoptosis Regulator); BMI1 (BMI1 Proto-Oncogene, Polycomb Ring Finger); BRCA1 (BRCA1 DNA Repair Associated); CDKN1A (Cyclin Dependent Kinase Inhibitor 1A); CXCL1 (C-X-C Motif Chemokine Ligand 1); CXCL2 (C-X-C Motif Chemokine Ligand 2); DLC1 (DLC1 Rho GTPase Activating Protein); DNMT1 (DNA Methyltransferase 1); DNMT3A (DNA Methyltransferase 3 Alpha); DNMT3B (DNA Methyltransferase 3 Beta); HDAC (Histone Deacetylase); KLF4 (Kruppel Like Factor 4); MDM2 (MDM2 Proto-Oncogene); MMPs (Matrix Metallopeptidases); PTEN (Phosphatase And Tensin Homolog PTEN); RASSF1 (Ras Association Domain Family Member 1; Tumor Suppressor Protein RDA32); SNCG (Synuclein, Gamma (Breast Cancer-Specific Protein 1)); SP1 (Sp1 Transcription Factor); TET1 (Tet Methylcytosine Dioxygenase 1); TP53 (Tumor Protein P53).

Furthermore, apart from that direct chemical mechanism of the reduction of DNMTs enzymatic activity in response to curcumin exposure, the second one is the biological inhibition of DNMTs synthesis. In many studies, in breast cancer cells incubated with curcumin, the DNMT1 protein level was significantly decreased [59,63–66] and it was consistent with the mRNA levels of DNMT1, also significantly downregulated in curcumin-exposed cells [59,63,65,66]. These curcumin-mediated effects may be associated with the disruption of binding of the NF-κB/SP1 complex to the *DNMT1* promoter region [66].

4.2. Curcumin and HDACs/HATs

Epigenetic alterations, which may occur as a part of the carcinogenesis process, modulate gene expression. In tumor cells, silencing of TSGs and activation of oncogenes are observed and usually, these aberrant methylation patterns are not caused by histone modification or DNA methylation alone. All those epigenetic processes are combined and interdependent [13].

Histone deacetylases (HDACs) and histone acetyltransferases (HATs) are the two major groups of enzymes that modulate chromatin structure via histone modifications. The balance between histone acetylation and deacetylation is important for the epigenetic regulation of gene function and dysregulation of these processes may contribute to cancer development. Curcumin has been reported to inhibit HDACs such as HDAC1 and HDAC2 in breast cancer cell lines, MCF-7 and MDA-MBA-231. The HDAC1 and HDAC2 expression levels were constitutively very high in these breast cancer cell lines, comparing to the normal breast epithelial cells MCF-12F. Treatment of cells with 50 μM curcumin for 24 h led to 84% inhibition of HDAC1 expression and 70% inhibition of HDAC2 expression in the case of MCF-7 cells. Similar exposure to curcumin in MDA-MB-231 cells downregulated HDACs, particularly HDAC2, but to a lesser extent (HDAC1 by 75% and HDAC2 by 45%). These changes were accompanied by cell cycle arrest via P21 upregulation in breast cancer MCF-7 cells, and by apoptosis induction via caspase-9 activation [59,67].

Similar observations were made by Mirza et al., 96 h exposure of MCF-7 and MDA-MB-231 cell to 10 μM curcumin led to a 3-fold decrease in HDAC1 level in both cell lines. A decrease in the HDAC1 (and DNMT1) expression caused an opposite effect on the P21 protein level. Curcumin exposures of the breast cancer cells resulted in 2- and 4-fold increases in the P21 expression in MDA-MB-231 and MCF-7, respectively [59].

Curcumin was also found to be highly potent direct HDAC inhibitor. Molecular docking studies showed that curcumin binds to HDAC8 (the other class I HDAC, apart from HDAC1 and 2) and makes hydrophobic contact with active site residues of the enzyme [72].

4.3. Curcumin and miRNAs

Curcumin has also been shown to modulate the expression of miRNAs (small non-coding RNA sequences containing about 22 nucleotides involved in post-transcriptional regulation of gene expression) in breast cancer [73]. Curcumin was able to affect the expression of oncogenic (miR-19a and miR-19b) [68] and tumor-suppressive miRNAs (miR-15a, miR-16, miR-29a, miR-34a, and miR-181b) in breast cancer cells [64,69–71]. The observed changes in miRNA expression led in consequence to the suppression of tumorigenesis and metastasis, and induction of apoptosis.

4.4. Curcumin Epigenetic Anti-Cancer Effects Revealed in In Vivo Studies

Du et al. revealed that curcumin in MCF-7 cells downregulates the mRNA and protein level of DNMT1, thereby decreasing the methylating activity of the nuclear extract and global DNA methylation in MCF-7 cells. Curcumin reactivates a silenced TSG ras-association domain family protein 1A (RASSF1A) at least partially due to its promoter hypomethylation in breast cancer MCF-7 and MDA-MB-231 cell lines. Since *RASSF1A* tumor suppressor silencing is associated with the deregulated proliferation activity of some cancer cells, the curcumin-mediated RASSF1A reactivation may be associated with its anti-proliferative activity in vitro and anti-tumor growth activity in vivo. The mRNA level of *RASSF1A* was found to be significantly higher not only in breast cancer cell lines but also in MCF-7 cell engrafted tumor tissue collected from tumor bearing nude mice treated with an intraperitoneal administration of 100 mg/kg curcumin in reference to those treated with the vehicle. Curcumin treatment caused also 65% decrease in tumor size in nude mice without any observed cytotoxicity [66].

The antitumor activity of curcumin in ER-negative human breast cancer was assessed in in vivo mouse model of breast cancer (MDA-MB-231 xenograft model in female Foxn1nu/nu mice). 16 mice were randomized into two groups: 8 controls (normal diet) and 8 curcumin-treated through 6 weeks after tumor cell implementation. Results obtained in the study indicated that curcumin inhibits tumor growth and angiogenesis without toxicity effect on mice. The data showed that curcumin represses the activation of NF-κB as far as NF-κB-regulated gene products such as cyclin D1, p65, and PECAM-1 (significant reductions in the expression of PECAM-1, cyclin D1, and p65 compared to the control group were observed) [74]. Similar observations were made in the other study, in which the effects of curcumin on the human breast cancer cell line MDA-MB-231 in vitro and in a mouse metastasis model (MDA-MB-231 cells were intracardiac injected in immunodeficient mice) were examined. Curcumin appeared to inhibit the expression and activity of AP-1 and NF-κB and consequently reduce the expression of major matrix metalloproteinases (MMPs). Curcumin caused also a diminution of IκB and p65 phosphorylation, reduced activation of survival pathway NF-κB and the number of metastases [75].

Curcumin alone and in combination with mitomycin C (MMC) inhibited MCF-7 breast cancer cell proliferation and viability in vitro and in vivo. In MCF-7 xenografts, combined administration of curcumin (100 mg/kg) and MMC (1–2 mg/kg) for 4 weeks produced significantly greater inhibition on tumor growth than either treatment alone. The combined treatment resulted in significantly greater G1 arrest than MMC or curcumin alone. Moreover, the cell cycle arrest was associated with inhibition of cyclin D1, cyclin E, cyclin A, cyclin-dependent kinase 2 (CDK2) and CDK4, along with the induction of the cell cycle inhibitor P21 and P27 both in MCF-7 cells and in MCF-7 xenografts. These proteins were regulated through the P38 MAPK pathway [76].

4.5. Curcumin Epigenetic Anti-Cancer Effects Revealed in Clinical Trials

Since there is a need to improve the efficacy of breast cancer chemotherapy especially with safe molecules, therefore the feasibility and tolerability of the combination of chemotherapeutics with natural compounds are investigated.

In a clinical trial evaluating curcumin in combination with docetaxel in advanced and metastatic breast cancer patients, it was revealed that the safety profile of the combination is consistent with that observed with monotherapy of docetaxel. Curcumin was given orally for seven consecutive days in 8000 mg/day dose, in combination with docetaxel 100 mg/m^2 administered every 3 weeks for six cycles. Curcumin/docetaxel combination demonstrated antitumor activity. Among the 14 patients enrolled in this study, nine patients were evaluated for tumor response. In seven patients, the biological response was documented with the decrease of tumor markers, which was up to 50% in four patients [77]

In the other clinical trial the efficacy and safety of curcumin, administering intravenously, in combination with paclitaxel in patients with advanced, metastatic breast cancer was explored. A total of 150 women with advanced and metastatic breast cancer were randomly assigned to receive either paclitaxel (80 mg/m^2) plus placebo or paclitaxel plus curcumin (CUC-1®, 300 mg solution, once per week) intravenously for 12 weeks with 3 months of follow-up (133 patients complete the study treatments). Objective response rate (percent of patients with complete and partial tumor reduction) of curcumin/paclitaxel combination was significantly higher (50.7%) than that of the placebo/paclitaxel (33.3%; $p < 0.05$) after 12 weeks of treatment and 4 weeks of follow-up. Intravenously administered curcumin caused no major safety issues and no reduction in quality of life, and it was assumed as slight beneficial in reducing fatigue [78].

Those findings confirmed the safety of curcumin and its potential in elevating antitumor activity of conventional chemotherapeutics applied in anti-cancer treatment, also in breast cancer therapy.

5. Insight into the Other Bioactive Components of Turmeric Rhizome as Potential Epigenetic Modifiers

The turmeric rhizome contains the numerous bioactive agents, including not only polyphenolic curcumin but also the B-group vitamins (riboflavin, pyridoxine and folic acid) antioxidative vitamins C and E, several minerals (i.e., zinc), and turmeric oil [19]. Folic acid as a dietary methyl donor, as well as vitamins B2 and B6 as cofactors of enzymes, are implicated in one-carbon metabolism involved in SAM synthesis and DNA methylation reaction. The bioactive compounds derived from turmeric rhizome may affect DNA methylation processes by changes in DNMTs expression and activity, as well as by alterations in SAM level, the ubiquitous donor of a methyl group in DNA methylation reaction (Figure 2).

DNA methylation is a SAM-dependent modulation in which SAM plays the role of methyl group donor. The process of SAM synthesis (from methionine) is a part of the methionine cycle [79]. In the cycle, SAM is converted to S-adenosylhomocysteine (SAH) that is transformed to homocysteine. Subsequently obtaining methyl group from 5-methyltetrahydrofolate (5-MTHF) enables the methylation of homocysteine to methionine and then the formation of SAM. Importantly, 5-methyltetrahydrofolate is created in the folate cycle that begins with the conversion of folic acid. The connection of folate and methionine cycles allows transferring a methyl group from 5-MTHF to methionine [80].

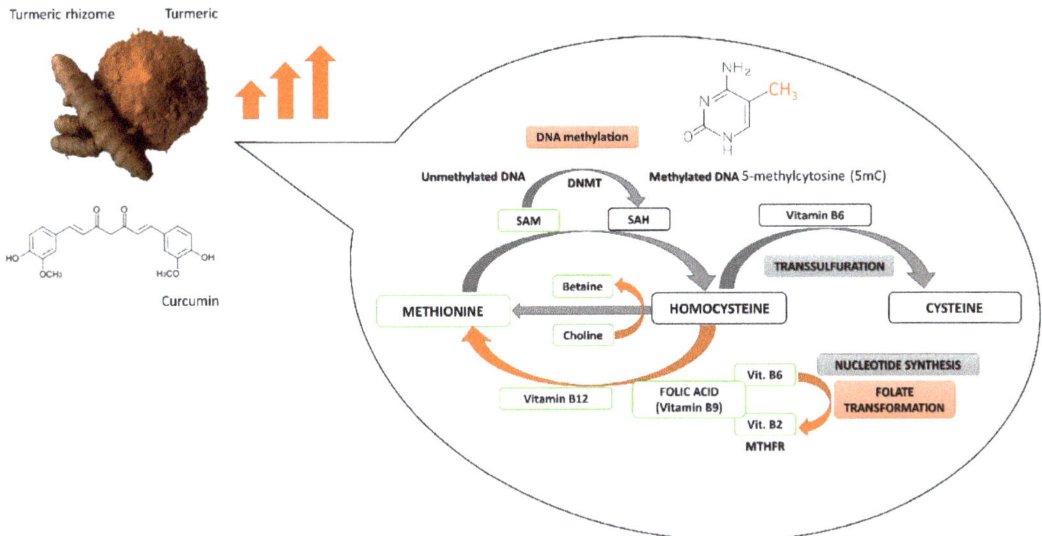

Figure 2. Potential interference of curcumin and other components of turmeric in DNA methylation reaction and one-carbon metabolism. MTHFR, Methylenetetrahydrofolate reductase; SAH, S-adenosyl-L-homocysteine; SAM, S-adenosyl-L-methionine.

After carrying out the genome methylation, SAM is converted to SAH that is a potent inhibitor of DNMTs, especially DNMT1. SAH can bind to the catalytic region of most SAM-dependent methyltransferases [79]. The ratio of SAM to SAH is called "methylation index" [81]. The proper ratio of SAM to SAH is very important for the stability of DNA methylation patterns. Methylation index can indicate the probability of DNA hyper or hypomethylation [82]. Cancer development is associated with SAM depletion, global DNA hypomethylation, activation of oncogenes as well as hypermethylation and silencing of TSGs [83–85].

As mentioned above, the methionine cycle is connected with a folate cycle. The folate cycle is regulated by B-group vitamins which, as cofactors of the cycle, indirectly affect the synthesis and concentration of SAM. One carbon metabolism pathway starts with folic acid (vitamin B9) which is transformed into dihydrofolate (DHF), tetrahydrofolate (THF), 5,10-methylenetetrahydrofolate (5,10-MTHF) and 5-MTHF [86]. Conversion of THF to 5-MTHF occurs in the presence of pyridoxine (vitamin B6), while riboflavin (vitamin B2) as a component of FAD, participates in the conversion of 5,10-MTHF to 5-MTHF. A low pyridoxine plasma concentration is linked to an increase in SAH level, a lower ratio of SAM to SAH and DNA hypomethylation, which can increase the risk of cancer. According to the literature, vitamins B9 and B6 take part in the reduction of breast cancer growth [13]. They regulate indirectly the level of SAM and SAH, while SAH is an inhibitor of DNMTs.

Moreover, folic acid can enhance curcumin inhibitory effect on DNMTs activity, probably by improving the solubility and bioavailability of curcumin, what was documented by Liu Z [87]. The important role of folates is confirmed by studies including Chinese breast cancer cases. The studies indicate that high folate intake may reduce cancer risk in premenopausal women. The protective role of folate in breast cancer, particularly in ER-negative cancer was shown in another prospective study [88]. On the other hand, some cases indicated that low methionine and B-vitamins intake may be related to a breast cancer carcinogenesis [89]. Other studies showed that folate deficiency leads to decreased SAM level and slight, statistically significant global hypomethylation which is an early epigenetic modification in many cancers [90].

As it was mentioned before, antioxidant and anti-inflammatory properties are also the two mechanisms of curcumin chemopreventive activity. Accumulation of free-radicals (reactive oxygen species (ROS) and reactive nitrogen species (RNS)) is responsible for peroxidation of membrane lipids and oxidative damage of DNA and proteins, and it is implicated in the development of chronic pathological complications, such as cancer. Curcumin has been shown to inhibit cyclooxygenase-2 (COX-2), lipoxygenase (LOX), inducible nitric oxide synthase (iNOS) and xanthine oxidoreductase (XOR) enzymes, generating ROS. That is the reason why curcumin is considered as a potent chemopreventive antioxidant agent [91].

Additionally, curcumin enhances cell antioxidant ability via increasing the activity of antioxidant enzymes such as glutathione peroxidase (GPx), glutathione reductase (GR), glutathione S-transferase (GST) and superoxide dismutase (SOD), glucose-6-phosphate dehydrogenase and catalase [92,93].

It was also reported that curcumin mediates nuclear transport of transcription factor NF2L2 (Nrf-2; Nuclear factor erythroid 2-related factor 2) which regulates antioxidant signaling pathway. NF2L2-targeted genes encode proteins that may be classified as the phase II xenobiotic-metabolizing antioxidant enzymes, playing a pivotal role in cancer prevention [94].

However, curcumin can also activate ROS formation and enhance oxidative stress in cancer cells. Curcumin-mediated rapid generation of ROS induces apoptosis via caspase activation and alterations in mitochondrial membrane potential followed by cytochrome c release. Thus, curcumin can initiate apoptosis and mediate chemosensitization of cancer cells [57,95].

In human breast cancer cells, a combination of arabinogalactan and curcumin significantly decreased cell growth without any significant effect on normal cells. This combination of compounds promoted apoptosis by increased ROS level, changed mitochondrial membrane potential and glutathione reduction. Moreover, in vivo mice studies indicated that the combination of curcumin and arabinogalactan inhibited the progression of breast tumors [96].

Interestingly, the turmeric rhizome also contains two potent antioxidants: vitamin C and vitamin E. The vitamins may intensify the antioxidative properties of curcumin by neutralizing free radicals of environmental carcinogens. Moreover, according to Minor's studies, vitamin C participates in the hydroxylation of 5-methylcytosine to 5-hydroxymethylcytosine in DNA and it probably takes part in the modulation of epigenetic control of genome activity leading to the active demethylation of DNA [97].

It is also necessary to mention about two important substances that are present in a turmeric rhizome: zinc ions and turmeric oil. Some studies show that zinc deficiency may result in reduced methyl group transfer from SAM to cytosine in the methylated gene. It can also complicate the binding of DNMT1 to DNA, which can lead to DNA demethylation and global DNA hypomethylation [98].

6. Discussion, Conclusions and Future Perspectives

Numerous data have shown that curcumin from turmeric rhizome can modulate our epigenome (Figure 3). It raises questions on its pharmacological applications in epigenetic chemoprevention and anti-cancer therapy. From the experimental evidence discussed in the present review, epigenetic modifications, mainly DNA methylation, are one of the mechanisms by which curcumin inhibits breast cancer cell growth.

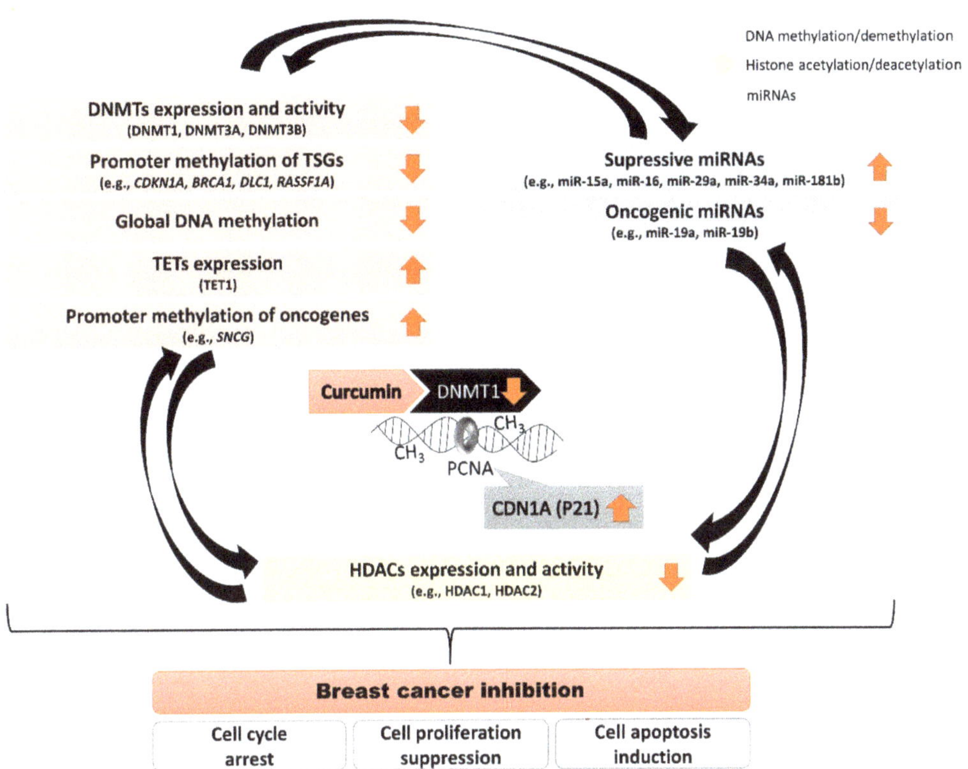

Figure 3. Scheme demonstrating mechanisms used by curcumin to drive changes in the epigenome in breast cancer inhibition. Curcumin binds to the DNMT1 catalytic domain and impairs its enzymatic activity. PCNA is crucial for DNMT1 activity during replication when DNA methylation pattern is copied from a parental to a daughter DNA strand. CDN1A (P21 encoded by *CDKN1A* gene) competes with DNMT1 for the same binding site on PCNA, which impairs DNMT1 activity. Curcumin that leads to an increase in P21 expression may affect DNA methylation. Interconnections between the components of the epigenome: DNA methylation, histone modifications and miRNAs. Curcumin driving changes in DNA methylation patterns in breast cancer cells may have indirect effects on other epigenetic components (histone modifications and miRNAs) and vice versa.

Aberrant epigenetic processes lead to dysregulated expression of many genes, with the upregulated oncogenes and silenced TSGs at the different stages of breast cancer development. Curcumin binds to the DNMT1 catalytic domain and impairs its enzymatic activity. Moreover, curcumin-mediated CDN1A (P21 protein encoded by *CDKN1A* gene) upregulation impairs DNMT1 activity. Since PCNA is crucial for DNMT1 activity during replication when DNA methylation pattern is copied from a parental to a daughter DNA strand, curcumin-reactivated CDN1A (P21) competes with DNMT1 for the same binding site on PCNA. Hence, curcumin that leads to a relevant increase in P21 expression may affect DNA methylation processes. The hitherto studies in different breast cancer models have shown that curcumin exposure caused relevant changes in the expression profile of genes encoding DNA methylating/demethylating enzymes (DNMTs (mostly DNMT1) downregulation and TET1 upregulation), and histone deacetylating enzymes (HDAC1 and HDAC2 downregulation), and alterations in the activity of numerous miRNAs. Therefore, the interconnections between the components of the epigenome: DNA methylation, histone modifications and miRNAs have to be highlighted. Curcumin driving changes in DNA methylation patterns in breast cancer cells may have indirect effects on other epigenetic components (histone modifications and miRNAs) and vice versa.

Recent studies have shown that, as compared to most single-target drugs, the majority of bioactive compounds, including curcumin do not target a single protein or pathway but affects multiple cell signaling pathways. It may allow phytochemicals such as curcumin to evade the development of resistance due to the activation of supporting alternative pathways. Curcumin via epigenetic anti-cancer activity may lead to re-activation of numerous DNA-methylation-silenced TSGs and downregulation of oncogenes via promoter hypermethylation, and the proteins encoded by these genes might be the negative or positive regulators of various intracellular oncogenic signaling pathways. Moreover, curcumin by modulating the activity of different transcription factors, such as AP-1, NF-κB/SP1 or TP53, may affect the transcriptional activity of *DNMT1*, as the AP-1, NF-κB/SP1 or TP53 binding sites have been identified within the DNMT1 regulatory region. It seems like the curcumin would have the potential to alter "cancer-specific" DNA methylation pattern and gene expression profile towards "normal-like". Cell-type-specific effects of curcumin are remarkable in many types of cancer, including breast cancer and only continued research can allow a better understanding of cell signaling pathways targeted by this potent anti-cancer agent.

Thus, curcumin showed promising chemopreventive effects in the laboratory experiments, but its clinical application is still limited because of low water solubility and low metabolic stability. Nevertheless, the numerous studies undergoing in various laboratories and clinics intensively working on evolving novel phytochemical-based treatment options for breast cancer implies that phytochemicals such as curcumin are the leading molecules for future anti-cancer drug development targeting breast cancer-related signaling pathways. Moreover, the effective doses of curcumin have not been shown to exert any toxicities or side effects making this bioactive compound ideal preventative and anti-cancer agent.

Thus, the main challenges in the investigation of curcumin as an epigenetic anti-cancer agent are new strategies of increasing its bioavailability, assessing the efficacy of the metabolites and determining the role of this phytochemical alone or in combination with other turmeric-derived compounds and existing drugs in improving anti-cancer efficacy. These important issues should be addressed in future nutriepigenomic studies. Whilst there is much more to be done, the available data so far indicate that curcumin can be a potential target of drug development against cancer.

Author Contributions: All the authors contributed substantially to the work reported. Conceptualization, K.F.-M., A.K.-S. and K.L.; methodology, K.F.-M., A.K.-S. and K.L.; writing—original draft preparation, K.F.-M., A.K.-S., A.S.-K., J.J. and K.L.; writing—review and editing, K.F.-M., A.K.-S. A.S.-K., J.J., G.M. and K.L.; visualization, A.K.-S., A.S.-K. and K.L.; supervision, K.F.-M., A.K.-S. and K.L.; funding acquisition, K.F.-M. and K.L. All authors have read and agreed to the published version of the manuscript.

Funding: This work was funded by the Medical University of Lodz, grant number 503/6-099-01/503-61-001-19-00.

Conflicts of Interest: The authors declare no conflict of interest.

References

1. Bray, F.; Ferlay, J.; Soerjomataram, I.; Siegel, R.L.; Torre, L.A.; Jemal, A. Global cancer statistics 2018: GLOBOCAN estimates of incidence and mortality worldwide for 36 cancers in 185 countries. *CA Cancer J. Clin.* **2018**, *68*, 394–424. [CrossRef] [PubMed]
2. Bray, F.; Ferlay, J.; Soerjomataram, I.; Siegel, R.L.; Torre, L.A.; Jemal, A. Erratum: Global cancer statistics 2018: GLOBOCAN estimates of incidence and mortality worldwide for 36 cancers in 185 countries. *CA. Cancer J. Clin.* **2020**. [CrossRef]
3. Sørlie, T.; Perou, C.M.; Tibshirani, R.; Aas, T.; Geisler, S.; Johnsen, H.; Hastie, T.; Eisen, M.B.; van de Rijn, M.; Jeffrey, S.S.; et al. Gene expression patterns of breast carcinomas distinguish tumor subclasses with clinical implications. *Proc. Natl. Acad. Sci. USA* **2001**, *98*, 10869–10874. [CrossRef]
4. Sotiriou, C.; Neo, S.Y.; McShane, L.M.; Korn, E.L.; Long, P.M.; Jazaeri, A.; Martiat, P.; Fox, S.B.; Harris, A.L.; Liu, E.T. Breast cancer classification and prognosis based on gene expression profiles from a population-based study. *Proc. Natl. Acad. Sci. USA* **2003**, *100*, 10393–10398. [CrossRef] [PubMed]

5. Moo, T.A.; Sanford, R.; Dang, C.; Morrow, M. Overview of breast cancer therapy. *PET Clin.* **2018**, *13*, 339–354. [CrossRef]
6. Singhal, S.K.; Usmani, N.; Michiels, S.; Metzger-Filho, O.; Saini, K.S.; Kovalchuk, O.; Parliament, M. Towards understanding the breast cancer epigenome: A comparison of genome-wide DNA methylation and gene expression data. *Oncotarget* **2016**, *7*, 3002–3017. [CrossRef]
7. Young, G.P.; Le Leu, R.K. Preventing cancer: Dietary lifestyle or clinical intervention? Preventing cancer: Diet or screening? *Asia Pac. J. Clin. Nutr.* **2002**, *11*, S618–S631. [CrossRef]
8. Chlebowski, R.T.; Chen, Z.; Anderson, G.L.; Rohan, T.; Aragaki, A.; Lane, D.; Dolan, N.C.; Paskett, E.D.; McTiernan, A.; Hubbell, F.A.; et al. Ethnicity and breast cancer: Factors influencing differences in incidence and outcome. *JNCI J. Natl. Cancer Inst.* **2005**, *97*, 439–448. [CrossRef]
9. Youlden, D.R.; Cramb, S.M.; Yip, C.H.; Baade, P.D. Incidence and mortality of female breast cancer in the Asia-Pacific region. *Cancer Biol. Med.* **2014**, *11*, 101–115. [CrossRef]
10. Hewlings, S.J.; Kalman, D.S. Curcumin: A review of its' effects on human health. *Foods* **2017**, *6*, 92. [CrossRef]
11. Dandawate, P.R.; Subramaniam, D.; Jensen, R.A.; Anant, S. Targeting cancer stem cells and signaling pathways by phytochemicals: Novel approach for breast cancer therapy. *Semin. Cancer Biol.* **2016**, *40–41*, 192–208. [CrossRef] [PubMed]
12. Boyanapalli, S.S.S.; Kong, A.-N.T. "Curcumin, the king of spices": Epigenetic regulatory mechanisms in the prevention of cancer, neurological, and inflammatory diseases. *Curr. Pharmacol. Rep.* **2015**, *1*, 129–139. [CrossRef] [PubMed]
13. Stefanska, B.; Karlic, H.; Varga, F.; Fabianowska-Majewska, K.; Haslberger, A. Epigenetic mechanisms in anti-cancer actions of bioactive food components—The implications in cancer prevention. *Br. J. Pharmacol.* **2012**, *167*, 279–297. [CrossRef] [PubMed]
14. Banik, U.; Parasuraman, S.; Adhikary, A.K.; Othman, N.H. Curcumin: The spicy modulator of breast carcinogenesis. *J. Exp. Clin. Cancer Res. CR* **2017**, *36*. [CrossRef] [PubMed]
15. Madhavi, D.; Kagan, D. Bioavailability of a sustained release formulation of curcumin. *Integr. Med.* **2014**, *13*, 7.
16. Thorat, B.; Jangle, R. Reversed-phase high-performance liquid chromatography method for analysis of curcuminoids and curcuminoid-loaded liposome formulation. *Indian J. Pharm. Sci.* **2013**, *75*, 60. [CrossRef]
17. Ganiger, S.; Malleshappa, H.N.; Krishnappa, H.; Rajashekhar, G.; Ramakrishna Rao, V.; Sullivan, F. A two generation reproductive toxicity study with curcumin, turmeric yellow, in Wistar rats. *Food Chem. Toxicol.* **2007**, *45*, 64–69. [CrossRef]
18. Joint FAO/WHO Expert Committee on Food Additives. *Safety Evaluation of Certain Food Additives and Contaminants*; World Health Organization (IPCS): Geneva, Switzerland, 2004; ISBN 978-92-4-166052-5.
19. FoodData Central. Available online: https://fdc.nal.usda.gov/fdc-app.html#/food-details/172231/nutrients (accessed on 26 November 2020).
20. Kurita, T.; Makino, Y. Novel curcumin oral delivery systems. *Anticancer Res.* **2013**, *33*, 2807–2821.
21. Schiborr, C.; Kocher, A.; Behnam, D.; Jandasek, J.; Toelstede, S.; Frank, J. The oral bioavailability of curcumin from micronized powder and liquid micelles is significantly increased in healthy humans and differs between sexes. *Mol. Nutr. Food Res.* **2014**, *58*, 516–527. [CrossRef]
22. Pan, M.-H.; Huang, T.-M.; Lin, J.-K. Biotransformation of curcumin through reduction and glucuronidation in mice. *Drug Metab. Dispos.* **1999**, *27*, 486–494.
23. Frank, J.; Schiborr, C.; Kocher, A.; Meins, J.; Behnam, D.; Schubert-Zsilavecz, M.; Abdel-Tawab, M. Transepithelial transport of curcumin in caco-2 cells is significantly enhanced by micellar solubilisation. *Plant Foods Hum. Nutr.* **2017**, *72*, 48–53. [CrossRef] [PubMed]
24. Bolat, Z.B.; Islek, Z.; Demir, B.N.; Yilmaz, E.N.; Sahin, F.; Ucisik, M.H. Curcumin- and piperine-loaded emulsomes as combinational treatment approach enhance the anticancer activity of curcumin on HCT116 colorectal cancer model. *Front. Bioeng. Biotechnol.* **2020**, *8*. [CrossRef] [PubMed]
25. Thulasidasan, A.K.T.; Retnakumari, A.P.; Shankar, M.; Vijayakurup, V.; Anwar, S.; Thankachan, S.; Pillai, K.S.; Pillai, J.J.; Nandan, C.D.; Alex, V.V.; et al. Folic acid conjugation improves the bioavailability and chemosensitizing efficacy of curcumin-encapsulated PLGA-PEG nanoparticles towards paclitaxel chemotherapy. *Oncotarget* **2017**, *8*, 107374–107389. [CrossRef] [PubMed]
26. Liu, W.; Zhai, Y.; Heng, X.; Che, F.Y.; Chen, W.; Sun, D.; Zhai, G. Oral bioavailability of curcumin: Problems and advancements. *J. Drug Target.* **2016**, *24*, 694–702. [CrossRef] [PubMed]
27. Baylin, S.B.; Esteller, M.; Rountree, M.R.; Bachman, K.E.; Schuebel, K.; Herman, J.G. Aberrant patterns of DNA methylation, chromatin formation and gene expression in cancer. *Hum. Mol. Genet.* **2001**, *10*, 687–692. [CrossRef]
28. Das, P.M.; Singal, R. DNA methylation and cancer. *J. Clin. Oncol. Off. J. Am. Soc. Clin. Oncol.* **2004**, *22*, 4632–4642. [CrossRef]
29. Ehrlich, M. DNA hypomethylation in cancer cells. *Epigenomics* **2009**, *1*, 239–259. [CrossRef]
30. Newell-Price, J.; Clark, A.J.; King, P. DNA methylation and silencing of gene expression. *Trends Endocrinol. Metab. TEM* **2000**, *11*, 142–148. [CrossRef]
31. Esteller, M. CpG island hypermethylation and tumor suppressor genes: A booming present, a brighter future. *Oncogene* **2002**, *21*, 5427–5440. [CrossRef]
32. Hermann, A.; Gowher, H.; Jeltsch, A. Biochemistry and biology of mammalian DNA methyltransferases. *Cell. Mol. Life Sci. CMLS* **2004**, *61*, 2571–2587. [CrossRef]
33. Jones, P.A.; Baylin, S.B. The fundamental role of epigenetic events in cancer. *Nat. Rev. Genet.* **2002**, *3*, 415–428. [CrossRef] [PubMed]

34. Kang, J.H.; Kim, S.J.; Noh, D.Y.; Park, I.A.; Choe, K.J.; Yoo, O.J.; Kang, H.S. Methylation in the p53 promoter is a supplementary route to breast carcinogenesis: Correlation between CpG methylation in the p53 promoter and the mutation of the p53 gene in the progression from ductal carcinoma in situ to invasive ductal carcinoma. *Lab. Investig. J. Tech. Methods Pathol.* **2001**, *81*, 573–579. [CrossRef]
35. Veland, N.; Lu, Y.; Hardikar, S.; Gaddis, S.; Zeng, Y.; Liu, B.; Estecio, M.R.; Takata, Y.; Lin, K.; Tomida, M.W.; et al. DNMT3L facilitates DNA methylation partly by maintaining DNMT3A stability in mouse embryonic stem cells. *Nucleic Acids Res.* **2019**, *47*, 152–167. [CrossRef] [PubMed]
36. Goll, M.G.; Kirpekar, F.; Maggert, K.A.; Yoder, J.A.; Hsieh, C.-L.; Zhang, X.; Golic, K.G.; Jacobsen, S.E.; Bestor, T.H. Methylation of tRNAAsp by the DNA methyltransferase homolog Dnmt2. *Science* **2006**, *311*, 395–398. [CrossRef] [PubMed]
37. Girault, I.; Tozlu, S.; Lidereau, R.; Bièche, I. Expression analysis of DNA methyltransferases 1, 3A, and 3B in sporadic breast carcinomas. *Clin. Cancer Res. Off. J. Am. Assoc. Cancer Res.* **2003**, *9*, 4415–4422.
38. Iida, T.; Suetake, I.; Tajima, S.; Morioka, H.; Ohta, S.; Obuse, C.; Tsurimoto, T. PCNA clamp facilitates action of DNA cytosine methyltransferase 1 on hemimethylated DNA. *Genes Cells Devoted Mol. Cell. Mech.* **2002**, *7*, 997–1007. [CrossRef] [PubMed]
39. Chuang, L.S.; Ian, H.I.; Koh, T.W.; Ng, H.H.; Xu, G.; Li, B.F. Human DNA-(cytosine-5) methyltransferase-PCNA complex as a target for p21WAF1. *Science* **1997**, *277*, 1996–2000. [CrossRef]
40. Milutinovic, S.; Knox, J.D.; Szyf, M. DNA methyltransferase inhibition induces the transcription of the tumor suppressor p21(WAF1/CIP1/sdi1). *J. Biol. Chem.* **2000**, *275*, 6353–6359. [CrossRef]
41. Pradhan, S.; Kim, G.-D. The retinoblastoma gene product interacts with maintenance human DNA (cytosine-5) methyltransferase and modulates its activity. *EMBO J.* **2002**, *21*, 779–788. [CrossRef]
42. Rountree, M.R.; Bachman, K.E.; Baylin, S.B. DNMT1 binds HDAC2 and a new co-repressor, DMAP1, to form a complex at replication foci. *Nat. Genet.* **2000**, *25*, 269–277. [CrossRef]
43. Kimura, H.; Shiota, K. Methyl-CpG-binding protein, MeCP2, is a target molecule for maintenance DNA methyltransferase, Dnmt1. *J. Biol. Chem.* **2003**, *278*, 4806–4812. [CrossRef] [PubMed]
44. Tatematsu, K.I.; Yamazaki, T.; Ishikawa, F. MBD2-MBD3 complex binds to hemi-methylated DNA and forms a complex containing DNMT1 at the replication foci in late S phase. *Genes Cells Devoted Mol. Cell. Mech.* **2000**, *5*, 677–688. [CrossRef] [PubMed]
45. Fuks, F.; Burgers, W.A.; Brehm, A.; Hughes-Davies, L.; Kouzarides, T. DNA methyltransferase Dnmt1 associates with histone deacetylase activity. *Nat. Genet.* **2000**, *24*, 88–91. [CrossRef] [PubMed]
46. Zardo, G.; Reale, A.; Passananti, C.; Pradhan, S.; Buontempo, S.; De Matteis, G.; Adams, R.L.P.; Caiafa, P. Inhibition of poly(ADP-ribosyl)ation induces DNA hypermethylation: A possible molecular mechanism. *FASEB J. Off. Publ. Fed. Am. Soc. Exp. Biol.* **2002**, *16*, 1319–1321. [CrossRef]
47. Wu, Y.; Sarkissyan, M.; Vadgama, J.V. Epigenetics in breast and prostate cancer. *Methods Mol. Biol.* **2015**, *1238*, 425–466. [CrossRef]
48. Rasmussen, K.D.; Helin, K. Role of TET enzymes in DNA methylation, development, and cancer. *Genes Dev.* **2016**, *30*, 733–750. [CrossRef]
49. Yang, L.; Yu, S.-J.; Hong, Q.; Yang, Y.; Shao, Z.-M. Reduced Expression of TET1, TET2, TET3 and TDG mRNAs are associated with poor prognosis of patients with early breast cancer. *PLoS ONE* **2015**, *10*, e0133896. [CrossRef]
50. Yang, H.; Liu, Y.; Bai, F.; Zhang, J.-Y.; Ma, S.-H.; Liu, J.; Xu, Z.-D.; Zhu, H.-G.; Ling, Z.-Q.; Ye, D.; et al. Tumor development is associated with decrease of TET gene expression and 5-methylcytosine hydroxylation. *Oncogene* **2013**, *32*, 663–669. [CrossRef]
51. Zheng, J.; Zhou, Y.; Li, Y.; Xu, D.-P.; Li, S.; Li, H.-B. Spices for prevention and treatment of cancers. *Nutrients* **2016**, *8*, 495. [CrossRef]
52. Hu, S.; Xu, Y.; Meng, L.; Huang, L.; Sun, H. Curcumin inhibits proliferation and promotes apoptosis of breast cancer cells. *Exp. Ther. Med.* **2018**, *16*, 1266–1272. [CrossRef]
53. Wang, Y.; Yu, J.; Cui, R.; Lin, J.; Ding, X. Curcumin in treating breast cancer: A review. *J. Lab. Autom.* **2016**, *21*, 723–731. [CrossRef] [PubMed]
54. Liu, J.-L.; Pan, Y.-Y.; Chen, O.; Luan, Y.; Xue, X.; Zhao, J.-J.; Liu, L.; Jia, H.-Y. Curcumin inhibits MCF-7 cells by modulating the NF-κB signaling pathway. *Oncol. Lett.* **2017**, *14*, 5581–5584. [CrossRef] [PubMed]
55. Kunwar, A.; Barik, A.; Mishra, B.; Rathinasamy, K.; Pandey, R.; Priyadarsini, K.I. Quantitative cellular uptake, localization and cytotoxicity of curcumin in normal and tumor cells. *Biochimica et Biophysica Acta* **2008**, *1780*, 673–679. [CrossRef]
56. Ravindran, J.; Prasad, S.; Aggarwal, B.B. Curcumin and cancer cells: How many ways can curry kill tumor cells selectively? *AAPS J.* **2009**, *11*, 495–510. [CrossRef]
57. Syng-Ai, C.; Kumari, A.L.; Khar, A. Effect of curcumin on normal and tumor cells: Role of glutathione and bcl-2. *Mol. Cancer Ther.* **2004**, *3*, 1101–1108. [PubMed]
58. Abramczyk, H.; Surmacki, J.; Kopeć, M.; Olejnik, A.K.; Lubecka-Pietruszewska, K.; Fabianowska-Majewska, K. The role of lipid droplets and adipocytes in cancer. Raman imaging of cell cultures: MCF10A, MCF7, and MDA-MB-231 compared to adipocytes in cancerous human breast tissue. *Analyst* **2015**, *140*, 2224–2235. [CrossRef] [PubMed]
59. Mirza, S.; Sharma, G.; Parshad, R.; Gupta, S.D.; Pandya, P.; Ralhan, R. Expression of DNA methyltransferases in breast cancer patients and to analyze the effect of natural compounds on DNA methyltransferases and associated proteins. *J. Breast Cancer* **2013**, *16*, 23–31. [CrossRef] [PubMed]

60. Yu, Z.; Xiao, Q.; Zhao, L.; Ren, J.; Bai, X.; Sun, M.; Wu, H.; Liu, X.; Song, Z.; Yan, Y.; et al. DNA methyltransferase 1/3a overexpression in sporadic breast cancer is associated with reduced expression of estrogen receptor-alpha/breast cancer susceptibility gene 1 and poor prognosis. *Mol. Carcinog.* **2015**, *54*, 707–719. [CrossRef] [PubMed]
61. Al-Kharashi, L.A.; Al-Mohanna, F.H.; Tulbah, A.; Aboussekhra, A. The DNA methyl-transferase protein DNMT1 enhances tumor-promoting properties of breast stromal fibroblasts. *Oncotarget* **2018**, *9*, 2329–2343. [CrossRef]
62. Liu, Z.; Xie, Z.; Jones, W.; Pavlovicz, R.E.; Liu, S.; Yu, J.; Li, P.; Lin, J.; Fuchs, J.R.; Marcucci, G.; et al. Curcumin is a potent DNA hypomethylation agent. *Bioorg. Med. Chem. Lett.* **2009**, *19*, 706–709. [CrossRef]
63. Chatterjee, B.; Ghosh, K.; Kanade, S.R. Curcumin-mediated demethylation of the proximal promoter CpG island enhances the KLF4 recruitment that leads to increased expression of p21Cip1 in vitro. *J. Cell. Biochem.* **2019**, *120*, 809–820. [CrossRef] [PubMed]
64. Al-Yousef, N.; Shinwari, Z.; Al-Shahrani, B.; Al-Showimi, M.; Al-Moghrabi, N. Curcumin induces re-expression of BRCA1 and suppression of γ synuclein by modulating DNA promoter methylation in breast cancer cell lines. *Oncol. Rep.* **2020**, *43*, 827–838. [CrossRef] [PubMed]
65. Liu, Y.; Zhou, J.; Hu, Y.; Wang, J.; Yuan, C. Curcumin inhibits growth of human breast cancer cells through demethylation of DLC1 promoter. *Mol. Cell. Biochem.* **2017**, *425*, 47–58. [CrossRef] [PubMed]
66. Du, L.; Xie, Z.; Wu, L.; Chiu, M.; Lin, J.; Chan, K.K.; Liu, S.; Liu, Z. Reactivation of RASSF1A in breast cancer cells by curcumin. *Nutr. Cancer* **2012**, *64*, 1228–1235. [CrossRef] [PubMed]
67. Mukherjee, S.; Sarkar, R.; Biswas, J.; Roy, M. Curcumin inhibits histone deacetylase leading to cell cycle arrest and apoptosis via upregulation of p21 in breast cancer cell lines. *Int. J. Green Nanotechnol.* **2012**, *4*, 183–197. [CrossRef]
68. Li, X.; Xie, W.; Xie, C.; Huang, C.; Zhu, J.; Liang, Z.; Deng, F.; Zhu, M.; Zhu, W.; Wu, R.; et al. Curcumin modulates miR-19/PTEN/AKT/p53 axis to suppress bisphenol A-induced MCF-7 breast cancer cell proliferation. *Phytother. Res. PTR* **2014**, *28*, 1553–1560. [CrossRef] [PubMed]
69. Guo, J.; Li, W.; Shi, H.; Xie, X.; Li, L.; Tang, H.; Wu, M.; Kong, Y.; Yang, L.; Gao, J.; et al. Synergistic effects of curcumin with emodin against the proliferation and invasion of breast cancer cells through upregulation of miR-34a. *Mol. Cell. Biochem.* **2013**, *382*, 103–111. [CrossRef]
70. Kronski, E.; Fiori, M.E.; Barbieri, O.; Astigiano, S.; Mirisola, V.; Killian, P.H.; Bruno, A.; Pagani, A.; Rovera, F.; Pfeffer, U.; et al. miR181b is induced by the chemopreventive polyphenol curcumin and inhibits breast cancer metastasis via down-regulation of the inflammatory cytokines CXCL1 and -2. *Mol. Oncol.* **2014**, *8*, 581–595. [CrossRef]
71. Yang, J.; Cao, Y.; Sun, J.; Zhang, Y. Curcumin reduces the expression of Bcl-2 by upregulating miR-15a and miR-16 in MCF-7 cells. *Med. Oncol.* **2010**, *27*, 1114–1118. [CrossRef]
72. Bora-Tatar, G.; Dayangaç-Erden, D.; Demir, A.S.; Dalkara, S.; Yelekçi, K.; Erdem-Yurter, H. Molecular modifications on carboxylic acid derivatives as potent histone deacetylase inhibitors: Activity and docking studies. *Bioorg. Med. Chem.* **2009**, *17*, 5219–5228. [CrossRef]
73. Norouzi, S.; Majeed, M.; Pirro, M.; Generali, D.; Sahebkar, A. Curcumin as an adjunct therapy and microRNA modulator in breast cancer. *Curr. Pharm. Des.* **2018**, *24*, 171–177. [CrossRef] [PubMed]
74. Bimonte, S.; Barbieri, A.; Palma, G.; Rea, D.; Luciano, A.; D'Aiuto, M.; Arra, C.; Izzo, F. Dissecting the role of curcumin in tumour growth and angiogenesis in mouse model of human breast cancer. *BioMed Res. Int.* **2015**, *2015*, 878134. [CrossRef] [PubMed]
75. Bachmeier, B.; Nerlich, A.G.; Iancu, C.M.; Cilli, M.; Schleicher, E.; Vené, R.; Dell'Eva, R.; Jochum, M.; Albini, A.; Pfeffer, U. The chemopreventive polyphenol curcumin prevents hematogenous breast cancer metastases in immunodeficient mice. *Cell. Physiol. Biochem. Int. J. Exp. Cell. Physiol. Biochem. Pharmacol.* **2007**, *19*, 137–152. [CrossRef]
76. Zhou, Q.; Wang, X.; Liu, X.; Zhang, H.; Lu, Y.; Su, S. Curcumin enhanced antiproliferative effect of mitomycin C in human breast cancer MCF-7 cells in vitro and in vivo. *Acta Pharmacologica Sinica* **2011**, *32*, 1402–1410. [CrossRef]
77. Bayet-Robert, M.; Kwiatkowski, F.; Leheurteur, M.; Gachon, F.; Planchat, E.; Abrial, C.; Mouret-Reynier, M.-A.; Durando, X.; Barthomeuf, C.; Chollet, P. Phase I dose escalation trial of docetaxel plus curcumin in patients with advanced and metastatic breast cancer. *Cancer Biol. Ther.* **2010**, *9*, 8–14. [CrossRef]
78. Saghatelyan, T.; Tananyan, A.; Janoyan, N.; Tadevosyan, A.; Petrosyan, H.; Hovhannisyan, A.; Hayrapetyan, L.; Arustamyan, M.; Arnhold, J.; Rotmann, A.-R.; et al. Efficacy and safety of curcumin in combination with paclitaxel in patients with advanced, metastatic breast cancer: A comparative, randomized, double-blind, placebo-controlled clinical trial. *Phytomedicine* **2020**, *70*, 153218. [CrossRef]
79. Soda, K. Polyamine metabolism and gene methylation in conjunction with one-carbon metabolism. *Int. J. Mol. Sci.* **2018**, *19*, 3106. [CrossRef]
80. Li, Y.; Tollefsbol, T.O. Impact on DNA methylation in cancer prevention and therapy by bioactive dietary components. *Curr. Med. Chem.* **2010**, *17*, 2141–2151. [CrossRef]
81. Yi, P.; Melnyk, S.; Pogribna, M.; Pogribny, I.P.; Hine, R.J.; James, S.J. Increase in plasma homocysteine associated with parallel increases in plasma S-adenosylhomocysteine and lymphocyte DNA hypomethylation. *J. Biol. Chem.* **2000**, *275*, 29318–29323. [CrossRef]
82. Waterland, R.A. Assessing the effects of high methionine intake on DNA methylation. *J. Nutr.* **2006**, *136*, 1706S–1710S. [CrossRef]
83. James, S.J.; Pogribny, I.P.; Pogribna, M.; Miller, B.J.; Jernigan, S.; Melnyk, S. Mechanisms of DNA damage, DNA hypomethylation, and tumor progression in the folate/methyl-deficient rat model of hepatocarcinogenesis. *J. Nutr.* **2003**, *133*, 3740S–3747S. [CrossRef] [PubMed]

84. Davis, C.D.; Uthus, E.O. DNA methylation, cancer susceptibility, and nutrient interactions. *Exp. Biol. Med.* **2004**, *229*, 988–995 [CrossRef]
85. Duthie, S.J. Folate and cancer: How DNA damage, repair and methylation impact on colon carcinogenesis. *J. Inherit. Metab. Dis* **2011**, *34*, 101–109. [CrossRef] [PubMed]
86. Crider, K.S.; Yang, T.P.; Berry, R.J.; Bailey, L.B. Folate and DNA methylation: A review of molecular mechanisms and the evidence for folate's role. *Adv. Nutr.* **2012**, *3*, 21–38. [CrossRef] [PubMed]
87. Liu, Z.; Huang, P.; Law, S.; Tian, H.; Leung, W.; Xu, C. Preventive effect of curcumin against chemotherapy-induced side-effects *Front. Pharmacol.* **2018**, *9*, 1374. [CrossRef] [PubMed]
88. Maruti, S.S.; Ulrich, C.M.; White, E. Folate and one-carbon metabolism nutrients from supplements and diet in relation to breast cancer risk. *Am. J. Clin. Nutr.* **2009**, *89*, 624–633. [CrossRef]
89. Shrubsole, M.J.; Shu, X.O.; Li, H.-L.; Cai, H.; Yang, G.; Gao, Y.-T.; Gao, J.; Zheng, W. Dietary B vitamin and methionine intakes and breast cancer risk among Chinese women. *Am. J. Epidemiol.* **2011**, *173*, 1171–1182. [CrossRef]
90. Liu, J.J.; Ward, R.L. Folate and one-carbon metabolism and its impact on aberrant DNA methylation in cancer. *Adv. Genet.* **2010** *71*, 79–121. [CrossRef]
91. Menon, V.P.; Sudheer, A.R. Antioxidant and anti-inflammatory properties of curcumin. *Adv. Exp. Med. Biol.* **2007**, *595*, 105–125 [CrossRef]
92. Liu, Y.; Wu, Y.; Zhang, P. Protective effects of curcumin and quercetin during benzo (a) pyrene induced lung carcinogenesis in mice. *Eur. Rev. Med. Pharmacol. Sci.* **2015**, *19*, 1736–1743.
93. Gupta, N.; Verma, K.; Nalla, S.; Kulshreshtha, A.; Lall, R.; Prasad, S. Free radicals as a double-edged sword: The cancer preventive and therapeutic roles of curcumin. *Molecules* **2020**, *25*, 5390. [CrossRef] [PubMed]
94. Tu, Z.-S.; Wang, Q.; Sun, D.-D.; Dai, F.; Zhou, B. Design, synthesis, and evaluation of curcumin derivatives as Nrf2 activators and cytoprotectors against oxidative death. *Eur. J. Med. Chem.* **2017**, *134*, 72–85. [CrossRef] [PubMed]
95. Khan, M.A.; Gahlot, S.; Majumdar, S. Oxidative stress induced by curcumin promotes the death of cutaneous T-cell lymphoma (HuT-78) by disrupting the function of several molecular targets. *Mol. Cancer Ther.* **2012**, *11*, 1873–1883. [CrossRef] [PubMed]
96. Moghtaderi, H.; Sepehri, H.; Attari, F. Combination of arabinogalactan and curcumin induces apoptosis in breast cancer cells in vitro and inhibits tumor growth via overexpression of p53 level in vivo. *Biomed. Pharmacother.* **2017**, *88*, 582–594. [CrossRef]
97. Minor, E.A.; Court, B.L.; Young, J.I.; Wang, G. Ascorbate induces ten-eleven translocation (Tet) methylcytosine dioxygenase mediated generation of 5-hydroxymethylcytosine. *J. Biol. Chem.* **2013**, *288*, 13669–13674. [CrossRef]
98. Kadayifci, F.Z.; Zheng, S.; Pan, Y.-X. Molecular mechanisms underlying the link between diet and DNA methylation. *Int. J. Mol Sci.* **2018**, *19*, 4055. [CrossRef]

Review

Curcumin: Could This Compound Be Useful in Pregnancy and Pregnancy-Related Complications?

Tiziana Filardi [1,*], Rosaria Varì [2], Elisabetta Ferretti [3], Alessandra Zicari [4], Susanna Morano [1] and Carmela Santangelo [2]

1 Department of Experimental Medicine, Section of Medical Pathophysiology, Food Science and Endocrinology, Sapienza University, Viale Regina Elena 324, 00161 Rome, Italy; susanna.morano@uniroma1.it
2 Center for Gender-Specific Medicine, Gender Specific Prevention and Health Unit, Istituto Superiore di Sanità, Viale Regina Elena 299, 00161 Rome, Italy; rosaria.vari@iss.it (R.V.); carmela.santangelo@iss.it (C.S.)
3 Oncogenomic Unit, Department of Experimental Medicine, Sapienza University, Viale Regina Elena 324, 00161 Rome, Italy; elisabetta.ferretti@uniroma1.it
4 Department of Experimental Medicine, 2nd Section of Cell Pathology, Sapienza University, Viale Regina Elena 324, 00161 Rome, Italy; alessandra.zicari@uniroma1.it
* Correspondence: tiziana.filardi@uniroma1.it

Received: 22 September 2020; Accepted: 12 October 2020; Published: 17 October 2020

Abstract: Curcumin, the main polyphenol contained in turmeric root (*Curcuma longa*), has played a significant role in medicine for centuries. The growing interest in plant-derived substances has led to increased consumption of them also in pregnancy. The pleiotropic and multi-targeting actions of curcumin have made it very attractive as a health-promoting compound. In spite of the beneficial effects observed in various chronic diseases in humans, limited and fragmentary information is currently available about curcumin's effects on pregnancy and pregnancy-related complications. It is known that immune-metabolic alterations occurring during pregnancy have consequences on both maternal and fetal tissues, leading to short- and long-term complications. The reported anti-inflammatory, antioxidant, antitoxicant, neuroprotective, immunomodulatory, antiapoptotic, antiangiogenic, anti-hypertensive, and antidiabetic properties of curcumin appear to be encouraging, not only for the management of pregnancy-related disorders, including gestational diabetes mellitus (GDM), preeclampsia (PE), depression, preterm birth, and fetal growth disorders but also to contrast damage induced by natural and chemical toxic agents. The current review summarizes the latest data, mostly obtained from animal models and in vitro studies, on the impact of curcumin on the molecular mechanisms involved in pregnancy pathophysiology, with the aim to shed light on the possible beneficial and/or adverse effects of curcumin on pregnancy outcomes.

Keywords: curcumin; pregnancy; pregnancy complications; postpartum depression; fetal development; preterm birth; adverse effects

1. Introduction

Maternal nutrition is an essential and modifiable environmental factor that deeply influences maternal and offspring health in the short and long-term [1–6]. Genetics, nutrition, and other environmental factors significantly contribute to the physiological immune and metabolic modifications occurring in pregnancy, to favor maternal adaptation to the growing and developing fetus. Maternal malnutrition adversely affects these dynamic processes by acting on the mechanisms related to the nutritional programming, including nutrition sensing signals, epigenetic regulation, gut microbiome, as well as on the nutrient-nutrient and nutrient-drug interactions, modulating maternal and fetal genes in a sex-specific manner [3,6–9].

Over the last decades, the advantages of a healthy diet, rich in fruit and vegetables, have been widely explored, highlighting that culinary herbs and spices might also effectively reduce the risk of developing chronic diseases [10]. Among them, curcumin, a compound extracted from the rhizome of *Curcuma longa*, has been extensively studied in light of a wide range of properties, including anti-inflammatory, antioxidant, anti-toxicant, antiapoptotic, immunomodulatory, neuroprotective, hepatoprotective, antiangiogenic, anti-hypertensive, and antidiabetic activities, emerging as a candidate therapeutic agent for several diseases [10–13]. Data from animal and in vitro studies provided evidence that curcumin might be effective in counteracting the adverse programming processes in pregnancy.

The known pathophysiological mechanisms underlying pregnancy and the most common pregnancy-related complications, such as gestational diabetes mellitus (GDM) [14], hypertension and preeclampsia [15], fetal growth disorders [16], as well as the damage induced by natural and chemical toxic agents [12] seem to be positively modulated by curcumin, although observed in in vitro and animal studies.

Additionally, promising results from preclinical studies on the use of curcumin in different neurological disorders [17] suggest a potential role in the treatment of post-partum depression (PPD) as well, a largely underestimated pregnancy-related disorder [18].

Harmful effects of curcumin on embryo development in the early stages of pregnancy have also been observed in animal studies [19]. Hence, the increasing consumption of natural products during pregnancy requires particular attention, considering the complexity of the largely unknown processes underlying maternal adaptation and fetal development.

We conducted a comprehensive literature search until 22 July 2020 using PubMed; and found a good number of articles in English, using the keywords "pregnancy", "pregnancy complications", "gestational diabetes", "preeclampsia", "reproductive toxicity", "post-partum depression", "placenta", "oocyte", "blastocyst", "embryo", "preterm labor", "fetal growth and development", in combination with the keywords "curcumin" and "dietary curcumin". The aim of this review is to provide an overview of both the potential health benefits and the possible adverse effects of curcumin in pregnancy and pregnancy-related complications.

2. Curcumin: Functions, Bioavailability, and Delivery

Curcumin, also called diferuloylmethane, is a lipophilic polyphenol extracted from the rhizome of *Curcuma Longa* (commonly known as turmeric). It has been widely used in traditional Indian and Chinese medicine for thousands of years [20]. The pharmacological effects of turmeric have been attributed mainly to curcuminoids, comprising curcumin and two related compounds, demethoxycurcumin and bisdemethoxycurcumin, which are contained in commercial curcumin [21]. Curcumin is a potent anti-inflammatory and antioxidant agent that exerts a myriad of biological activities by influencing multiple signaling pathways [10,11,13,22]. Curcumin is able to interact with a large number of molecular and cellular targets (as summarized in this recent review [13]) and regulates gene expression also by modulating epigenetic modifications (i.e., DNA methylation, histone modification, and microRNA expression) [23,24]. This compound, by mutually interacting with intestinal microflora, ameliorates gut microbiome dysbiosis, and influences the "gut–brain–microflora axis" to preserve and favor brain health [25,26]. The overall result of these different activities is the improvement in several disease states, including inflammatory, metabolic, endocrine, cardiovascular, gastrointestinal, neurological, respiratory, viral, skin diseases, and cancer, as highlighted by the impressive number of in vitro and in vivo studies summarized in recent papers [13,24,27,28]. Numerous clinical trials have shown good tolerability, safety, and efficacy of curcumin in the treatment of multiple chronic diseases—including cardiovascular diseases, diabetes, neurodegeneration, arthritis, and cancer—at doses up to 6–12 g/day [10,11,13]. In light of this, the United States Food and Drug Administration (FDA) has "Generally Recognized As Safe" (GRAS) curcumin as an ingredient in various food categories (0.5–100 mg/100 g) [29]; and the European Food Safety Authority (EFSA) Panel on Food Additives and Nutrient Sources added to Food (ANS), defined the Allowable Daily Intake (ADI) value of 0–3 mg/kg bw/day of curcumin as a food additive [30]. However, despite its potential therapeutic benefits, curcumin is poorly bioavailable due to its rapid metabolism, and the small portion of substance that is absorbed is extensively bio-transformed into its water-soluble metabolites, glucuronides, and sulfates [10]. Therefore, several strategies have been developed to enhance its bioavailability and efficacy, to increase oral and gastro-intestinal absorption, and to reduce the clearance from the body [31–33]. For this purpose, taking into consideration that curcumin is fat-soluble, several delivery systems have been developed to obtain a number of formulations by mixing curcumin with different materials, including adjuvants, such as piperine [32,33]. Micelles, liposomes, phospholipid complexes, phytosomes, emulsions, microemulsions, nano-emulsions, solid lipid nanoparticles, nanostructured lipid carriers, biopolymer nanoparticles, and microgels represent different and recent technical approaches to encapsulate curcumin [32–34], although further studies are needed to evaluate their effectiveness and safety as potential health-promoting compounds in humans.

3. Role of Curcumin in Pregnancy

3.1. Altered Glucose Metabolism

It is well known that dynamic changes in insulin sensitivity take place during healthy pregnancy to allow adequate supply to the growing fetus [35]. In pregnancy, several players, including hormones, cytokines, and metabolic factors, contribute to the development of insulin resistance through complex mechanisms, not yet completely understood [36,37]. Maternal obesity, related to unhealthy diet and lifestyle, can negatively affect insulin sensitivity leading to the development of GDM and type 2 diabetes (T2D), with serious short and long-term health consequences for both the mother and the offspring [38,39]. Recent evidence emphasized the anti-hyperglycemic activity of curcumin, both in animals and humans [40]. Specifically, this compound had the capability to improve glucose uptake, insulin sensitivity, and pancreatic β-cell function, as well as liver and kidney function, and to reduce glucose and lipid levels, oxidative stress, and inflammation [41], by interacting with almost all the players involved in these processes, as demonstrated in in vitro studies [13,41].

As regards human studies, the effects of curcumin supplementation have been evaluated in several randomized controlled trials. A recent intervention study showed that 1500 mg/day curcumin supplementation (500 mg capsules: 347 mg of curcumin, 84 mg of demethoxycurcumin, and 9 mg of bisdemethoxycurcumin) for 10 weeks reduced triglycerides (TG) and C-reactive protein (CRP), and increased adiponectin levels [42], whereas 500 mg/day curcumin co-administered with piperine 5 mg/day for three months was able to reduce blood glucose, C-peptide, glycated hemoglobin (HbA1c), alanine aminotransferase (ALT) and aspartate aminotransferase (AST), in patients with T2D [43]. Another study showed that the daily ingestion of 2100 mg turmeric powder for eight weeks resulted in a reduction in body weight, low density lipoprotein-cholesterol (LDL-c), and TG levels, with no significant effects on glycemia, CRP, and HbA1c, in hyperlipidemic T2D patients [44]. In obese women with polycystic ovary syndrome (PCOS), 1000 mg/day curcumin supplementation (500 mg twice daily: 70–80% curcumin, 15–20% demethoxycurcumin and 2.5–6.5% bisdemethoxycurcumin) for six weeks improved serum insulin and the Quantitative Insulin Sensitivity Check Index [45]. A recent meta-analysis reported that curcumin intake was associated with reduced body mass index (BMI), body weight, body fat, leptin value, and increased adiponectin levels in patients with metabolic syndrome and related disorders [46]. Overall, the dosage and duration of curcumin supplementation appear to differently modulate glucose metabolism in humans.

A recent promising approach to treat hyperglycemia consists of combining the effects of curcumin and the ongoing antidiabetic agents, as observed in diabetic rats treated with a combination of curcumin and metformin. Specifically, this association improved hyperglycemia, dyslipidemia, and oxidative stress, increasing the activity of the antioxidant enzyme paraoxonase 1 (PON1), in diabetic rats [47].

Dietary bioactive compounds might have beneficial effects on GDM [5,48]. In particular, curcumin appeared to improve GDM and GDM-related complications in a recent study in a mouse model. Specifically, C57 BL/KsJdb/+ diabetic pregnant mice were supplemented with different curcumin dosages: 50 mg/kg and 100 mg/kg/day, from gestational day zero (GD 0) to GD20. Results showed that 100 mg/kg curcumin significantly reduced blood glucose and insulin levels, increased hepatic glycogen content, and improved oxidative stress by reducing thiobarbituric acid reactive substance (TBARS) and increasing glutathione (GSH) levels, superoxide dismutase (SOD), and catalase (CAT) activities in the liver of diabetic pregnant mice at gestational day 20. The reduced 5′ adenosine monophosphate-activated protein kinase (AMPK) and increased Histone Deacetylase 4 HDAC4 activities observed in GDM liver were reverted by curcumin treatment. Furthermore, curcumin positively influenced the offspring of mothers with GDM, restoring litter size and birth weight, and inducing the reduction of glucose-6-phosphatase (G6Pase) expression and activity in the liver [14] (Table 1). Congenital birth defects, including neural tube defects (NTD), occur more often in the offspring of diabetic mothers. In a recent study, mouse embryos (at E8.5 of development) were cultured for 24 h with 100 mg/dL glucose, in the absence or presence of curcumin (10 and 20 μM). Remarkably, 20 μM curcumin was able to reduce the rate of embryos with NTD induced by high glucose. Curcumin reduced high glucose-induced oxidative and nitrosative stress [i.e., decreased 4-hydroxynonenal (4-HNE), nitrotyrosine levels, and lipid hydroperoxide (LPO)], as well as endoplasmic reticulum (ER) stress (i.e., decreased expression of ER-markers stress such as phosphorylated protein kinase-like endoplasmic reticulum kinase (p-PERK), phosphorylated inositol-requiring protein-1α (p-IRE1α), phosphorylated eukaryotic initiation factor 2α (p-eIF2α), C/EBP-homologous protein (CHOP), binding immunoglobulin protein (BiP), and x-box binding protein 1 (XBP1). Moreover, 20 μM curcumin inhibited the cleavage of pro-apoptotic caspases (i.e., casp-3 and -8) [49]. Although the results from preclinical studies are overall promising, further research is needed to better understand the molecular mechanisms underlying diabetic complications, as well as the pharmacodynamics and pharmacokinetics of curcumin in pregnancy, to conceivably employ this compound as a therapeutic agent for human pregnancy complications.

Table 1. Effects of curcumin on pregnancy and pregnancy-related disorders.

Curcumin	Experimental Model	Outcomes	References
Altered glucose metabolism			
100 mg/kg/day (from 0 to 20 GD)	Mouse model of GDM	↓Maternal glucose and insulin levels; improved oxidative stress (↑ GSH, SOD, CAT), and ↑AMPK and ↓HDAC4, in the liver; restored offspring litter size and body weight	Lu, X., 2019 [14]
20 μM for 24 h	Mouse embryos (E8.5 of development) cultured for 24 h with 100 mg/dL glucose	↓Neural tube defects by reducing oxidative stress (↓4-HNE, ↓LPO, ER stress (↓p-PERK, p-IRE1α, p-eIF2α, CHOP, BiP and XBP1 expression), and apoptosis (↓caspase-3 and -8 cleavage)	Wu, Y., 2015 [49]
Cardiovascular disorders			
0.36 mg/kg/day (from 0 to GD18)	Rat model of PE (LPS-induced)	Improved hypertension, proteinuria, and renal damage; ↓serum levels of IL-6 and MCP-1; ↓ placental TLR4, IL-6, and NFkB expression; improved trophoblast invasion and spiral artery remodeling	Gong, P., 2016 [50]
0.36 mg/kg/day (from 0.5 to GD18)	Mouse model of PE (LPS-induced)	↑Number of live pups, and fetal and placental weight; ↓inflammation (↓TNF-α, IL-1β, IL-6, MCP-1, and MIP-1 placental expression), ↑ Akt activation	Zhou, J., 2017 [51]
5–10 μM for 24 h	HTR8/SVneotrophoblast cells (model for human first-trimester placenta)	↑Proliferation associated with Akt activation, ↑tube formation; ↑proangiogenic factors VEGF, VEGFR2, and FABP4 expression; ↑ expression of NOTCH-signaling pathway mediators; ↑promoter hypomethylation of oxidative and metabolic stress genes	Basak, H., 2020 [15]

Table 1. *Cont.*

Curcumin	Experimental Model	Outcomes	References
5 μM for 24 h	HTR8/SVneo trophoblast cells (H_2O_2-treated)	↑Cells viability; ↓oxidative stress (↑CAT, GSH-Px activities); ↑Nrf2 activation and ↓ caspase-3 activation	Qi, L., 2020 [52]
60 μM for 24 h	Human placental and fetal membranes, LPS-treated	↓IL-6, IL-8, and COX-2 mRNA expression; ↓PGE2 and PGF2a release; ↓MMP-9 expression and NFkB activation	Lim, R., 2013 [53]
100 mg (single dose)	47 pregnant women with PE	No differences in serum level of COX-2 and IL-10	Fadinie, W., 2019 [54]
Fetal growth and development			
100 mg/kg/day (from 1.5 to 19.5 GD)	Mouse model of FGR (low-protein diet)	↓Placental apoptosis and ↑ placental blood sinusoids area; ↑GSH-Px activity, Nfr2 mRNA expression; ↑antioxidant genes expression (SOD1, SOD2, CAT, Nrf2, and HO-1), in fetal liver	Qi, L., 2020 [16]
400 mg/kg/day at 6 weeks of age for 6 weeks	FGR newborn rats	↓TNF-α, IL-1β and IL-6 levels, ↓activity of AST, ALT, and MDA, ↑Gpx and GSH activity, in serum;↓NF-kB and JAK2 expression, ↑antioxidant genes (Nqo1, Hmox1, Gst, Gpx1 and Sod1), an Nfr2 activation, in the liver	He, J., 2018 [55]
400 mg/kg/day at 6 weeks of age for 6 weeks	FGR newborn rats	↓Glucose levels and IR; ↓TAG, NEFA, total cholesterol, ↑glycogen (↓IRS-1 and Akt phosphorylation, CD36, SREBP-1, and FASN expression, ↑PPARα), in the liver	Niu, Y., 2019 [56]
100 mg/kg (single dose)	Mouse model of PTB, LPS-induced	↓TNF-α, IL-8, MDA, and ↑SOD serum levels; ↓NFkB activation in placenta	Guo, Y.Z., 2017 [57]
Toxicant agents			
200 mg/kg/day (from 7 to PND28)	Pregnant rats, BPA-treated	Neuroprotective; ↑proliferation and differentiation of neuronal stem cells (↑neurogenin and neuroD1 expression); ↓apoptosis (↓Bax, ↑Bcl-2 expression); improvement in learning and memory	Tiwari, S.K., 2019 [58]
150/300 ppm/day (from GD1 to 15PND)	Pregnant mice, $HgCl_2$-treated	↑Neurodevelopment and ↓anxiety (↑levels of DA, 5-HT, AChE, and GSH)	Abu-Taweel, G.M., 2019 [59]
150/300 ppm/day (from GD1 to 15PND)	Pregnant mice, $HgCl_2$-treated	↑Pups body weight; ↑male genitalia weight, testosterone, and FSH levels; ↑ovary weight and progesterone, FSH and LH levels; improved sexual behavior in both sexes	Abu-Taweel, G.M., 2020 [60]
16 g/kg/day during pregnancy and lactation	Pregnant rats, Pb-treated	Prevented central nervous system dysfunction allowing normal locomotor behavior	Benammi, H., 2017 [61]
Pretreatment with curcumin 500 nmol/kg/day (from ED 13.5 to E16.5)	Pregnant mice, celecoxib-treated	↑Neurogenesis in fetal frontal cortex (↑Cyclin D1 expression, and activation of Wnt/βcatenin signaling in neural progenitor cells)	Wang, R., 2017 [62]
Single-dose curcumin (1 g/kg) in neonatal rats	Pregnant rats, VPA-treated	↑Body and brain weight in pups; ↓IL-6, IFN-γ, and ↑GSH, CYP450 expression, in brain pups	Al-Askar, M., 2017 [63]
Offsprings 100 mg/kg/day (from 28 to 35 PND	PLAE-pregnant mice (offspring peri-adolescence period)	Improved offspring anxiety and memory deficits; ↓Neuroinflammation (↓IL-6, TNF-α, and NF-kB expression)	Cantacorps, L., 2020 [64]
Embryos 25 μM for 24 h	PAE-pregnant mice (embryos E17.5)	Improved offspring anxiety and memory deficits; ↓neuroinflammation (↓IL-6, TNF-α, and NF-kB expression)	Yan, X., 2017 [65]
Adverse effects on embryos			
24 μM for 24 h	Mouse blastocysts	↑Apoptosis (↑Bax and ↓Bcl-2 expression); ↓ implantation rate and development	Chen, C.C., 2010 [66]
24 μM for 24 h	Mouse oocytes	↑Apoptosis; ↓ oocytes fertilization; ↓implantation rate and development	Chen, C.C., 2012 [67]

Table 1. *Cont.*

Curcumin	Experimental Model	Outcomes	References
6–24 μM for 24 h	Mouse blastocysts (at implantation stage and during the early post-implantation stage)	Dose-dependent damage, 24 μM lethal for all blastocysts	Huang, F.J., 2013 [68]
Curcuma longa extract (7.80–125 μg/mL) for 5 days	Zebrafish embryos and larvae at different hours of post-fertilization (24–120 h)	Dose-dependent toxic effects: malformations above 62.50 μg /mL, and mortality at 125.0 μg/mL	Alafiatayo, A.A., 2019 [19]

Abbreviations: ↑ Increases; ↓ Decreases; GDM, gestational diabetes mellitus; GD, gestational day; GSH, glutathione; SOD, superoxide dismutase; CAT, catalase; AMPK, 5′ AMP-activated protein chinasi; HDAC4, histone deacetylase 4; 4-HNE, 4-hydroxynonenal; LPO, lipid peroxidation; ER, endoplasmic reticulum; p-PERK, phospho-protein kinase-like endoplasmic reticulum kinase; p-IRE1α, phospho-inositol-requiring kinase 1α; p-eIF2α, phospho-eukaryotic Initiation Factor 2α; CHOP, C/EBP homologous protein; BiP, binding immunoglobulin protein; XBP1, X-box-binding protein-1; PE, preeclampsia; LPS, lipopolysaccharides; IL6, interleukin-6; MCP-1, monocyte chemoattractant protein-1; TLR4, toll-like Receptor 4; NFkB, nuclear transcriptor factor kappa B; TNFα, tumor necrosis factor α; IL1β, interleukin-1β; MIP-1, macrophage inflammatory protein-1; Akt, protein kinase B; VEGF, vascular endothelial growth; VEGFR2, vascular endothelial growth factor receptor 2; FABP4, fatty acid binding protein 4; GSH-Px, glutathione peroxidase; Nrf2, nuclear factor erythroid-2-related factor-2; IL-8, interleukin-8; COX-2, cyclooxigenase-2; PGE2, prostaglandin E2; PGF2a, prostaglandin F2α; MMP-9, metalloproteinase-9; IL-10, interleukin-10; FGR, fetal growth restriction; HO-1, heme oxygenase-1(enzyme); AST, aspartate aminotransferase; ALT, aminotransferase; MDA, malondialdehyde; JAK2, Janus kinase 2; Nqo1, quinone dehydrogenase; Hmox1, heme oxygenase 1 (gene); Gst, glutathione S-transferase; Gpx1, glutathione peroxidase; IR, insulin resistance; TAG, triglycerides; NEFA, Non-Esterified Fatty Acids; IRS-1, insulin receptor substrate-1; PTB, preterm birth; CD36, cluster of differentiation 36; SREBP-1, stearoyl CoA desaturase-1; FASN, Fatty acid synthase; PPARα, Peroxisome Proliferator Activated Receptors-α; PND, postnatal day; BPA, bisphenol-A; DA, dopamine; 5-HT, serotonin; AChE, acetylcholinesterase; FSH, follicle stimulating hormone; LH, luteinizing hormone; ED, embrionic day; Pb, plumbum (lead); VPA, valproic acid; IFN-γ, interferon γ; CYP450, cytochromes P450; PLAE, prenatal and lactational alcohol exposure; PAE, prenatal alcohol exposure; PND, postnatal day; B-cell lymphoma protein 2 (Bcl-2)-associated X (Bax); B-cell lymphoma protein 2 (Bcl-2).

3.2. Cardiovascular Disorders

Critical changes in the cardiovascular system occur in physiological pregnancy, to ensure maternal and fetal adaptation to the increased metabolic demand and to guarantee adequate uteroplacental circulation for fetal growth. A healthy pregnancy is hallmarked by systemic vasodilatation, significantly related to the high levels of estrogen and progesterone. Cardiac output and heart rate rise during gestation and the activation of the renin-angiotensin-aldosterone system leads to a significant increase in total blood volume. Alterations in these processes are associated with maternal and fetal morbidity and mortality [69]. Obesity, older maternal age, and diabetes mellitus increase the risk of cardiovascular diseases in pregnancy (1–4%), with a higher prevalence when including hypertensive disorders—chronic hypertension, pregnancy-induced hypertension, pre-eclampsia, and HELLP syndrome (hemolysis, elevated liver enzymes, and low platelet count) [70]. Considering the anti-inflammatory, antioxidant, and antiangiogenic activities observed in several studies, curcumin is a potential therapeutic compound in cardiovascular disorders [71].

Preeclampsia (PE) is a systemic syndrome characterized by hypertension and proteinuria, which begins after 20 weeks of gestation; it occurs in 2–8% of pregnancies, and it is a leading cause of maternal and fetal morbidity and mortality [72]. Although the pathophysiology of PE remains to be elucidated, alterations in maternal vascular physiology have been described, leading to a generalized vasoconstrictive state, systemic oxidative stress, inflammation, and endothelial cell dysfunction, with severe adverse effects on the placenta, one of the major organs that develops after conception [73,74]. Strategies to reverse or arrest the pathological processes of PE are aimed at reducing excessive inflammatory response, micro-emboli formation, and vasoconstriction by using specific drugs or natural products [75]. For this purpose, studies in animal models have been performed. It has been observed that in lipopolysaccharides (LPS)-treated pregnant rats to create a PE model (LPS 0.5 μg/kg on gestational day 5), the administration of curcumin (0.36 mg/kg, from GD 0 to GD18) improved hypertension, proteinuria, and renal damage, and reduced serum levels of IL-6 and monocyte chemoattractant protein-1 (MCP-1). Curcumin treatment ameliorated inadequate trophoblast invasion and spiral artery remodeling, significant histopathological alterations observed in PE. Analysis of placental tissue showed that curcumin administration decreased the LPS-induced expression of the inflammatory molecules toll-like receptor (TLR)-4, IL-6, and the proinflammatory transcription factor NF-kB. According to the obtained data, the authors hypothesized that curcumin may positively modulate the cascade of different signaling pathways involved in PE development [50]. Similar results were obtained in a mouse model of LPS-induced PE. In this study, in addition to blood pressure and proteinuria reduction, curcumin increased the number of live pups, fetal and placental weight, and decreased fetal desorption. These effects were associated with the inhibition of placental expression of TNF-α, IL-1β, IL-6 cytokines, and MCP-1 and MIP-1 chemokines, and with a reduction in macrophage infiltration. The reduced inflammatory status was accompanied by increased activation of the serine/threonine-specific protein kinase Akt, involved in cellular proliferation [51]. Neo-vascularization is a critical event mediated by several angiogenic factors—including the vascular endothelial growth factor (VEGF), fibroblast growth factors (FGFs), matrix metalloproteinases (MMPs)—and inflammatory factors such as Cyclooxygenase (COX)-2 and NF-kB, occurring not only in tumor progression but also in early placentation [76,77]. Curcumin appears to modulate the above-mentioned factors, influencing vessel formation by acting either as a proangiogenic or as an antiangiogenic molecule, depending on the concentration and the cell type [77]. A recent study investigated the effect of curcumin in HTR8/SVneo trophoblasts cells, a model of the human first-trimester placenta. Incubation with curcumin at low concentration (5–10 μM for 24 h) stimulated (i) proliferation with concomitant activation of Akt, (ii) tube formation of placental trophoblast HTR8/SVneo cells, (iii) and increased the expression of the proangiogenic factors VEGF, VEGFR2, and FABP4. In addition, curcumin treatment strongly increased the mRNA and protein expression of HLA-G, involved in the immune regulation during trophoblast invasion; and mRNA expression of a relevant number of genes related to the NOTCH-signaling pathway, which regulates angiogenesis. The authors examined the promoter methylation of genes involved in metabolic and oxidative stress and observed that curcumin induced hypomethylation in genes involved in the protection against oxidative stress and DNA damage. Altogether these data indicate that curcumin is able to promote angiogenesis and to activate protective pathways in the first trimester of pregnancy, and supports the development of the placental trophoblast [15]. Moreover, HTR8/SVneo trophoblast cells were used to evaluate the protective effects of curcumin against oxidative stress induced by H_2O_2 (400 μM for 24 h). Results showed that pretreatment with curcumin (5 μM for 24 h) increased cell viability, upregulated the activities of the antioxidant enzymes CAT and glutathione peroxidase (GSH-Px), reduced the H_2O_2-induced ROS accumulation and the apoptotic rate. At molecular levels, these data were associated with an increased nuclear translocation of the antioxidant transcription factor Nrf2, and reduced expression of cleaved-caspase 3 [52].

The anti-inflammatory activity of curcumin has been also observed in vitro in human gestational tissues treated with LPS. Specifically, incubation with curcumin (60 μM for 24 h) reduced IL-6 release, and IL-6 and IL-8 mRNA expression induced by LPS, in both placenta and fetal membranes. Moreover, curcumin decreased placental COX-2 mRNA expression, prostaglandin PGE2 and PGF2a release, and the expression and activity of the matrix-degrading enzyme MMP-9, in association with reduced activation of NF-kB [53].

Although several clinical trials emphasized the benefits of curcumin in different pathological contexts [10,11,32,78], there are few data on curcumin supplementation in human pregnancy. Recently, a double-blind randomized clinical trial involving 47 pregnant women with preeclampsia was conducted to evaluate the possible effect of curcumin on the expression of COX-2 and IL-10, thought to have a role in the pathogenesis of PE. The enrolled patients were randomized to receive either curcumin 100 mg/d (n = 23) or placebo (n = 24) [54]. The authors analyzed the circulating levels of IL-10 and COX-2, at T0, 90 min after curcumin ingestion, and 12 h after delivery. Results showed that curcumin did not modify the expression of the analyzed molecules at any tested time. The authors hypothesized that the absence of effect might be due to the low dose of curcumin, taking into account that in non-pregnant subjects doses can reach more than 1 g/day [54].

3.3. Postpartum Depression

During the antenatal and postpartum periods, women are particularly prone to develop mental disorders, including depression. Postpartum depression (PPD) occurs in 10–20% of women, leading to significant health consequences for both mother and offspring [79]. This condition has been largely underestimated and understudied so far. Hence, its prevalence is supposed to be higher, conceivably reaching 50% of women. Symptoms of depression begin during pregnancy in about 30% of women and numerous environmental, genetic, biochemical, and epigenetic factors likely contribute to the onset of PPD [79–81], although the exact mechanisms responsible for this condition are not yet completely known. Several pharmacological and psychological approaches are currently adopted to treat PPD, even though complementary and alternative medicine have also been taken into consideration. Increasing data have suggested the neuroprotective roles of a healthy diet, rich in fruit and vegetables, highlighting its positive influence on mental health [82]. On the contrary, an unhealthy dietary pattern increases the risk of systemic low-grade inflammation and neuroinflammation, known to be associated with PPD [18]. The neuroprotective and antidepressant benefits of curcumin have been known for a long time [83–85]. Several preclinical studies have suggested potential positive effects of curcumin in treating neurological disorders, such as Alzheimer's disease, Parkinson's disease, multiple sclerosis, migraine, epilepsy, brain and spinal cord injury, and depression [17,86,87]. Lopresti and colleagues have investigated the effects of curcumin on depression outcomes in humans. They observed that eight-week curcumin supplementation (500 mg twice a day) in subjects with major depressive disorder (MDD) was effective in reducing depressive and anxiety symptoms, as demonstrated by the reduction in total depressive symptoms (total IDS score), mood/cognitive depressive symptoms (IDSm), arousal-related symptoms (IDSa), and trait anxiety (STAIt) [88]. This supplementation resulted in an increase in urinary levels of both the arachidonic acid metabolite thromboxane B2 (Tbx-B2) and the neuropeptide substance P (SUB-P), potentially involved in depression mechanisms. Moreover, although curcumin did not modify plasma levels of endothelin-1 and leptin, a greater antidepressant benefit was observed in subjects with the highest baseline levels of these molecules. The authors hypothesized that curcumin might act by increasing endothelin and leptin receptor activities [89]. Similarly, in another trial, 1000 mg/day curcumin ingestion for six weeks or the administration of the antidepressant drug fluoxetine showed comparable efficacy in subjects with MDD [90]. A recent meta-analysis provided relevant information about curcumin use in depression. Specifically, this analysis revealed that curcumin administration (i) appears to be more effective in reducing depression symptoms at a higher dosage (1 g/day) and for six weeks or more; (ii) can enhance the action of antidepressants; and (iii) has more effects on subjects with major depression and without other comorbidities [86]. These results indicate the need for further study to better comprehend the mechanisms of action of curcumin in depression treatment.

Data obtained from animal and in vitro studies have indicated that curcumin might exert antidepressant activity by acting on different signaling pathways involved in mental disorders. Specifically, this compound is able to ameliorate the hypothalamic-pituitary-adrenal (HPA) axis disturbances [91]. Curcumin can influence the unbalanced release of monoamine neurotransmitters—such as serotonin (5-HT), dopamine (DA), noradrenaline, and glutamate—the expression of monoamine oxidase (MAO), the expression of neurotrophic factors such as brain-derived neurotrophic factor (BDNF) and neurogenesis, as well as the dysregulated immune system function and oxidative and nitrosative stress. Thus, curcumin appears to promote neurogenesis and inhibit neuronal cell apoptosis [83,84,92,93]. Despite the consistent evidence of efficacy and safety of curcumin treatment in other pathological conditions, to date, data on its effects on depression in pregnancy are completely lacking. However, in the last years, there has been a growing awareness of the possible role of anti-inflammatory micronutrients in improving PPD symptoms [18].

3.4. Fetal Growth and Development

According to the theory of the fetal origin of adult diseases (FOAD) hypothesized by David Barker, the intrauterine environment has a relevant role in fetal growth and development and influences disease susceptibility in the offspring in the short and long term [94]. The physiological processes of pregnancy require immune and metabolic modifications to accommodate the growing fetus; maternal malnutrition negatively influences this dynamic equilibrium, leading to tissue-specific impairment, with serious adverse outcomes for both mother and child [3,6]. Taking into consideration the importance of nutrition in human development, there is a need for better understanding the nutritional programming and the related mechanisms and players acting during pregnancy.

The placenta has the fundamental role of transferring nutrients to the fetus, and alterations in placental function have severe effects on fetal growth. Placental insufficiency is the most common cause of fetal growth restriction (FGR), a serious condition that affects 3–7% of all newborns [95]. Although the pathophysiology of FGR is not completely known, excessive oxidative stress and inflammation, as well as the activation of a complex network of several signaling pathways, appear to be involved [95,96]. The antioxidant and anti-inflammatory effects exerted by curcumin on the placenta [53] were confirmed in a mouse model of FGR fed with a low-protein (LP) diet [16]. The authors showed that maternal supplementation with curcumin (100 mg/kg day, from 1.5 to 19.5 GD) induced a potent antioxidant response in LP-fed pregnant mice; specifically, curcumin (i) increased GSH-Px activity, Nfr2 mRNA expression, and the blood sinusoids area; (ii) reduced malondialdehyde (MDA) content and apoptosis in the placenta, leading to increased placental efficiency; and (iii) elevated the expression of the antioxidant genes SOD1, SOD2, and CAT, and protein expression of Nrf2 and eme oxygenase-1(HO-1) in the liver. Overall, curcumin supplementation during pregnancy was able to revert tissue damage and contrast the decrease in fetal weight induced by a LP diet [16]. Curcumin appeared to improve birth weight, inflammation, and oxidative damage also in FGR newborn rats. Indeed, FGR rats supplemented with 400 mg/kg curcumin (at six weeks of age for six weeks) displayed reduced levels of the inflammatory cytokines TNF-α, IL-1β, and IL-6, reduced activity of AST, ALT, and MDA enzymes, and increased Gpx and GSH activity in serum. Antioxidant defense in the liver was significantly improved as well. The attenuation of the inflammatory status induced by curcumin was associated with (i) reduced activation of NF-kB and JAK2; (ii) increased expression of the antioxidant genes (Nqo1, Hmox1, Gst, Gpx1, and Sod1), and activation of their regulatory transcription factor Nfr2, in the liver [55]. Successively, the same authors investigated the effects of curcumin on insulin resistance (IR) and hepatic lipid accumulation in FGR newborn rats. Specifically, supplementation with 400 mg/kg curcumin (at six weeks of age for six weeks) attenuated IR by reducing serum insulin, glycemia, and homeostasis model assessment of insulin resistance (HOMA-IR). Furthermore, in the liver, curcumin diminished total cholesterol, TG, and non-esterified fatty acids (NEFA); increased glycogen concentration and induced the activation of lipolytic enzymes, together with a reduction in IRS-1 and Akt phosphorylation, a decrease in CD36, SREBP-1, and FASN expression, and an increase in PPARα levels. Overall, these data showed that curcumin could improve IR and lipid accumulation in the liver by regulating insulin signaling pathways, and promoting lipolysis and fatty acid oxidation in FGR rats [56].

Of note, curcumin alleviated also jejunum damage in FGR growing pigs. Indeed, the addition of 200 mg/kg curcumin to diet improved antioxidant defense (i.e., increased SOD and decreased MDA activity), immune-related gene expression (reduced mRNA of TNFα, IL-6, and IFNγ, and increased IL-2), and decreased apoptotic genes, such as caspase3 and Bax in the jejunum. Moreover, curcumin supplementation increased mRNA expression of the tight junction-related gene ocln [97].

Preterm birth (PTB) is a pregnancy complication that affects about 11% of births worldwide and is associated with increased maternal and neonate morbidity and mortality [98]. An altered inflammatory status appears to be associated with PTB. Thus the anti-inflammatory activity of curcumin has been evaluated in a mouse model of PTB, obtained through LPS injection in the abdominal cavity [57]. The injection of 100 mg/kg curcumin into the abdominal cavity, one day before (preventative group) or one day after (treatment group) LPS treatment, significantly reduced serum levels of TNF-α, IL-8, and MDA, and increased SOD levels, in both the experimental conditions, in pregnant mice. The staining intensity of NF-κB p65 showed that curcumin was able to reduce the LPS-induced expression of this inflammatory transcription factor in placental tissue both in the preventative and in the treatment group [57].

3.5. Toxicant Agents

Besides maternal nutrition, many other factors, including exposure to chemical and natural toxic agents, drugs, alcohol, smoking, and maternal stress influence fetal growth and development [99].

Among the myriad of properties, curcumin appears to be able to reduce toxicity induced by several environmental agents in different organs and tissues, including the brain and liver [12].

Bisphenol-A (BPA) is a chemical substance adapted to produce plastic. It has been considered an endocrine disruptor by the European Chemicals Agency (ECHA 2017) [100] due to its estrogenic activity. BPA exposure in pregnancy is associated with negative outcomes, including impaired fetal growth and childhood adiposity [101].

Remarkably, this synthetic compound affected the processes of neurogenesis in the hippocampus of the developing rat brain, and curcumin treatment showed neuroprotective activity by reverting BPA-induced effects. Specifically, pups from a pregnant rat receiving BPA (40 μg/kg body weight/day from GD6 to PND28) were treated with curcumin (200 mg/kg body weight/day from PND7 to PND28). The authors performed accurate experiments on embryo and pup brains and examined the expression of genes and pathways involved in neurogenesis. They observed that curcumin attenuated the BPA-induced reduction in neuronal stem cells (NSC) proliferation and differentiation. At molecular levels, the improvement in neurogenesis was associated with the enhanced expression of the proneural transcription factors neurogenin and neuroD1, the reduced expression of the proapoptotic molecule Bax, the increased expression of the antiapoptotic molecule Bcl-2, and the activation of Wnt/βcatenin signaling that regulates NSC proliferation and differentiation. Of note, the benefits of curcumin resulted in improved learning and memory in BPA-treated pups [58].

Mercury (Hg) is a widely diffused toxic heavy metal that occurs naturally in three forms, namely metallic Hg, organic Hg, and inorganic Hg. Human exposure to Hg occurs mainly through the environment (e.g., mercury-contaminated sea fish, dental amalgam). Of note, occupation (e.g., mining) is another important source of exposure for humans and is associated with possible multi-organ toxicity [102]. As for the influence of Hg on neurodevelopment, a cross-sectional study, involving healthy Saudi mothers and their infants (age 3–12 months), showed an association between Hg exposure and neurodevelopmental delay, with possible negative effects persisting also in adulthood [103]. Interestingly, curcumin appeared to mitigate Hg toxicity in animal models [102]. Specifically, pregnant mice were exposed (from 1GD to 15PND) to 10 ppm mercuric chloride ($HgCl_2$) in the presence or absence of 150 and 300 ppm curcumin. Hg exposure induced serious damage to the development of neuromotors, and increased anxiety behavior in pups. Curcumin administration improved neurodevelopment and reduced anxiety, by restoring the levels of neurotransmitters DA, 5-HT, and acetylcholinesterase (AChE), and of the antioxidant GSH, decreased by Hg exposure, in forebrain pups [59]. Moreover, by using the same experimental conditions, the authors analyzed changes in body weight, sexual behavior, and fertility in male and female pups. The obtained data showed that curcumin counteracted the perinatal effects of Hg exposure by increasing (i) body weight, liver and brain weight in male and female pups; (ii) epididymis, seminal vesicle, testis weight in males; and (iii) ovary weight in females; also sexual behavior was improved in both sexes. Moreover, curcumin increased testosterone and FSH levels, and sperm motility in males, as well as FSH, LH, and progesterone in females, reduced by Hg exposure [60].

Lead (Pb) is a heavy metal widely spread in the environment. It is extremely dangerous for both animals and humans. Lead exposure occurs mainly through food and water contamination, and air pollution. Lead can cross the placental and blood-brain barrier, inducing neurotoxicity. Curcumin exerted neuroprotective effects contrasting lead-induced damage in rats. The concomitant exposure of rat mothers to Pb (3 g/L) and curcumin (16 g/kg) during pregnancy and lactation resulted in the recovery of the Pb-induced altered sensorimotor functions in neonatal rats. Pb neurotoxicity produced alterations in locomotor neuronal network development and curcumin treatment reversed these anomalies, allowing normal locomotor behavior. These findings indicate that curcumin has the capability to prevent central nervous system dysfunction induced by lead during the earlier stages of development [61].

Celecoxib is a selective inhibitor of COX-2 that is able to reduce pain and inflammation caused by several inflammatory conditions [62]. Since recent data have shown that the inhibition of COX-2 reduced adult neural cell proliferation and differentiation [104], Wang et al., investigated the neuroprotective action of curcumin on fetal brain development in pregnant mice treated with celecoxib [62]. Specifically, pregnant mice were pretreated with curcumin (500 nmol/kg body weight) from embryonic day (E) 13.5 to E16.5, and then with celecoxib (300 mg/kg body weight) from E16.5 to E17.5. Results showed that curcumin counteracted the celecoxib-induced inhibition of neurogenesis in the fetal frontal cortex, by increasing proliferation and Cyclin D1 expression in neural progenitor cells, and by activating Wnt/βcatenin signaling (i.e., decreased expression of glycogen synthase kinase 3 beta (GSK-3β), and increased expression of βcatenin) [62].

Valproic acid (VPA), a branched short-chain fatty acid, is an antiepileptic agent that has been associated with congenital malformations, including alterations in fetal brain development, and consequent intellectual disabilities and autistic spectrum disorders in the offspring [105]. Curcumin appears to attenuate the VPA-induced brain damage, as observed in a rodent model of autism. Neonatal rats, born to mothers treated with VPA from 12.5 gestational day, received a single dose of curcumin (1 g/kg day), and their brains were analyzed 28 days after birth. Curcumin was able to ameliorate body and brain weight, and the altered expression of IL-6, IFN-γ, GSH, CYP450, in the brain of VPA-exposed pups [63].

Prenatal alcohol exposure (PAE) has dramatic effects on fetal growth and development (fetal alcohol spectrum disorders: FASD) and is responsible for neurodevelopmental disorders (i.e., neurocognitive and behavioral deficits, and increased susceptibility to mental health disorders) and birth defects (growth deficits and physical abnormalities). PAE induces chromosomal rearrangements and epigenetic alterations, therefore leading to altered gene-environment interactions that are responsible for alcohol-induced disorders [106]. Curcumin (100 mg/kg body weight), administered during the peri-adolescence period (PND 28–35), appeared to counteract fetal brain damage induced by prenatal and lactational alcohol exposure (PLAE; 20% (*v*/*v*) alcohol solution) in mice. The authors showed that curcumin improved anxiety and memory deficits caused by PLAE, and these improvements were associated with reduced microglia activation and astrogliosis. At molecular levels, curcumin reduced protein expression of IL-6, TNF-α, and NF-kB. These data showed that curcumin may act against cognitive deficits and neuroinflammation induced by alcohol exposure in pregnancy [64].

Curcumin can counteract the deleterious effects of PAE on cardiac development, as demonstrated in a mouse model. Pregnant mice were daily exposed to ethanol (56% *v*/*v* in saline) between embryonic days 7.5 to 15.5; at embryonic day 17.5, mice were euthanized and embryonic hearts were removed. Results showed that PAE treatment increased apoptosis in pup hearts; this finding was associated with higher levels of caspase-3 and -8 mRNA expression, and reduced Bcl-2 mRNA expression, due to a different modulation of histone H3K9 acetylation near the promoter regions of caspase-3, caspase-8 (hyperacetylation), and Bcl-2 (hypoacetylation). In vitro, curcumin (25 μM for 24 h) treatment abolished apoptosis and reverted the expression of caspases and Bcl-2, induced by alcohol (200 mM), in cardiac progenitor cells. These results highlighted the capability of curcumin to prevent congenital heart diseases induced by PAE in pregnancy, by acting as an epigenetic modulator [65].

3.6. Adverse Effects on Embryos

Embryonic development is a complex process that is finely regulated and highly susceptible to environmental influences. Therefore, it is reasonable to hypothesize that the anti-inflammatory, antioxidative, antiproliferative, and antiangiogenic properties of curcumin could interfere with the blastocyst stage, implantation and post-implantation development of embryos [66].

Chen and colleagues evaluated the possible embryotoxicity of curcumin in mouse blastocysts both in vitro and in vivo. They observed that curcumin (24 μM for 24 h) induced apoptosis in mouse blastocysts, and reduced implantation rate and development, in vitro. Then, embryos treated with curcumin were transferred in vivo; results confirmed a significant reduction in implantation ratio, and, among the implanted embryos, a higher rate of failure to develop normally. The authors evaluated the possible mechanisms responsible for these effects and found that curcumin-induced apoptosis was associated with the modulation of pro- and anti-apoptotic molecules (i.e., increased Bax and reduced Bcl-2 expression), ROS generation, and caspase-3 activation [66]. Additionally, the same authors showed that curcumin (24 μM) adversely affected oocytes maturation, in vitro. This effect resulted in a reduced ability of oocytes to be fertilized, increased blastocyst apoptosis, and reduced blastocyst implantation ratio and development. These results were confirmed in oocytes collected from female mice after feeding them with curcumin supplementation (40 μM) for four days [67]. Another in vitro study highlighted that the degree of damage induced by curcumin (6, 12, or 24 μM curcumin for 24 h) on mouse blastocyst at the implantation stage and during the early post-implantation stage is dose-dependent. Specifically, 6 μM and 12 μM curcumin inhibited cell proliferation of the blastocyst but increased the formation of trophoblastic giant cells, whereas 24 μM curcumin exposure was lethal to all blastocysts, and induced severe damage to the implanted blastocysts [68].

Further evidence on these effects comes from a recent study in zebrafish. The exposure of zebrafish embryos and larvae to different concentrations of Curcuma Longa extract (7.80, 15.63, 31.25, 62.50, 125.0, and 250.0 μg/mL) at different hours post fertilization (hpf: 24, 48, 72, 96, 120 h) showed that a dosage above 62.50 μg/mL had toxic effects, and a dosage of 125.0 μg/mL increased embryo mortality and induced morphological deformities in larvae [19]. Despite the potential benefits of curcumin described in different pathological conditions, all these data indicate that dosage and time of exposure throughout pregnancy should be carefully evaluated to avoid serious damage to embryo development.

4. Conclusions and Future Perspectives

The use of the natural product curcumin to treat medical conditions is spreading around the world. There is an increasing public interest in the potential health benefits of this compound, as evidenced by the large number of currently available curcumin formulations, aimed at increasing its bioavailability and efficacy, and by the considerable number of scientific papers published over the last years.

This review has drawn attention towards the effects of curcumin on pregnancy and pregnancy complications, considering that during gestation, mother and fetus undergo significant (patho-) physiological changes.

Almost all data emphasizing the numerous biological activities of curcumin have been obtained from pregnant rodents and in vitro studies. Curcumin appeared to ameliorate diabetes in a GDM mouse model, as well as PE in a PE rat model, and was found to be neuroprotective against environmental toxic agents. The antidepressant activity of curcumin has also been tested in humans. However, to date, studies on the possible beneficial effects of curcumin on PPD, a largely underestimated and understudied condition, are completely lacking. As regards fetal growth and development, curcumin counteracted the modifications associated with FGR and PTB in rodent models but negatively affected blastocyst stage, implantation and post-implantation embryo development in healthy animals.

Altogether, these results indicate that the use of curcumin in pregnancy must be carefully evaluated. The growing use of curcumin as self-medication along with the misleading perception that "natural" is the equivalent of "safe" are additional issues of concern.

Further studies are needed to clarify whether pregnancy might benefit from curcumin's properties; for this purpose, the collaboration between multidisciplinary scientific teams is essential to provide a holistic view of the complex networks between natural products and human physiology. Systems biology and the recently developed network pharmacology represent new strategies to better comprehend the mechanisms underlying curcumin activities in the human body.

Author Contributions: Conceptualization, C.S. and S.M.; PubMed search, T.F. and R.V.; writing—draft preparation, T.F. and C.S.; writing—review and editing, S.M., E.F. and A.Z.; supervision and critical revision S.M., R.V. and C.S. All authors have read and agreed to the published version of the manuscript.

Funding: This review received no external funding.

Conflicts of Interest: The authors declare no conflict of interest.

References

1. Abu-Saad, K.; Fraser, D. Maternal Nutrition and Birth Outcomes. *Epidemiol. Rev.* **2010**, *32*, 5–25. [CrossRef]
2. Beulen, Y.; Super, S.; De Vries, J.; Koelen, M.; Feskens, E.J.; Wagemakers, A. Dietary Interventions for Healthy Pregnant Women: A Systematic Review of Tools to Promote a Healthy Antenatal Dietary Intake. *Nutrients* **2020**, *12*, 1981. [CrossRef]
3. Hsu, C.-N.; Tain, Y.-L. Tain the Good, the Bad, and the Ugly of Pregnancy Nutrients and Developmental Programming of Adult Disease. *Nutrients* **2019**, *11*, 894. [CrossRef]

4. Santangelo, C.; Varì, R.; Scazzocchio, B.; Filesi, C.; Masella, R. Management of Reproduction and Pregnancy Complications in Maternal Obesity: Which Role for Dietary Polyphenols? *BioFactors* **2013**, *40*, 79–102. [CrossRef]
5. Santangelo, C.; Zicari, A.; Mandosi, E.; Scazzocchio, B.; Mari, E.; Morano, S.; Masella, R. Could Gestational Diabetes Mellitus Be Managed Through Dietary Bioactive Compounds? Current Knowledge and Future Perspectives. *Br. J. Nutr.* **2016**, *115*, 1129–1144. [CrossRef]
6. Segovia, S.A.; Vickers, M.H.; Gray, C.; Reynolds, C.M. Maternal Obesity, Inflammation, and Developmental Programming. *BioMed Res. Int.* **2014**, *2014*, 1–14. [CrossRef] [PubMed]
7. Filardi, T.; Catanzaro, G.; Mardente, S.; Zicari, A.; Santangelo, C.; Lenzi, A.; Morano, S.; Ferretti, E. Non-Coding RNA: Role in Gestational Diabetes Pathophysiology and Complications. *Int. J. Mol. Sci.* **2020**, *21*, 4020. [CrossRef] [PubMed]
8. Franzago, M.; La Rovere, M.; Franchi, P.G.; Vitacolonna, E.; Stuppia, L. Epigenetics and Human Reproduction: The Primary Prevention of the Noncommunicable Diseases. *Epigenomics* **2019**, *11*, 1441–1460. [CrossRef] [PubMed]
9. Franzago, M.; Santurbano, D.; Vitacolonna, E.; Stuppia, L. Genes and Diet in the Prevention of Chronic Diseases in Future Generations. *Int. J. Mol. Sci.* **2020**, *21*, 2633. [CrossRef]
10. Vázquez-Fresno, R.; Rosana, A.R.R.; Sajed, T.; Onookome-Okome, T.; Wishart, N.A.; Wishart, D.S. Herbs and Spices-Biomarkers of Intake Based on Human Intervention Studies–A Systematic Review. *Genes Nutr.* **2019**, *14*, 18. [CrossRef]
11. Hewlings, S.; Kalman, D. Curcumin: A Review of Its Effects on Human Health. *Foods* **2017**, *6*, 92. [CrossRef]
12. Hosseini, A.; Hosseinzadeh, H. Antidotal or Protective Effects of Curcuma Longa (Turmeric) and Its Active Ingredient, Curcumin, Against Natural and Chemical Toxicities: A Review. *Biomed. Pharmacother.* **2018**, *99*, 411–421. [CrossRef] [PubMed]
13. Patel, S.S.; Acharya, A.; Ray, R.S.; Agrawal, R.; Raghuwanshi, R.; Jain, P. Cellular and Molecular Mechanisms of Curcumin in Prevention and Treatment of Disease. *Crit. Rev. Food Sci. Nutr.* **2019**, *60*, 887–939. [CrossRef] [PubMed]
14. Lu, X.; Wu, F.; Jiang, M.; Sun, X.; Tian, G. Curcumin Ameliorates Gestational Diabetes in Mice Partly Through Activating AMPK. *Pharm. Biol.* **2019**, *57*, 250–254. [CrossRef] [PubMed]
15. Basak, S.; Srinivas, V.; Mallepogu, A.; Duttaroy, A.K. Curcumin Stimulates Angiogenesis through VEGF and Expression of HLA-G in First-Trimester Human Placental Trophoblasts. *Cell Biol. Int.* **2020**, *44*, 1237–1251. [CrossRef] [PubMed]
16. Qi, L.; Jiang, J.; Zhang, J.; Zhang, L.; Wang, T. Maternal Curcumin Supplementation Ameliorates Placental Function and Fetal Growth in Mice with Intrauterine Growth Retardation. *Biol. Reprod.* **2020**, *102*, 1090–1101. [CrossRef]
17. Salehi, B.; Calina, D.; Docea, A.O.; Koirala, N.; Aryal, S.; Lombardo, D.; Pasqua, L.; Taheri, Y.; Castillo, C.M.S.; Martorell, M.; et al. Curcumin's Nanomedicine Formulations for Therapeutic Application in Neurological Diseases. *J. Clin. Med.* **2020**, *9*, 430. [CrossRef] [PubMed]
18. Matrisciano, F.; Epinna, G. PPAR and Functional Foods: Rationale for Natural Neurosteroid-Based Interventions for Postpartum Depression. *Neurobiol. Stress* **2020**, *12*, 100222. [CrossRef] [PubMed]
19. Alafiatayo, A.A.; Lai, K.-S.; Syahida, A.; Maziah, M.; Shaharudin, N. Phytochemical Evaluation, Embryotoxicity, and Teratogenic Effects of Curcuma longa Extract on Zebrafish (Danio rerio). *Evid.-Based Complementary Altern. Med.* **2019**, *2019*, 3807207–3807210. [CrossRef] [PubMed]
20. Goel, A.; Kunnumakkara, A.B.; Aggarwal, B.B. Curcumin as "Curecumin": From Kitchen to Clinic. *Biochem. Pharmacol.* **2008**, *75*, 787–809. [CrossRef]
21. Paramasivam, M.; Poi, R.; Banerjee, H.; Bandyopadhyay, A. High- Performance Thin Layer Chromatographic Method for Quantitative Determination of Curcuminoids in Curcuma Longa Germplasm. *Food Chem.* **2009**, *113*, 640–644. [CrossRef]
22. Kahkhaie, K.R.; Mirhosseini, A.; Aliabadi, A.; Mohammadi, A.; Mousavi, M.J.; Haftcheshmeh, S.M.; Sathyapalan, T.; Sahebkar, A. Curcumin: A Modulator of Inflammatory Signaling Pathways in the Immune System. *Inflammopharmacology* **2019**, *27*, 885–900. [CrossRef]
23. Cheng, D.; Li, W.; Wang, L.; Lin, T.; Poiani, G.; Wassef, A.; Hudlikar, R.; Ondar, P.; Brunetti, L.; Kong, A.-N.T. Pharmacokinetics, Pharmacodynamics, and PKPD Modeling of Curcumin in Regulating Antioxidant and Epigenetic Gene Expression in Healthy Human Volunteers. *Mol. Pharm.* **2019**, *16*, 1881–1889. [CrossRef] [PubMed]

24. Hassan, F.-U.; Rehman, M.S.-U.; Khan, M.S.; Ali, M.A.; Javed, A.; Nawaz, A.; Yang, C. Curcumin as an Alternative Epigenetic Modulator: Mechanism of Action and Potential Effects. *Front. Genet.* **2019**, *10*, 514. [CrossRef] [PubMed]
25. Di Meo, F.; Margarucci, S.; Galderisi, U.; Crispi, S.; Peluso, G. Curcumin, Gut Microbiota, and Neuroprotection. *Nutrients* **2019**, *11*, 2426. [CrossRef]
26. Pluta, R.; Januszewski, S.; Ułamek-Kozioł, M. Mutual Two-Way Interactions of Curcumin and Gut Microbiota. *Int. J. Mol. Sci.* **2020**, *21*, 1055. [CrossRef] [PubMed]
27. Kotha, R.R.; Luthria, D.L. Curcumin: Biological, Pharmaceutical, Nutraceutical, and Analytical Aspects. *Molecules* **2019**, *24*, 2930. [CrossRef]
28. Zahedipour, F.; Hosseini, S.A.; Sathyapalan, T.; Majeed, M.; Jamialahmadi, T.; Al-Rasadi, K.; Banach, M.; Sahebkar, A. Potential Effects of Curcumin in the Treatment of COVID-19 Infection. *Phytother. Res.* **2020**. [CrossRef]
29. U.S. Food & Drug Administration. GRAS Notice Inventory. 2018. Available online: https://www.fda.gov/food/generally-recognized-safe-gras/gras-notice-inventory (accessed on 9 September 2020).
30. European Food Safety Authority. Scientific Opinion on the Re-Evaluation of Curcumin (E 100) as a Food Additive. Available online: https://efsa.onlinelibrary.wiley.com/doi/pdf/10.2903/j.efsa.2010.1679 (accessed on 7 October 2020).
31. Del Prado-Audelo, M.L.; Caballero-Florán, I.H.; Meza-Toledo, J.A.; Mendoza-Muñoz, N.; Torres, M.G.; Florán, B.; Cortés, H.; Leyva-Gómez, G. Formulations of Curcumin Nanoparticles for Brain Diseases. *Biomolecules* **2019**, *9*, 56. [CrossRef]
32. Mirzaei, H.; Shakeri, A.; Rashidi, B.; Jalili, A.; Banikazemi, Z.; Sahebkar, A. Phytosomal Curcumin: A Review of Pharmacokinetic, Experimental and Clinical Studies. *Biomed. Pharmacother.* **2017**, *85*, 102–112. [CrossRef]
33. Stohs, S.J.; Chen, O.; Ray, S.D.; Ji, J.; Bucci, L.R.; Preuss, H.G. Highly Bioavailable Forms of Curcumin and Promising Avenues for Curcumin-Based Research and Application: A Review. *Molecules* **2020**, *25*, 1397. [CrossRef] [PubMed]
34. Nasery, M.M.; Abadi, B.; Poormoghadam, D.; Zarrabi, A.; Keyhanvar, P.; Khanbabaei, H.; Ashrafizadeh, M.; Mohammadinejad, R.; Tavakol, S.; Sethi, G. Curcumin Delivery Mediated by Bio-Based Nanoparticles: A Review. *Molecules* **2020**, *25*, 689. [CrossRef] [PubMed]
35. Kampmann, U.; Knorr, S.; Fuglsang, J.; Ovesen, P.G. Determinants of Maternal Insulin Resistance during Pregnancy: An Updated Overview. *J. Diabetes Res.* **2019**, *2019*, 5320156. [CrossRef]
36. Filardi, T.; Panimolle, F.; Crescioli, C.; Lenzi, A.; Morano, S. Gestational Diabetes Mellitus: The Impact of Carbohydrate Quality in Diet. *Nutrients* **2019**, *11*, 1549. [CrossRef] [PubMed]
37. Santangelo, C.; Filardi, T.; Perrone, G.; Mariani, M.; Mari, E.; Scazzocchio, B.; Masella, R.; Brunelli, R.; Lenzi, A.; Zicari, A.; et al. Cross- Talk Between Fetal Membranes and Visceral Adipose Tissue Involves HMGB1–RAGE and VIP–VPAC2 Pathways in Human Gestational Diabetes Mellitus. *Acta Diabetol.* **2019**, *56*, 681–689. [CrossRef]
38. Pintaudi, B.; Fresa, R.; Dalfrà, M.; Dodesini, A.R.; Vitacolonna, E.; Tumminia, A.; Sciacca, L.; Lencioni, C.; Marcone, T. The Risk Stratification of Adverse Neonatal Outcomes in Women with Gestational Diabetes (STRONG) Study. *Acta Diabetol.* **2018**, *55*, 1261–1273. [CrossRef]
39. Filardi, T.; Tavaglione, F.; Di Stasio, M.; Fazio, V.; Lenzi, A.; Morano, S. Impact of Risk Factors for Gestational Diabetes (GDM) on Pregnancy Outcomes in Women with GDM. *J. Endocrinol. Investig.* **2017**, *41*, 671–676. [CrossRef]
40. Pivari, F.; Mingione, A.; Brasacchio, C.; Soldati, L. Curcumin and Type 2 Diabetes Mellitus: Prevention and Treatment. *Nutrients* **2019**, *11*, 1837. [CrossRef]
41. Den Hartogh, D.J.; Gabriel, A.; Tsiani, E. Antidiabetic Properties of Curcumin II: Evidence from In Vivo Studies. *Nutrients* **2019**, *12*, 58. [CrossRef]
42. Adibian, M.; Hodaei, H.; Nikpayam, O.; Sohrab, G.; Hekmatdoost, A.; Hedayati, M. The Effects of Curcumin Supplementation on High-Sensitivity C-Reactive Protein, Serum Adiponectin, and Lipid Profile in Patients with Type 2 Diabetes: A Randomized, Double-Blind, Placebo-Controlled Trial. *Phytother. Res.* **2019**, *33*, 1374–1383. [CrossRef]
43. Panahi, Y.; Khalili, N.; Sahebi, E.; Namazi, S.E.; Simental-Mendía, L.; Majeed, M.; Sahebkar, A.; Simental-Mendía, L. Effects of Curcuminoids Plus Piperine on Glycemic, Hepatic and Inflammatory Biomarkers in Patients with Type 2 Diabetes Mellitus: A Randomized Double-Blind Placebo-Controlled Trial. *Drug Res.* **2018**, *68*, 403–409. [CrossRef]

44. Adab, Z.; Eghtesadi, S.; Vafa, M.; Heydari, I.; Shojaii, A.; Haqqani, H.; Arablou, T.; Eghtesadi, M. Effect of Turmeric on Glycemic Status, Lipid Profile, Hs-CRP, and Total Antioxidant Capacity in Hyperlipidemic Type 2 Diabetes Mellitus Patients. *Phytother. Res.* **2019**, *33*, 1173–1181. [CrossRef] [PubMed]
45. Sohaei, S.; Amani, R.; Tarrahi, M.J.; Ghasemi-Tehrani, H. The Effects of Curcumin Supplementation on Glycemic Status, Lipid Profile and Hs-CRP Levels in Overweight/Obese Women with Polycystic Ovary Syndrome: A Randomized, Double-Blind, Placebo-Controlled Clinical Trial. *Complementary Ther. Med.* **2019**, *47*, 102201. [CrossRef] [PubMed]
46. Akbari, M.; Lankarani, K.B.; Tabrizi, R.; Ghayour-Mobarhan, M.; Peymani, P.; Ferns, G.; Ghaderi, A.; Asemi, Z. The Effects of Curcumin on Weight Loss among Patients with Metabolic Syndrome and Related Disorders: A Systematic Review and Meta-Analysis of Randomized Controlled Trials. *Front. Pharmacol.* **2019**, *10*, 649. [CrossRef]
47. Roxo, D.F.; Arcaro, C.A.; Gutierres, V.O.; Costa, M.C.; Oliveira, J.O.; Lima, T.F.O.; Assis, R.P.; Brunetti, I.L.; Baviera, A.M. Curcumin Combined with Metformin Decreases Glycemia and Dyslipidemia, and Increases Paraoxonase Activity in Diabetic Rats. *Diabetol. Metab. Syndr.* **2019**, *11*, 33. [CrossRef] [PubMed]
48. Franzago, M.; Fraticelli, F.; Stuppia, L.; Vitacolonna, E. Nutrigenetics, Epigenetics and Gestational Diabetes: Consequences in Mother and Child. *Epigenetics* **2019**, *14*, 215–235. [CrossRef]
49. Wu, Y.; Wang, F.; Reece, E.A.; Yang, P. Curcumin Ameliorates High Glucose-Induced Neural Tube Defects by Suppressing Cellular Stress and Apoptosis. *Am. J. Obstet. Gynecol.* **2015**, *212*, e801–e808. [CrossRef]
50. Gong, P.; Liu, M.; Hong, G.; Li, Y.; Xue, P.; Zheng, M.; Wu, M.; Shen, L.; Yang, M.; Diao, Z.; et al. Curcumin Improves LPS-Induced Preeclampsia-Like Phenotype in Rat by Inhibiting the TLR4 Signaling Pathway. *Placenta* **2016**, *41*, 45–52. [CrossRef]
51. Zhou, J.; Miao, H.; Li, X.; Hu, Y.; Sun, H.; Hou, Y. Curcumin Inhibits Placental Inflammation to Ameliorate LPS-Induced Adverse Pregnancy Outcomes in Mice via Upregulation of Phosphorylated Akt. *Inflamm. Res.* **2016**, *66*, 177–185. [CrossRef] [PubMed]
52. Qi, L.; Jiang, J.; Zhang, J.; Zhang, L.; Wang, T. Curcumin Protects Human Trophoblast HTR8/SVneo Cells from H2O2-Induced Oxidative Stress by Activating Nrf2 Signaling Pathway. *Antioxidants* **2020**, *9*, 121. [CrossRef]
53. Lim, R.; Barker, G.; Wall, C.A.; Lappas, M. Dietary Phytophenols Curcumin, Naringenin and Apigenin Reduce Infection-Induced Inflammatory and Contractile Pathways in Human Placenta, Foetal Membranes and Myometrium. *Mol. Hum. Reprod.* **2013**, *19*, 451–462. [CrossRef] [PubMed]
54. Fadinie, W.; Lelo, A.; Wijaya, D.W.; Lumbanraja, S.N. Curcumin's Effect on COX-2 and IL-10 Serum in Preeclampsia's Patient Undergo Sectio Caesarea with Spinal Anesthesia. *Open Access Maced. J. Med. Sci.* **2019**, *7*, 3376–3379. [CrossRef]
55. He, J.; Niu, Y.; Wang, F.; Wang, C.; Cui, T.; Bai, K.; Zhang, J.; Zhong, X.; Zhang, L.; Wang, T. Dietary Curcumin Supplementation Attenuates Inflammation, Hepatic Injury and Oxidative Damage in a Rat Model of Intra-Uterine Growth Retardation. *Br. J. Nutr.* **2018**, *120*, 537–548. [CrossRef] [PubMed]
56. Niu, Y.; He, J.; Ahmad, H.; Wang, C.; Zhong, X.; Zhang, L.; Cui, T.; Zhang, J.; Wang, T. Curcumin Attenuates Insulin Resistance and Hepatic Lipid Accumulation in a Rat Model of Intra-Uterine Growth Restriction Through Insulin Signalling Pathway and Sterol Regulatory Element Binding Proteins. *Br. J. Nutr.* **2019**, *122*, 1–9. [CrossRef] [PubMed]
57. Guo, Y.-Z.; Feng, A.-M.; He, P. Effect of Curcumin on Expressions of NF-κBp65, TNF-α and IL-8 in Placental Tissue of Premature Birth of Infected Mice. *Asian Pac. J. Trop. Med.* **2017**, *10*, 175–178. [CrossRef]
58. Tiwari, S.K.; Agarwal, S.; Tripathi, A.; Chaturvedi, R.K. Correction to: Bisphenol-A Mediated Inhibition of Hippocampal Neurogenesis Attenuated by Curcumin via Canonical Wnt Pathway. *Mol. Neurobiol.* **2019**, *56*, 6660–6662. [CrossRef] [PubMed]
59. Abu-Taweel, G.M. Neurobehavioral Protective Properties of Curcumin against the Mercury Chloride Treated Mice Offspring. *Saudi J. Biol. Sci.* **2019**, *26*, 736–743. [CrossRef]
60. Abu-Taweel, G.M. Curcumin Palliative Effects on Sexual Behavior, Fertility and Reproductive Hormones Disorders in Mercuric Chloride Intoxicated Mice Offspring. *J. King Saud Univ. Sci.* **2020**, *32*, 1293–1299. [CrossRef]
61. Benammi, H.; Erazi, H.; El Hiba, O.; Vinay, L.; Bras, H.; Viemari, J.-C.; Gamrani, H. Disturbed Sensorimotor and Electrophysiological Patterns in Lead Intoxicated Rats During Development Are Restored by Curcumin I. *PLoS ONE* **2017**, *12*, e0172715. [CrossRef]
62. Wang, R.; Tian, S.; Yang, X.; Liu, J.; Wang, Y.; Sun, K. Celecoxib- Induced Inhibition of Neurogenesis in Fetal Frontal Cortex Is Attenuated by Curcumin via Wnt/β-Catenin Pathway. *Life Sci.* **2017**, *185*, 95–102. [CrossRef]

63. Al-Askar, M.; Bhat, R.S.; Selim, M.E.; Al-Ayadhi, L.; El-Ansary, A. Postnatal Treatment Using Curcumin Supplements to Amend the Damage in VPA-Induced Rodent Models of Autism. *BMC Complement. Altern. Med.* **2017**, *17*, 259. [CrossRef]
64. Cantacorps, L.; Montagud-Romero, S.; Valverde, O. Curcumin Treatment Attenuates Alcohol-Induced Alterations in a Mouse Model of Foetal Alcohol Spectrum Disorders. *Prog. Neuro-Psychopharmacol. Biol. Psychiatry* **2020**, *100*, 109899. [CrossRef] [PubMed]
65. Yan, X.; Pan, B.; Lu, T.; Liu, L.; Zhu, J.; Shen, W.; Huang, X.; Tian, J. Inhibition of Histone Acetylation by Curcumin Reduces Alcohol-Induced Fetal Cardiac Apoptosis. *J. Biomed. Sci.* **2017**, *24*, 1. [CrossRef] [PubMed]
66. Chen, C.-C.; Hsieh, M.-S.; Hsuuw, Y.-D.; Huang, F.-J.; Chan, W.-H. Hazardous Effects of Curcumin on Mouse Embryonic Development through a Mitochondria-Dependent Apoptotic Signaling Pathway. *Int. J. Mol. Sci.* **2010**, *11*, 2839–2855. [CrossRef] [PubMed]
67. Chen, C.-C.; Chan, W.-H. Injurious Effects of Curcumin on Maturation of Mouse Oocytes, Fertilization and Fetal Development via Apoptosis. *Int. J. Mol. Sci.* **2012**, *13*, 4655–4672. [CrossRef] [PubMed]
68. Huang, F.-J.; Lan, K.-C.; Kang, H.-Y.; Liu, Y.-C.; Hsuuw, Y.-D.; Chan, W.-H.; Huang, K.-E. Effect of Curcumin on in Vitro Early Post-Implantation Stages of Mouse Embryo Development. *Eur. J. Obstet. Gynecol. Reprod. Biol.* **2013**, *166*, 47–51. [CrossRef] [PubMed]
69. Sanghavi, M.; Rutherford, J.D. Cardiovascular Physiology of Pregnancy. *Circulation* **2014**, *130*, 1003–1008. [CrossRef]
70. Ramlakhan, K.P.; Johnson, M.R.; Roos-Hesselink, J.W. Pregnancy and Cardiovascular Disease. *Nat. Rev. Cardiol.* **2020**, 1–14. [CrossRef]
71. Banez, M.J.; Geluz, M.I.; Chandra, A.; Hamdan, T.; Biswas, O.S.; Bryan, N.S.; Von Schwarz, E.R. A Systemic Review on the Antioxidant and Anti-Inflammatory Effects of Resveratrol, Curcumin, and Dietary Nitric Oxide Supplementation on Human Cardiovascular Health. *Nutr. Res.* **2020**, *78*, 11–26. [CrossRef]
72. Khan, K.S.; Wojdyla, D.; Say, L.; Gülmezoglu, A.M.; Van Look, P.F. WHO Analysis of Causes of Maternal Death: A Systematic Review. *Lancet* **2006**, *367*, 1066–1074. [CrossRef]
73. Nakashima, A.; Shima, T.; Tsuda, S.; Aoki, A.; Kawaguchi, M.; Yoneda, S.; Yamaki-Ushijima, A.; Cheng, S.-B.; Sharma, S.; Saito, S. Disruption of Placental Homeostasis Leads to Preeclampsia. *Int. J. Mol. Sci.* **2020**, *21*, 3298. [CrossRef] [PubMed]
74. Zenerino, C.; Nuzzo, A.M.; Giuffrida, D.; Biolcati, M.; Zicari, A.; Todros, T.; Rolfo, A. The HMGB1/RAGE Pro-Inflammatory Axis in the Human Placenta: Modulating Effect of Low Molecular Weight Heparin. *Molecules* **2017**, *22*, 1997. [CrossRef] [PubMed]
75. El-Sayed, A.A.F. Preeclampsia: A Review of the Pathogenesis and Possible Management Strategies Based on Its Pathophysiological Derangements. *Taiwan J. Obstet. Gynecol.* **2017**, *56*, 593–598. [CrossRef] [PubMed]
76. Duzyj, C.; Buhimschi, I.A.; Laky, C.A.; Cozzini, G.; Zhao, G.; Wehrum, M.; Buhimschi, C.S. Extravillous Trophoblast Invasion in Placenta Accreta Is Associated with Differential Local Expression of Angiogenic and Growth Factors: A Cross-Sectional Study. *BJOG* **2018**, *125*, 1441–1448. [CrossRef]
77. Wang, T.-Y.; Chen, J.-X. Effects of Curcumin on Vessel Formation Insight into the Pro- and Antiangiogenesis of Curcumin. *Evid.-Based Complementary Altern. Med.* **2019**, *2019*, 1–9. [CrossRef] [PubMed]
78. Kunnumakkara, A.B.; Bordoloi, D.; Padmavathi, G.; Monisha, J.; Roy, N.K.; Prasad, S.; Aggarwal, B.B. Curcumin, the Golden Nutraceutical: Multitargeting for Multiple Chronic Diseases. *Br. J. Pharmacol.* **2016**, *174*, 1325–1348. [CrossRef] [PubMed]
79. Payne, J.L.; Maguire, J.L. Pathophysiological Mechanisms Implicated in Postpartum Depression. *Front. Neuroendocrinol.* **2019**, *52*, 165–180. [CrossRef]
80. Dadi, A.F.; Miller, E.R.; Mwanri, L. Antenatal Depression and Its Association with Adverse Birth Outcomes in Low and Middle-Income Countries: A Systematic Review and Meta-Analysis. *PLoS ONE* **2020**, *15*, e0227323. [CrossRef]
81. Lahti-Pulkkinen, M.; Girchenko, P.; Robinson, R.; Lehto, S.M.; Toffol, E.; Heinonen, K.; Reynolds, R.M.; Kajantie, E.; Laivuori, H.; Villa, P.M.; et al. Maternal Depression and Inflammation During Pregnancy. *Psychol. Med.* **2019**, *50*, 1839–1851. [CrossRef]
82. Głąbska, D.; Guzek, D.; Groele, B.; Gutkowska, K. Fruit and Vegetable Intake and Mental Health in Adults: A Systematic Review. *Nutrients* **2020**, *12*, 115. [CrossRef]
83. Lopresti, A.L.; Hood, S.; Drummond, P.D. Multiple Antidepressant Potential Modes of Action of Curcumin: A Review of Its Anti-Inflammatory, Monoaminergic, Antioxidant, Immune-Modulating and Neuroprotective Effects. *J. Psychopharmacol.* **2012**, *26*, 1512–1524. [CrossRef]

84. Huang, Q.; Liu, H.; Suzuki, K.; Ma, S.; Liu, C. Linking What We Eat to Our Mood: A Review of Diet, Dietary Antioxidants, and Depression. *Antioxidants* **2019**, *8*, 376. [CrossRef] [PubMed]
85. Fusar-Poli, L.; Vozza, L.; Gabbiadini, A.; Vanella, A.; Concas, I.; Tinacci, S.; Petralia, A.; Signorelli, M.S.; Aguglia, E. Curcumin for Depression: A Meta-Analysis. *Crit. Rev. Food Sci. Nutr.* **2019**, *60*, 1–11. [CrossRef]
86. Al-Karawi, D.; Al Mamoori, D.A.; Tayyar, Y. The Role of Curcumin Administration in Patients with Major Depressive Disorder: Mini Meta-Analysis of Clinical Trials. *Phytother. Res.* **2015**, *30*, 175–183. [CrossRef] [PubMed]
87. Bhat, A.; Mahalakshmi, A.M.; Ray, B.; Tuladhar, S.; Hediyal, T.A.; Manthiannem, E.; Padamati, J.; Chandra, R.; Chidambaram, S.B.; Sakharkar, M.K. Benefits of Curcumin in Brain Disorders. *BioFactors* **2019**, *45*, 666–689. [CrossRef] [PubMed]
88. Lopresti, A.L.; Maes, M.; Maker, G.L.; Hood, S.; Drummond, P.D. Curcumin for the Treatment of Major Depression: A Randomised, Double-Blind, Placebo Controlled Study. *J. Affect. Disord.* **2014**, *167*, 368–375. [CrossRef]
89. Lopresti, A.L.; Maes, M.; Meddens, M.J.; Maker, G.L.; Arnoldussen, E.; Drummond, P.D. Curcumin and Major Depression: A Randomised, Double-Blind, Placebo-Controlled Trial Investigating the Potential of Peripheral Biomarkers to Predict Treatment Response and Antidepressant Mechanisms of Change. *Eur. Neuropsychopharmacol.* **2015**, *25*, 38–50. [CrossRef] [PubMed]
90. Sanmukhani, J.; Satodia, V.; Trivedi, J.; Patel, T.; Tiwari, D.; Panchal, B.; Goel, A.; Tripathi, C.B. Efficacy and Safety of Curcumin in Major Depressive Disorder: A Randomized Controlled Trial. *Phytother. Res.* **2013**, *28*, 579–585. [CrossRef] [PubMed]
91. Xu, Y.; Ku, B.; Tie, L.; Yao, H.; Jiang, W.; Ma, X.; Li, X. Curcumin Reverses the Effects of Chronic Stress on Behavior, the HPA Axis, BDNF Expression and Phosphorylation of CREB. *Brain Res.* **2006**, *1122*, 56–64. [CrossRef] [PubMed]
92. Szałach, Ł.P.; Lisowska, K.A.; Cubała, W.J. The Influence of Antidepressants on the Immune System. *Arch. Immunol. Ther. Exp.* **2019**, *67*, 143–151. [CrossRef]
93. Zhang, Y.; Li, L. Curcumin in Antidepressant Treatments: An Overview of Potential Mechanisms, Pre-Clinical/ Clinical Trials and Ongoing Challenges. *Basic Clin. Pharmacol. Toxicol.* **2020**. [CrossRef] [PubMed]
94. Barker, D.J.P. Developmental Origins of Adult Health and Disease. *J. Epidemiol. Community Health* **2004**, *58*, 114–115. [CrossRef] [PubMed]
95. Schlembach, D. Fetal Growth Restriction—Diagnostic Work-up, Management and Delivery. *Geburtshilfe Frauenheilkd* **2020**, *80*, 1016–1025. [CrossRef]
96. Chassen, S.; Jansson, T. Complex, Coordinated and Highly Regulated Changes in Placental Signaling and Nutrient Transport Capacity in IUGR. *Biochim. Biophys. Acta Mol. Basis Dis.* **2020**, *1866*, 165373. [CrossRef]
97. Yan, E.; Zhang, J.; Han, H.; Wu, J.; Gan, Z.; Wei, C.; Zhang, L.; Wang, C.; Wang, T. Curcumin Alleviates IUGR Jejunum Damage by Increasing Antioxidant Capacity through Nrf2/Keap1 Pathway in Growing Pigs. *Animals* **2019**, *10*, 41. [CrossRef]
98. Walani, S.R. Global Burden of Preterm Birth. *Int. J. Gynecol. Obstet.* **2020**, *150*, 31–33. [CrossRef]
99. Reynolds, L.P.; Borowicz, P.P.; Caton, J.S.; Crouse, M.S.; Dahlen, C.R.; Ward, A.K. Developmental Programming of Fetal Growth and Development. *Vet. Clin. N. Am. Food Anim. Pract.* **2019**, *35*, 229–247. [CrossRef]
100. European Chemicals Agency (ECHA). MSC Unanimously Agrees That Bisphenol a Is an Endocrine Disruptor. 2017. Available online: https://echa.europa.eu/-/msc-unanimously-agrees-that-bisphenol-a-is-an-endocrine-disruptor (accessed on 13 September 2020).
101. Filardi, T.; Panimolle, F.; Lenzi, A.; Morano, S. Bisphenol A and Phthalates in Diet: An Emerging Link with Pregnancy Complications. *Nutrients* **2020**, *12*, 525. [CrossRef]
102. Bhattacharya, S. Medicinal Plants and Natural Products Can Play a Significant Role in Mitigation of Mercury Toxicity. *Interdiscip. Toxicol.* **2018**, *11*, 247–254. [CrossRef]
103. Al-Saleh, I.; Nester, M.; Abduljabbar, M.; Al-Rouqi, R.; Eltabache, C.; Al-Rajudi, T.; Elkhatib, R. Mercury (Hg) Exposure and Its Effects on Saudi Breastfed Infant's Neurodevelopment. *Int. J. Hyg. Environ. Health* **2016**, *219*, 129–141. [CrossRef]
104. Nam, S.M.; Kim, J.W.; Yoo, D.Y.; Choi, J.H.; Kim, W.; Jung, H.Y.; Won, M.-H.; Hwang, I.K.; Seong, J.K.; Yoon, Y.S. Comparison of Pharmacological and Genetic Inhibition of Cyclooxygenase-2: Effects on Adult Neurogenesis in the Hippocampal Dentate Gyrus. *J. Vet. Sci.* **2015**, *16*, 245–251. [CrossRef] [PubMed]

105. Fujimura, K.; Mitsuhashi, T.; Takahashi, T. Adverse Effects of Prenatal and Early Postnatal Exposure to Antiepileptic Drugs: Validation from Clinical and Basic Researches. *Brain Dev.* **2017**, *39*, 635–643. [CrossRef] [PubMed]
106. Kaminen-Ahola, N. Fetal Alcohol Spectrum Disorders: Genetic and Epigenetic Mechanisms. *Prenat. Diagn.* **2020**, *40*, 1185–1192. [CrossRef]

Publisher's Note: MDPI stays neutral with regard to jurisdictional claims in published maps and institutional affiliations.

© 2020 by the authors. Licensee MDPI, Basel, Switzerland. This article is an open access article distributed under the terms and conditions of the Creative Commons Attribution (CC BY) license (http://creativecommons.org/licenses/by/4.0/).

 nutrients

Review

Interaction between Gut Microbiota and Curcumin: A New Key of Understanding for the Health Effects of Curcumin

Beatrice Scazzocchio [1], Luisa Minghetti [2] and Massimo D'Archivio [1,*]

1. Center for Gender-Specific Medicine, Istituto Superiore di Sanità, 00161 Rome, Italy; beatrice.scazzocchio@iss.it
2. Research Coordination and Support Service, Istituto Superiore di Sanità, 00161 Rome, Italy; luisa.minghetti@iss.it

* Correspondence: massimo.darchivio@iss.it

Received: 23 July 2020; Accepted: 17 August 2020; Published: 19 August 2020

Abstract: Curcumin, a lipophilic polyphenol contained in the rhizome of *Curcuma longa* (turmeric), has been used for centuries in traditional Asian medicine, and nowadays it is widely used in food as dietary spice worldwide. It has received considerable attention for its pharmacological activities, which appear to act primarily through anti-inflammatory and antioxidant mechanisms. For this reason, it has been proposed as a tool for the management of many diseases, among which are gastrointestinal and neurological diseases, diabetes, and several types of cancer. However, the pharmacology of curcumin remains to be elucidated; indeed, a discrepancy exists between the well-documented in vitro and in vivo activities of curcumin and its poor bioavailability and chemical instability that should limit any therapeutic effect. Recently, it has been hypothesized that curcumin could exert direct regulative effects primarily in the gastrointestinal tract, where high concentrations of this polyphenol have been detected after oral administration. Consequently, it might be hypothesized that curcumin directly exerts its regulatory effects on the gut microbiota, thus explaining the paradox between its low systemic bioavailability and its wide pharmacological activities. It is well known that the microbiota has several important roles in human physiology, and its composition can be influenced by a multitude of environmental and lifestyle factors. Accordingly, any perturbations in gut microbiome profile or dysbiosis can have a key role in human disease progression. Interestingly, curcumin and its metabolites have been shown to influence the microbiota. It is worth noting that from the interaction between curcumin and microbiota two different phenomena arise: the regulation of intestinal microflora by curcumin and the biotransformation of curcumin by gut microbiota, both of them potentially crucial for curcumin activity. This review summarizes the most recent studies on this topic, highlighting the strong connection between curcumin and gut microbiota, with the final aim of adding new insight into the potential mechanisms by which curcumin exerts its effects.

Keywords: gut microbiota; curcumin; polyphenols; health

---

1. Introduction

Curcumin, one of the major curcuminoids contained in the rhizome of *Curcuma longa* (turmeric), is a lipophilic polyphenol that has been used for centuries as an essential tool of traditional medicine in Asia [1]. Nowadays, it is widely used as dietary spice, but also in cosmetic and pharmaceutical industries [2].

Curcumin has received considerable attention in the last years for its pharmacological activities. Due to the presence of conjugated double bonds in its chemical structure, this polyphenol serves as an effective electron donor to counteract the production of reactive oxygen species (ROS) in many redox reactions [3], acting as a potent antioxidant. In addition, it has other important biological functions, such as anti-inflammatory, antitumor, antimicrobial, and antiviral ones [4–7].

Different studies highlighted that curcumin, like other dietary polyphenols, counteracts the effects of toxic damage in different tissues [8,9] and, in addition, it is able to interfere with key cancer-associated signaling pathways by directly targeting proteins or regulating gene expression [10,11]. According to its biological activities, curcumin has been proposed as a potential treatment for many diseases, among which are gastrointestinal, cardiovascular, and neurological disorders, diabetes, and several types of cancer [12,13].

Unfortunately, these findings have not been consistently supported through human clinical trials, except for the treatment of arthritis, pain, and major depressive disorder [14–16]. Consequently, the real biological activities of curcumin remain to be better elucidated; indeed, a discrepancy exists between the well-detailed in vitro and in vivo activities of curcumin and its poor bioavailability and chemical instability that should limit any healthy therapeutic outcome.

High concentrations of curcumin have been detected in the gastrointestinal tract after oral administration, and this has led to the hypothesis that the polyphenol directly exerts its regulatory effects on gut microbiota, explaining in this way the paradox between its low systemic bioavailability and its wide pharmacological activities that would be mediated by the gut microbiota.

It is well known that the gut microbiota has several important roles in normal human physiology, and its composition can be influenced by a multitude of environmental and lifestyle factors [17–19]. Accordingly, any perturbations in gut microbiome profile, that is, dysbiosis, can have a key role in human disease progression. Interestingly, curcumin and its metabolites have been shown to influence the gut microbiota [20,21]. It is worth noting that the interaction between curcumin and gut microbiota gives rise to two different phenomena: the first is the direct regulation of intestinal microflora by curcumin and the second is the biotransformation of curcumin by gut microbiota, yielding active metabolites [22,23]; both these phenomena seem to be crucial for the activity of curcumin.

This review summarizes the most recent studies on the reciprocal interaction between curcumin and gut microbiota, with the final aim to provide novel insight for defining future effective preventive strategies and microbiota-targeted therapies using curcumin. The observed high concentrations of curcumin in the GI tract after oral administration can lead to two major effects: an altered gut microbiota and the modulation of intestinal functions. We focused our literature search on the altered gut microbiota. The search was conducted in PubMed on 29 May, 2020. The search syntax was "curcumin", "microbiota", and "microbiome". The scientific literatures were searched for in vivo, experimental and clinical studies, and human randomized controlled trials, reporting results on the interaction between curcumin and gut microbiota and vice versa.

The publication date was considered from 29 May, 2015 to 29 May, 2020 (five years).

In total, 89 titles were found through database search: after excluding review articles, studies in animal models and in humans were identified and discussed more in detail.

2. Curcumin: Metabolism and Bioavailability

Curcumin, [1,7-bis(4-hydroxy-3-methoxyphenyl)-1,6-heptadiene-3,5-dione], is the most representative polyphenol component extracted from the rhizome of *Curcuma longa* (turmeric). It is almost completely insoluble in water but it is easily soluble in organic solvents such as acetone and ethanol [24], and it is quite stable in the acidic pH of the stomach [25]. From a chemical viewpoint, the molecule is symmetric with two similar aromatic rings, and presents conjugate double bonds utilized as effective electron donor to hinder ROS formation.

Curcumin is widely consumed, particularly in Asia, as one of the culinary ingredients in food recipes. In the recent years, this polyphenol has increasingly received worldwide attention for its multiple pharmacological activities, primarily anti-inflammatory and antioxidant ones [26–29]. A recent meta-analysis has evidenced curcumin efficacy as a free radical scavenger and an inhibitor of malondialdehyde production, showing its ability in improving levels of antioxidants in diseased individuals susceptible to oxidative stress. The reduction of oxidative stress by curcumin supplementation was dependent on the dose and the duration of treatment [30].

Curcumin and the whole turmeric rhizome have some beneficial effects in the treatment of chronic diseases such as gastrointestinal, cardiovascular, and neurological disorders, diabetes, and several types of cancer [31–35]. Clinical trials based on curcumin administration have been published or are currently in progress, pointing out the expanding interest of the scientific community on the therapeutic potential of curcumin [36–39].

The safety of orally administered curcumin has been clearly demonstrated: the US Food and Drug Administration has approved curcumin as a compound "generally recognized as safe" and also JECFA (The Joint FAO/WHO Expert Committee on Food Additives) and EFSA (European Food Safety Authority) reported the ADI (acceptable daily intake) value of 0–3 $mg \cdot kg^{-1}$ for curcumin [40]. However, it has to been taken into account that very few reports on the potential adverse effects of curcumin exist: recently, a case report showed a liver injury attributed to a curcumin supplement in a woman with jaundice [41]. Curcumin could also interfere with systemic iron metabolism, suggesting limited application of this compound in patients with chronic disease or anemia [42].

In spite of its therapeutic potential against a wide spectrum of human pathologies, curcumin is known for its poor gastrointestinal absorption and low bioavailability, mainly attributed to water insolubility, and rapid metabolism and excretion [43].

In humans, after curcumin oral administration, glucuronide conjugates and sulfate conjugates are detected in blood, while intact curcumin is poorly detected [44]. As a first step, ingested curcumin passes through the stomach, where practically no absorption takes place. Due to its resistance to low pH, curcumin, without any chemical modifications, reaches the large intestine and undergoes extensive phase I and II metabolism. Firstly, it is metabolized by phase I enzymes: different reductases introduce reactive and polar groups in their substrates, yielding active metabolites, namely, dihydrocurcumin, tetrahydrocurcumin, and hexahydrocurcumin [45–47] (Figure 1). This reductive metabolic reaction of curcumin occurs extensively in enterocytes and hepatocytes [46,48,49].

Then, these phase I metabolites undergo phase II metabolism: in vitro and in vivo study have previously demonstrated that curcumin and its reductive metabolites are easily conjugated [45,46,50]. Glucuronidases and sulfotrasferases are capable of conjugating glucuronic acid or sulphate molecule, respectively, to any of the hydroxyl groups [51], to produce the corresponding glucuronide and sulfate O-conjugated metabolites [22] (Figure 1). The conjugation process typically involves the addition of a single moiety, although double glucuronidation has been reported in isolated liver microsomes [52], and diglutathionylated curcumin has been found in isolated reaction systems [53]. The predominating pathway of conjugation is represented by glucuronidation; indeed the glucuronide of curcumin is usually found as the major metabolite of curcumin in body fluids, organs, and cells [47,54] even though, due to the increase of molecular weight, these metabolites are less active than their substrates [54,55].

The very low concentrations of curcumin in blood plasma and urine after oral administration have been demonstrated in both animal and human studies. It is important to underline that this could be due to the fact that curcumin derivatives are not always assayed, thus underestimating its absorption. In rats administered with an oral dose of 1000 mg/kg of curcumin, about 75% was excreted in feces, and a very low amount was detected in the urine [56]. An oral dose of 0.1 g/kg administered to mice yielded a peak plasma concentration of free curcumin that was only 2.25 mg/mL [47]. Even after high oral doses (up to 8 g/day), serum levels of curcumin were undetectable in humans [57,58]. In another clinical trial with an oral dose of 3.6 g of curcumin, a plasma level as low as 11.1 nmol/L was detected an hour after oral dosing [59].

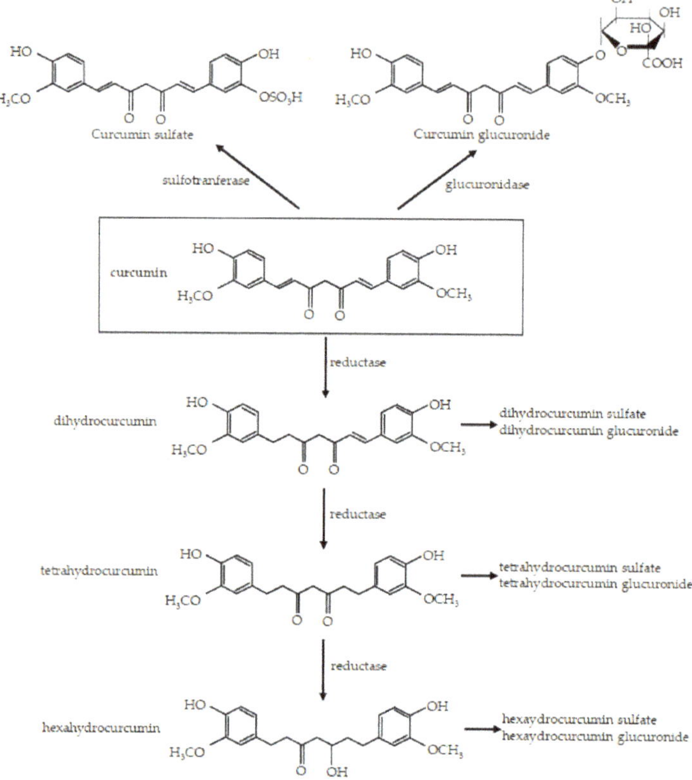

Figure 1. Metabolism of curcumin. Reduction pathway and the two main conjugation pathways, glucuronidation and sulfatation, are shown.

This low bioavailability of curcumin after oral administration could largely restrict its pharmacological potential and consequently, its clinical application [60,61]. As a result, different delivery systems including micelles, liposomes, phospholipid complexes, microemulsions, nanostructured lipid carriers, and biopolymer nanoparticles have been developed to increase curcumin bioavailability. Specifically, Kato et al., by using a new formulation where curcumin was dispersed with colloidal nanoparticles, succeeded in improving hyperglycemia via stimulation of GLP-1 (glucagon-like peptide 1) secretion and the subsequent insulin secretion [62], suggesting a possible use of curcumin formulation in diabetes treatment. Such formulations may also be effective against inflammatory status and osteoarthritis [63,64], even if the dosage represents a critical point because it should remain quite low to avoid toxicity. Very recently, Chen et al. [65] clearly demonstrated that the supplementation of nanobubble curcumin extract in mice had a beneficial effect on health and exercise performance, helping mice to overcome physical fatigue. Moreover, several natural agents have been used to improve curcumin bioavailability, most of which work by blocking curcumin metabolism in order to increase its absorption [66]. Among these agents, piperine, the major active component of black pepper [67], probably represents the most utilized one [68–70].

Some recent papers have also showed the importance of food matrix in curcumin absorption [71,72], highlighting an enhanced bioavailability when it is consumed as fresh or powdered turmeric with respect to supplements, which could be due to the synergic activity with other turmeric compounds and/or to a turmeric matrix effect [71].

However, as previously stated, curcumin could exert its main regulative effects primarily in the gut, where high concentrations are present after oral administration [73]. Actually, curcumin is able to modulate directly intestinal barrier function as well as dysregulated signaling pathways. On the other hand, it might act at intestinal level by promoting changes in the composition and diversity of the gut microbiota [74]. The possible role of the gut microbiota in the mechanisms responsible for the biological activities of curcumin represents an interesting and attractive area of research and will be discussed in detail in the following paragraphs.

3. Gut Microbiota

There is considerable attention given to the substantial discrepancy between the strong biological effects of some functional foods and the poor bioavailability of these substances. For orally administered drugs and functional food, the bioavailability is defined as "the quantity or the proportion of the ingested dose that is directly absorbed in the small intestine to enter into circulation". However, since the gut microbiota is actually considered as an effective bioreactor in the human intestinal tract and considering the emerging interactions between functional food and gut microbiota, probably we should think of a redefinition of the concept of bioavailability.

Recently, high concentrations of curcumin have been detected in the gastrointestinal tract after oral administration [20], thus suggesting that this polyphenol could directly interact with the gut microbiota exerting its regulatory effects. Before analyzing in more detail the mutual interaction between curcumin and gut microbiota, discussing the more recent findings on this topic, we will give a brief overview on human gut microbiota.

In the last decades, human microbiota has emerged as an area of utmost interest, as many studies have highlighted its impact on health and diseases [75]. It develops together with the host and fulfils essential physiological functions for the host, such as preventing infection, promoting the immune system maturation [76], participating in the regulation of nutritional absorption and metabolism [77], producing soluble B-vitamins (cobalamin, thiamine, pyridoxine, biotin, folate, nicotinic acid, pantothenic acid) and vitamin K lactic acid [78,79]. The colonization of newborn microbiota begins in utero [80] and changes suddenly during the first year of life.

The gut microbiota is a hugely complex ecology of organisms, primarily comprising many classes of bacteria (50 bacterial phyla and about a thousand of bacterial species), fungi, viruses, and a few other species [81,82]. Collectively, the whole of all microbiota genes, the microbiome, is 150 times larger than the human genome [83,84]. The gut microenvironment mainly favors the growth of bacteria from seven predominant phyla: Bacteroidetes, Firmicutes, Actinobacteria, Fusobacteria, Proteobacteria, Verrucomicrobia, and Cyanobacteria [85], with the first two phyla constituting more than 90% of the total gut population. Most of the species under Firmicutes, such as *Clostridium*, *Eubacterium*, and *Ruminococcus*, are the most representative in the gut [86].

The composition of microbiota varies greatly within each individual, in which about 150 bacterial species can predominate, getting benefits from the nutrient-rich environment of the gut and performing protective, metabolic, and structural functions. Understanding this variability in the "healthy microbiome" represents one of the major challenge in microbiota research.

The gut microbial community is very dynamic and has specific properties that allowed it to colonize the gut, among which are the possession of enzymes able to utilize the available nutrients, the right cell-surface molecular pattern to attach at the "right" habitat, and the ability to evade bacteriophages [86]. Usually, these microbes are mainly involved in nutrient metabolism through fermentation of complex carbohydrates. This leads to the synthesis of short-chain fatty acids, well known for their anti-inflammatory and anticancer properties, which represent an important energy source for colonocytes. Intestinal microorganisms might also influence lipid and protein metabolism [87].

There is growing evidence that any perturbation in gut microbiota composition (dysbiosis), associated with a reduced diversity and the predominance of a few pathogenic taxa, is closely linked to many human diseases [88–90]. In particular, dysbiosis has been related to pathological gastrointestinal conditions, such as inflammatory bowel disease and colorectal cancer (CRC) [91,92], but also to obesity, diabetes, asthma, and allergies [93,94].

Microbiota composition can be influenced by a multitude of environmental and lifestyle factors, among which dietary habits have a great impact on gut microbiome diversity. Del Bas et al. [95]

have highlighted that the correct balance between fibers, simple carbohydrates, and fats is crucial in determining the abundance of different gut microbial populations. It has also been shown that unbalanced diets cause alterations in gut microbiota composition, resulting in modification of gut permeability and in gut low-grade inflammation [96].

In view of the above, it is understandable the increasing interest in defining the effective interplay between curcumin and gut microbiota.

4. Curcumin Modifies Gut Microbiota

As stated above, curcumin preferentially accumulates in the gastrointestinal tract after oral or intraperitoneal administration, and therefore it is reasonable to hypothesize that this polyphenol may exert its regulatory effect by modulating the microbial richness, diversity, and composition of the intestinal microflora [97]. Many in vivo studies have confirmed this hypothesis, and the most recent and interesting are discussed below.

A research has been carried out on adult healthy volunteers [98] asked to consume daily for 28 days a dried *Curcuma longa* extract containing a standardized amount of curcuminoids. The product was formulated in tablets, each one containing 500 mg of *Curcuma longa* (equivalent to 100 mg of curcuminoids). Metabolome analysis was performed to better understand the changes of 24-h urinary metabolome composition following the extract consumption. The analysis revealed that curcumin induced changes in urinary metabolites. In particular, metabolites related to fatty acid metabolism, involved in energy production, and compounds related to inflammation were detected, suggesting a key role of curcumin on the regulation of metabolic and anti-inflammatory pathways. Furthermore, changes of several microbial metabolites clearly revealed, although indirectly, intestinal absorption of curcuma constituents and gut microbiota metabolic activity, thus demonstrating an interaction between curcumin and gut microbiota.

4.1. Curcumin Favors Beneficial Bacterial Strains in Gut Microbiota

Recently, an increasing number of studies have suggested that gut dysbiosis is linked with many metabolic diseases, and curcumin seems to have beneficial effects on gut microbiota, favoring the growth of beneficial bacteria strains.

Indeed, Zhai et al. [99] explored the effect of curcumin on ochratoxin-induced liver oxidative injury in an animal model of liver disease. A total of 720 ducks were randomly assigned into four different groups: control, ochhratoxin, curcumin (ducks fed a diet with 400 mg/kg curcumin), and ochhratoxin plus curcumin and treated for 21 days. The ducks were provided with the different pelleted diets and ad libitum access to feed and water. The authors demonstrated that curcumin counteracted ochhratoxin-induced oxidative injury and lipid metabolism disruption. By 16S rRNA gene sequencing of gut microbiota it was shown that curcumin supplementation was also able to neutralize the decrease in butyric acid-producing bacteria induced by ochratoxin, and increased the richness and diversity of gut microbiota [99]. Thus, the authors hypothesized that curcumin could alleviate liver oxidative injury by modulating the gut microbiota.

Rats fed a high-fat diet show an altered hepatic metabolism accompanied by modified gut microbiota composition and increased intestinal permeability. In a nonalcoholic fatty liver disease (NAFLD) rat model induced by high-fat diet [20], rats were randomly divided into three groups fed standard diet, high-fat diet, or high-fat diet plus curcumin (200 mg/kg of curcumin by gastric gavage, daily for four weeks), respectively. The addition of curcumin to the diet significantly shifted the composition of the microbiota toward that of lean control rats fed a standard diet. In particular, curcumin was able to significantly counteract the high-fat-diet-induced abundance of several genera that have previously been associated to diabetes and inflammation, such as *Ruminococcus* [100]. Moreover, the treatment with curcumin succeeded to decrease thirty-six potentially harmful bacterial strains positively correlated with hepatic steatosis [20]. These data suggest that curcumin may have the gut microbiota as target in the treatment of liver steatosis induced by high-fat diet.

Other studies confirmed that oral curcumin administration was able to remarkably shift the ratio between beneficial and harmful bacteria in gut microbiota community in favor of beneficial bacteria strains, such as *Bifidobacteria*, *Lactobacilli*, and butyrate-producing bacteria, and reduces the abundance of the pathogenic ones, such as *Prevotellaceae*, *Coriobacterales*, *Enterobacteria*, and *Rikenellaceae*, often associated to the onset of systemic diseases [101–103]. In particular, Shen et al. [101] investigated the regulative effects of oral curcumin administration on the gut microbiota of C57BL/6 mice. After receiving daily curcumin gavage in a dose of 100 mg/kg body weight for 15 days, the gut microbial composition was significantly modified, affecting the abundance of several representative pathogenic families such as *Prevotellaceae*, *Bacteroidaceae*, and *Rikenellaceae*.

In APP/PS1 mice, a model of Alzheimer's disease, it was found that curcumin administration improved the spatial learning and memory abilities, also reducing the amyloid plaque burden in the hippocampus [104]. Concomitantly, curcumin altered significantly the relative abundances of bacterial strains such as *Bacteroidaceae*, *Prevotellaceae*, and *Lactobacillaceae*, which have been reported to be key bacterial species associated with Alzheimer's disease development [23]. In another study [105], estrogen deficiency induced in rats by ovariectomy gave rise to changes in the structure and distribution of intestinal microflora. Curcumin, administered at 100 mg/kg/day by oral gavage to ovariectomised rats, was able to partially reverse changes in the diversity of gut microbiota after 12 weeks of treatment [105].

It is important to underline that curcumin treatment decreases the microbial abundance of cancer-related species [106], such as *Prevotella* that were found to be greater in stool from CRC patients than in that from cancer-free patients [107]. Mice with colon cancer were fed different pelleted diets, with a calculated human equivalency dose of curcumin ranging from 8/mg/kg/day to 162 mg/kg/day [102]. Curcumin administration, at the highest dose, reduced or eliminated colon tumor burden, increasing *Lactobacilli* and reducing *Coriobacterales*. It has also been clearly demonstrated that curcumin treatment reduces several *Ruminococcus* species [108]; this represents an interesting finding because increased population of *Ruminococcus* species has been linked to CRC occurrence [90,109], even if the pathogenic role of *Ruminococcus* in cancer development has not been yet fully clarified. Moreover, in mice treated with a mutagenic compound, dietary curcumin was able to restore to control levels the amount of *Lactobacilli* [102], which have been shown to possess antitumoral function [110]. All these results support the potential anticancer activity of curcumin, at least against CRC, and have prompted researchers to start clinical trials focused to define this issue.

Peterson et al., in a human randomized placebo-controlled trial [108], investigated the effects of turmeric and curcumin dietary supplementation vs. placebo on 30 healthy subjects (10 for each group) previously advised not to consume any other curcumin-containing food or supplements for the entire study period. The turmeric tablets contained 1000 mg *Curcuma longa* plus 1.25 mg extract of piperine; the curcumin tablets contained 1000 mg curcumin plus 1.25 mg extract of piperine; the subjects were instructed to take three tablets orally with food, twice a day (total 6000 mg daily). Microbiota analyses were performed at baseline and after 8 weeks of treatment. All the subjects showed both significant variations of microbiota composition over the time and an individualized response to treatment. The intestinal microflora varied significantly from person to person, and the responses to the treatment were not uniform across individuals. However, comparing the number of bacterial species present in each group before and after the treatments, the placebo group showed an overall reduction in species by 15%, whereas the turmeric- and curcumin-treated groups displayed increases by 7% and 69%, respectively.

All these studies strongly suggest a protective effect of curcumin most likely based on its ability to promote an evident shift from pathogenic to beneficial bacteria strains in the gut.

However, it must be highlighted that these studies provide data hardly comparable because they use different doses and formulations. Moreover, the prebiotic effect of curcumin on gut microbiota is probably due to an indirect effect. Indeed, it is unlikely that curcumin metabolism provides a "direct" fitness advantage to any bacterial species: probably its prebiotic effect is the result of the induced host changes that in turn alter the gut microbiota.

In the light of the increasing data supporting a role of gut microbiota in the pathogenesis of many diseases, the research findings defining the ability of curcumin to positively modulate gut microbiota may help us to better understand its therapeutic benefits.

4.2. Curcumin Acts on Intestinal Barrier Function

Curcumin not only modifies the composition of the microbiota but might also enhance the function of the intestinal barrier. The intestinal barrier primarily is composed of four different layers. In the first one, the presence of alkaline phosphatase can detoxify bacterial endotoxin lipopolysaccharide. The second layer (mucosa) inhibits the entry of pathogenic bacteria. The third layer consists of tight junctions between intestinal epithelial cells, which form a barrier against bacterial endotoxin. Antibacterial proteins, which do not allow bacteria to cross the intestinal barrier, constitute the final layer [111]. Obviously, any defects in the intestinal barrier integrity can provoke an invasion of bacteria into normal colonic tissue, giving rise to a dysregulation of intestinal epithelial cells [112] and a subsequent local inflammation. Chronic inflammation underlies the development of western-induced metabolic diseases, such as diabetes or atherosclerosis, but it is also believed to be one of the primary reasons for the initiation of CRC.

In vitro studies have demonstrated that curcumin represents a potential compound to restore disrupted intestinal permeability. Indeed, in CaCo2 cells, curcumin is able to attenuate the disruption of intestinal epithelial barrier function, counteracting LPS-induced IL-1β secretion and preventing tight junction protein disruption [113,114]. Furthermore, curcumin was also able to decrease p38 MAPK activation, induced by IL-1β, and the subsequent raise in the phosphorylation of tight junction proteins and resulting disruption of their normal arrangement [114].

These results have been confirmed in animal models; in rats fed a high-fat diet for 16 weeks, curcumin treatment (200 mg/kg by daily oral gavage) improved the structure of intestinal tight junctions, also reducing serum concentrations of TNF-α, and LPS and upregulating the expression of occludin in the intestinal mucosa [115]. In a similar study, mice fed a Western diet for 16 weeks, and supplemented with curcumin (100 mg/kg by daily oral gavage), significantly improved intestinal barrier function, restoring the intestinal alkaline phosphatase activity and the expression of tight junction proteins ZO-1 and claudin-1 [116].

It is well known that the decreased expression of tight junction proteins, such as ZO-1 and occludin, plays a key role also in the pathophysiology of NAFLD [117]. Starting from this assumption, Feng et al. [20] demonstrated the beneficial effect of curcumin on the intestinal barrier integrity in NAFLD rats. Immunohistochemical data and western blot analysis showed that protein expression levels of ZO-1 and occludin were reduced in distal ileal tissues from NAFLD rats but were restored following curcumin administration (200 mg/kg of curcumin by gastric gavage, daily for four weeks). This work clearly highlighted that curcumin, by improving intestinal barrier integrity in vivo, might have a role in a novel approach addressed to NAFLD therapy.

All together, these results provide convincing evidence that curcumin contributes to the maintenance of intestinal barrier integrity, and thus may represent a new tool in preventive/therapeutic strategies against intestinal pathologies. As previously stated in the introduction, the studies included in this review were obtained using as keywords "curcumin and microbiome/microbiota". This could represent a limitation of this section.

4.3. Curcumin Effects on Gut Inflammation

In a randomized placebo-controlled human trial, fifty-eight NAFLD patients were randomly allocated into two groups, which either received 250 mg curcumin–phospholipid delivery system, which was equivalent to 50 mg/day pure curcumin, or placebo [118]. Metabolomic analysis showed the beneficial effects of curcumin on biomarkers of oxidative stress and inflammation, which are considered two features of NAFLD. The authors suggested that curcumin treatment counteracted the increase of some bacterial strains that were changed during NAFLD progression.

An in vivo animal study [119] reported that a newly developed nanoparticle curcumin actively improves inflammation in mice with DSS (dextrane sulfate sodium)-colitis by inhibiting the expression of proinflammatory mediators and inducing Treg expansion which was also accompanied by the increase of fecal butyrate levels. Curcumin was mixed with the powder form of a normal rodent diet (containing 0.2% (*w*/*w*) nanoparticle curcumin): this compound was able to suppress the activation of NF-κB and the expression of proinflammatory mediators in the colonic epithelial cells of treated mice.

Alternatively, curcumin could act by attenuating LPS-induced inflammation by inhibiting the activation of TLR4/MyD88/NF-κB signaling pathways [120,121]. Moreover, curcumin has been shown to inhibit NF-κB nuclear translocation and to mitigate the expression of other pro-inflammatory genes that are overactivated in cancer [122].

In Weaned piglets fed a diet supplemented with curcumin (300 mg/kg of curcumin, mixed with the normal diet) for 28 days, Gan et al. demonstrated that this polyphenol was able to alleviate inflammation downregulating the expression of TLR4 by inhibiting *Escherichia coli* proliferation [123].

Unfortunately, only limited studies have been performed on human subjects up to now and, despite all the beneficial effects of curcumin described so far in in vivo studies, these results must be consistently supported through larger human clinical trials.

5. Gut Microbiota Metabolizes Curcumin

It is worth noting that the interaction between curcumin and microbiota is bidirectional (Figure 2). Consequently, an important effect of gut microbiota on curcumin has been evidenced. The metabolic transformation of curcumin does not occur only in the enterocytes and in hepatocytes, but it is also carried out by enzymes produced by the gut microbiota that generate many active metabolites [44]. The biological activity of curcumin metabolites may differ from that of the native curcumin, and specific biological properties attributed to curcumin actually depend on bioactive metabolites produced by gut microbiota digestion [124].

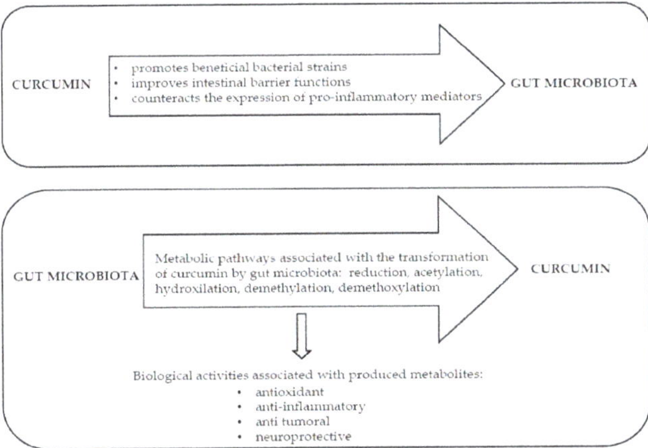

Figure 2. Reciprocal interaction between curcumin and gut microbiota.

Like for other dietary polyphenols such as anthocyanins, the biological activities of curcumin are related not only to the absorption rate, but also to the digestion by intestinal flora that leads to active curcumin metabolites. Moreover, the possible beneficial effects of curcumin depend not only on the dietary intake of curcumin, but also on the individual capacity of metabolizing it, that is, ultimately, on the composition of the gut microbiota of each person.

Several enteric bacteria capable of modifying curcumin have been identified: analyses of microorganisms isolated from human feces have previously shown that *Escherichia coli* represents the bacteria with the highest curcumin-metabolizing activity, via the NADPH-dependent curcumin/dihydrocurcumin reductase. This enzyme is able to convert curcumin into dihydrocurcumin, and then in the final product, tetrahydrocurcumin [125]. Other microorganisms such as *Bifidobacteria longum, Bifidobacteria pseudocatenulaum, Enterococcus faecalis, Lactobacillus acidophilus,* and *Lactobacillus casei* represent relevant bacterial strains able to metabolize curcumin, with a reduction of the parent compound higher than 50% [126].

Lou et al. [22] assessed whether curcumin was metabolized in vitro by human intestinal microorganisms, performing an ultra-performance liquid chromatography analysis, coupled with quadrupole time-of-flight mass. Curcumin (100 μM) was added to the intestinal bacteria culture obtained from fresh human fecal samples. The results clearly indicate that curcumin was extensively biotransformed by intestinal microflora, yielding 23 different metabolites also revealing different metabolic pathways, such as acetylation, hydroxylation, reduction, demethylation, or a combination of them, by which curcumin was metabolized by human intestinal microflora. The predominant metabolites produced by this intestinal microflora system derive from the reduced curcumin.

In addition to the known reductive metabolism of curcuminoids, alternative biotransformation pathways by human gut microbiota have been highlighted [127]. Fresh fecal samples from three healthy volunteers were taken for the preparation of mixed-cell cultures, and a curcuminoid mixture (10 mM), composed of curcumin, demethoxycurcumin, and bisdemethoxycurcumin, was added to begin the biotransformation. The human intestinal bacterium *Blautia* sp. (MRG-PMF1), through a demethylation process, was able to convert curcumin into two metabolites, bis-demethylcurcumin and demethylcurcumin [127]. Interestingly, the demethylated curcuminoid metabolites were present only in the culture where *Blautia* sp. was added, thus confirming that curcuminoids are differently metabolized depending on the individual fecal microflora.

In an in vitro model containing human fecal starters to investigate the colonic metabolism of curcuminoids [128], it was demonstrated that after 24 h of fermentation, up to 24% of curcumin, 61% of demethoxycurcumin, and 87% of bisdemethoxycurcumin were degraded by the human fecal microbiota. Three main metabolites were detected in the fermentation cultures: tetrahydrocurcumin, dihydroferulic acid, and 1-(4-hydroxy-3-methoxyphenyl)-2-propanol. There is evidence that curcumin metabolites have properties and potency similar to curcumin: tetrahydrocurcumin exhibits the same physiological and pharmacological properties of the parental compound, probably by means of the beta-diketone moiety, as well as phenolic hydroxyl groups [129]. Moreover, tetrahydrocurcumin is able to prevent oxidative stress and neuroinflammation, exhibiting also anticancer effects, probably due to inhibition of significant cytokine release, such as IL-6 and TNFα [130]. Consequently, the bacterial breakdown products should be considered in further studies on curcumin since they could be responsible for beneficial effects.

In transgenic mice with Alzheimer disease [23], the biotransformation of curcumin induced by gut microbiota was studied. The mice were divided in three groups (high-level curcumin group, low-level curcumin group, and control). Curcumin was administered daily at 200 mg/kg body weight (high) or 50 mg/kg (low) by oral gavage for 3 months, and the metabolites of curcumin from the feces of mice were identified by HPLC-Q-TOF/MS spectroscopy analysis. The authors shown that administered curcumin was transformed by gut microbiota through reduction, demethoxylation, demethylation, and hydroxylation processes. Eight metabolites of curcumin were identified (bisdemethylated hexahydrocurcumin, demethylated and dehydroxylated hexahydrocurcumin, demethylcurcumin, demethylated and demethoxylated curcumin, hydroxylatedcurcumin, dihydrocurcumin, hexahydrocurcumin, and demethylated hexahydrocurcumin). It is important to highlight that many of these metabolites have been reported to exhibit neuroprotective ability [131–133].

These findings not only could explain the paradox between the pharmacological effect of curcumin and its poor bioavailability, but also suggest that curcumin transformed by gut microbiota might act as an useful tool for microbiome-targeting therapies for Alzheimer disease [23].

Overall, all these results demonstrated that gut microbiota had a profound impact on the biotransformation of curcumin, also showing the huge potential of curcumin metabolites produced by the intestinal microflora as promising substances for the prevention or treatment of many diseases. The bidirectional interaction between curcumin and gut microbiota is reported in Figure 2.

These results clearly highlight how the different composition of the microbiota among the individuals will cause different biotransformation of dietary curcumin. Accordingly, the beneficial effects depend not only on the curcumin taken from the diet, but also on the type of microbial population of the individual. Therefore, future researches on human volunteers are needed to provide a basis for gut microbiota-based therapeutic applications of curcumin.

6. Conclusions and Perspectives

This review highlighted the strong connection between curcumin and gut microbiota with the final aim of adding novel insight for defining additional mechanism of action of curcumin. More knowledge available about the bidirectional interaction between gut microbiota and curcumin seems to clarify the paradox of the low-bioavailability curcumin and its wide impact on health.

However, in vivo human studies on this topic are almost lacking. Additional human trials, which also take into account an accurate dietary assessment to investigate better the relation between diet and microbiota, are needed. This could allow us to understand the complex interactions between gut microflora and curcumin, providing a better comprehension of its therapeutic efficacy.

We would expect that, in the near future, extensive research will allow to define the gut microbiota as a biomarker for many diseases and the use of curcumin and other probiotics as possible agents to treat dysbiosis and associated diseases. Considering that more than two-thirds of patients do not disclose supplement use to their health care providers nowadays [134], it would be relevant to reinforce the need to consume curcumin, like any other supplement, exclusively under the supervision of health care medical providers.

Author Contributions: B.S., L.M., and M.D. provided substantial contributions to the conception of the work as well as manuscript writing. All authors have read and agreed to the published version of the manuscript.

Funding: This research was funded by Italian National Institute of Health, in the context of JPI HDHL (Knowledge Platform on food, diet, intestinal microbiomics and human health).

Conflicts of Interest: The authors declare no conflict of interest.

References

1. Willenbacher, E.; Khan, S.Z.; Mujica, S.C.A.; Trapani, D.; Hussain, S.; Wolf, D.; Willenbacher, W.; Spizzo, G.; Seeber, A. Curcumin: New Insights into an Ancient Ingredient against Cancer. *Int. J. Mol. Sci.* **2019**, *20*, 1808. [CrossRef]
2. Ayati, Z.; Ramezani, M.; Amiri, M.S.; Moghadam, A.T.; Rahimi, H.; Abdollahzade, A.; Sahebkar, A.; Emami, S.A. Ethnobotany, phytochemistry and traditional uses of Curcuma spp. and pharmacological profile of two important species (*C. longa* and *C. zedoaria*): A review. *Curr. Pharm. Des.* **2019**, *25*, 871–935. [CrossRef]
3. Chikara, S.; Nagaprashantha, L.D.; Singhal, J.; Horne, D.; Awasthi, S.; Singhal, S.S. Oxidative stress and dietary phytochemicals: Role in cancer chemoprevention and treatment. *Cancer Lett.* **2018**, *413*, 122–134. [CrossRef]
4. Chen, C.Y.; Kao, C.L.; Liu, C.M. The cancer prevention, anti-inflammatory and anti-oxidation of bioactive phytochemicals targeting the TLR4 signaling pathway. *Int. J. Mol. Sci.* **2018**, *19*, 2729. [CrossRef]
5. Eke-Okoro, U.J.; Raffa, R.B.; Pergolizzi, J.V., Jr.; Breve, F.; Taylor, R., Jr.; Group, N.R. Curcumin in turmeric: Basic and clinical evidence for a potential role in analgesia. *J. Clin. Pharm. Ther.* **2018**, *43*, 460–466. [CrossRef]
6. Kunnumakkara, A.B.; Bordoloi, D.; Padmavathi, G.; Monisha, J.; Roy, N.K.; Prasad, S.; Aggarwal, B.B. Curcumin, the golden nutraceutical: Multitargeting for multiple chronic diseases. *Br. J. Pharmacol.* **2017**, *174*, 1325–1348. [CrossRef]
7. Naeini, M.B.; Momtazi, A.A.; Jaafari, M.R.; Johnston, T.P.; Barreto, G.; Banach, M.; Sahebkar, A. Antitumor effects of curcumin: A lipid perspective. *J. Cell Physiol.* **2019**, *234*, 14743–14758. [CrossRef]
8. Del Rio, D.; Rodriguez-Mateos, A.; Spencer, J.P.; Tognolini, M.; Borges, G.; Crozier, A. Dietary (poly)phenolics in human health: Structures, bioavailability, and evidence of protective effects against chronic diseases. *Antioxid. Redox. Signal.* **2013**, *18*, 1818–1892. [CrossRef]
9. Di Meo, F.; Margarucci, S.; Galderisi, U.; Crispi, S.; Peluso, G. Curcumin, gut microbiota, and neuroprotection. *Nutrients* **2019**, *11*, 2426. [CrossRef]

10. Bahrami, A.; Amerizadeh, F.; ShahidSales, S.; Khazaei, M.; Ghayour-Mobarhan, M.; Sadeghnia, H.R.; Maftouh, M.; Hassanian, S.M.; Avan, A. Therapeutic potential of targeting wnt/beta-catenin pathway in treatment of colorectal cancer: Rational and progress. *J. Cell Biochem.* **2017**, *118*, 1979–1983. [CrossRef]
11. Carlos-Reyes, A.; Lopez-Gonzalez, J.S.; Meneses-Flores, M.; Gallardo-Rincon, D.; Ruiz-Garcia, E.; Marchat, L.A.; Astudillo-de la Vega, H.; Hernandez de la Cruz, O.N.; Lopez-Camarillo, C. Dietary compounds as epigenetic modulating agents in cancer. *Front. Genet.* **2019**, *10*, 79. [CrossRef]
12. Kunwar, A.; Priyadarsini, K.I. Curcumin and its role in chronic diseases. *Adv. Exp. Med. Biol.* **2016**, *928*, 1–25.
13. Pulido-Moran, M.; Moreno-Fernandez, J.; Ramirez-Tortosa, C.; Ramirez-Tortosa, M. Curcumin and health. *Molecules* **2016**, *21*, 264. [CrossRef]
14. Al-Karawi, D.; Al Mamoori, D.A.; Tayyar, Y. The Role of Curcumin administration in patients with major depressive disorder: Mini meta-analysis of clinical trials. *Phytother. Res.* **2016**, *30*, 175–183. [CrossRef]
15. Daily, J.W.; Yang, M.; Park, S. Efficacy of turmeric extracts and Curcumin for alleviating the symptoms of joint arthritis: A systematic review and meta-analysis of randomized clinical trials. *J. Med. Food* **2016**, *19*, 717–729. [CrossRef]
16. Sahebkar, A.; Henrotin, Y. Analgesic efficacy and safety of Curcuminoids in clinical practice: A systematic review and meta-analysis of randomized controlled trials. *Pain Med.* **2016**, *17*, 1192–1202. [CrossRef]
17. Flandroy, L.; Poutahidis, T.; Berg, G.; Clarke, G.; Dao, M.C.; Decaestecker, E.; Furman, E.; Haahtela, T.; Massart, S.; Plovier, H.; et al. The impact of human activities and lifestyles on the interlinked microbiota and health of humans and of ecosystems. *Sci. Total Environ.* **2018**, *627*, 1018–1038. [CrossRef]
18. Gentile, C.L.; Weir, T.L. The gut microbiota at the intersection of diet and human health. *Science* **2018**, *362*, 776–780. [CrossRef]
19. Rothschild, D.; Weissbrod, O.; Barkan, E.; Kurilshikov, A.; Korem, T.; Zeevi, D.; Costea, P.I.; Godneva, A.; Kalka, I.N.; Bar, N.; et al. Environment dominates over host genetics in shaping human gut microbiota. *Nature* **2018**, *555*, 210–215. [CrossRef]
20. Feng, W.; Wang, H.; Zhang, P.; Gao, C.; Tao, J.; Ge, Z.; Zhu, D.; Bi, Y. Modulation of gut microbiota contributes to curcumin-mediated attenuation of hepatic steatosis in rats. *Biochim. Biophys. Acta Gen. Subj.* **2017**, *1861*, 1801–1812. [CrossRef]
21. Cotillard, A.; Kennedy, S.P.; Kong, L.C.; Prifti, E.; Pons, N.; Le Chatelier, E.; Almeida, M.; Quinquis, B.; Levenez, F.; Galleron, N.; et al. Dietary intervention impact on gut microbial gene richness. *Nature* **2013**, *500*, 585–588. [CrossRef] [PubMed]
22. Lou, Y.; Zheng, J.; Hu, H.; Lee, J.; Zeng, S. Application of ultra-performance liquid chromatography coupled with quadrupole time-of-flight mass spectrometry to identify curcumin metabolites produced by human intestinal bacteria. *J. Chromatogr. B Analyt. Technol. Biomed. Life Sci.* **2015**, *985*, 38–47. [CrossRef] [PubMed]
23. Sun, Z.Z.; Li, X.Y.; Wang, S.; Shen, L.; Ji, H.F. Bidirectional interactions between curcumin and gut microbiota in transgenic mice with Alzheimer's disease. *Appl. Microbiol. Biotechnol.* **2020**, *104*, 3507–3515. [CrossRef]
24. Wang, Y.J.; Pan, M.H.; Cheng, A.L.; Lin, L.I.; Ho, Y.S.; Hsieh, C.Y.; Lin, J.K. Stability of curcumin in buffer solutions and characterization of its degradation products. *J. Pharm. Biomed. Anal.* **1997**, *15*, 1867–1876. [CrossRef]
25. Aggarwal, B.B.; Kumar, A.; Bharti, A.C. Anticancer potential of curcumin: Preclinical and clinical studies. *Anticancer Res.* **2003**, *23*, 363–398. [PubMed]
26. Esatbeyoglu, T.; Huebbe, P.; Ernst, I.M.; Chin, D.; Wagner, A.E.; Rimbach, G. Curcumin—From molecule to biological function. *Angew. Chem. Int. Ed. Engl.* **2012**, *51*, 5308–5332. [CrossRef]
27. Gupta, S.C.; Sung, B.; Kim, J.H.; Prasad, S.; Li, S.; Aggarwal, B.B. Multitargeting by turmeric, the golden spice: From kitchen to clinic. *Mol. Nutr. Food Res.* **2013**, *57*, 1510–1528. [CrossRef]
28. Jacob, J.N.; Badyal, D.K.; Bala, S.; Toloue, M. Evaluation of the in vivo anti-inflammatory and analgesic and in vitro anti-cancer activities of curcumin and its derivatives. *Nat. Prod. Commun.* **2013**, *8*, 359–362. [CrossRef]
29. Mansuri, M.L.; Parihar, P.; Solanki, I.; Parihar, M.S. Flavonoids in modulation of cell survival signalling pathways. *Genes Nutr.* **2014**, *9*, 400. [CrossRef]
30. Alizadeh, M.; Kheirouri, S. Curcumin reduces malondialdehyde and improves antioxidants in humans with diseased conditions: A comprehensive meta-analysis of randomized controlled trials. *Biomedicine* **2019**, *9*, 23. [CrossRef]

31. Di Meo, F.; Filosa, S.; Madonna, M.; Giello, G.; Di Pardo, A.; Maglione, V.; Baldi, A.; Crispi, S. Curcumin C3 complex(R)/Bioperine(R) has antineoplastic activity in mesothelioma: An in vitro and in vivo analysis. *J. Exp. Clin. Cancer Res.* **2019**, *38*, 360. [CrossRef]
32. Gupta, S.C.; Patchva, S.; Aggarwal, B.B. Therapeutic roles of curcumin: Lessons learned from clinical trials. *AAPS J.* **2013**, *15*, 195–218. [CrossRef]
33. Mantzorou, M.; Pavlidou, E.; Vasios, G.; Tsagalioti, E.; Giaginis, C. Effects of curcumin consumption on human chronic diseases: A narrative review of the most recent clinical data. *Phytother Res.* **2018**, *32*, 957–975. [CrossRef] [PubMed]
34. Masoodi, M.; Mahdiabadi, M.A.; Mokhtare, M.; Agah, S.; Kashani, A.H.F.; Rezadoost, A.M.; Sabzikarian, M.; Talebi, A.; Sahebkar, A. The efficacy of curcuminoids in improvement of ulcerative colitis symptoms and patients' self-reported well-being: A randomized double-blind controlled trial. *J. Cell Biochem.* **2018**, *119*, 9552–9559. [CrossRef] [PubMed]
35. Tong, W.; Wang, Q.; Sun, D.; Suo, J. Curcumin suppresses colon cancer cell invasion via AMPK-induced inhibition of NF-kappaB, uPA activator and MMP9. *Oncol. Lett.* **2016**, *12*, 4139–4146. [CrossRef] [PubMed]
36. Aslanabadi, N.; Entezari-Maleki, T.; Rezaee, H.; Jafarzadeh, H.R.; Vahedpour, R. Curcumin for the prevention of myocardial injury following elective percutaneous coronary intervention; a pilot randomized clinical trial. *Eur. J. Pharmacol.* **2019**, *858*, 172471. [CrossRef]
37. Hatcher, H.; Planalp, R.; Cho, J.; Torti, F.M.; Torti, S.V. Curcumin: From ancient medicine to current clinical trials. *Cell Mol. Life Sci.* **2008**, *65*, 1631–1652. [CrossRef]
38. Irving, G.R.; Iwuji, C.O.; Morgan, B.; Berry, D.P.; Steward, W.P.; Thomas, A.; Brown, K.; Howells, L.M. Combining curcumin (C3-complex, Sabinsa) with standard care FOLFOX chemotherapy in patients with inoperable colorectal cancer (CUFOX): Study protocol for a randomised control trial. *Trials* **2015**, *16*, 110. [CrossRef]
39. Nouri-Vaskeh, M.; Malek Mahdavi, A.; Afshan, H.; Alizadeh, L.; Zarei, M. Effect of curcumin supplementation on disease severity in patients with liver cirrhosis: A randomized controlled trial. *Phytother. Res.* **2020**, *34*, 1446–1454. [CrossRef]
40. Kocaadam, B.; Sanlier, N. Curcumin, an active component of turmeric (*Curcuma longa*), and its effects on health. *Crit. Rev. Food Sci. Nutr.* **2017**, *57*, 2889–2895. [CrossRef]
41. Imam, Z.; Khasawneh, M.; Jomaa, D.; Iftikhar, H.; Sayedahmad, Z. Drug induced liver injury attributed to a Curcumin supplement. *Case Rep. Gastrointest. Med.* **2019**, *2019*, 6029403. [CrossRef] [PubMed]
42. Jiao, Y.; Wilkinson, J.T.; Di, X.; Wang, W.; Hatcher, H.; Kock, N.D.; D'Agostino, R., Jr.; Knovich, M.A.; Torti, F.M.; Torti, S.V. Curcumin, a cancer chemopreventive and chemotherapeutic agent, is a biologically active iron chelator. *Blood* **2009**, *113*, 462–469. [CrossRef] [PubMed]
43. Stohs, S.J.; Chen, O.; Ray, S.D.; Ji, J.; Bucci, L.R.; Preuss, H.G. Highly bioavailable forms of Curcumin and promising avenues for Curcumin-based research and application: A review. *Molecules* **2020**, *25*, 1397. [CrossRef] [PubMed]
44. Tsuda, T. Curcumin as a functional food-derived factor: Degradation products, metabolites, bioactivity, and future perspectives. *Food Funct.* **2018**, *9*, 705–714. [CrossRef]
45. Asai, A.; Miyazawa, T. Occurrence of orally administered curcuminoid as glucuronide and glucuronide/sulfate conjugates in rat plasma. *Life Sci.* **2000**, *67*, 2785–2793. [CrossRef]
46. Ireson, C.R.; Jones, D.J.; Orr, S.; Coughtrie, M.W.; Boocock, D.J.; Williams, M.L.; Farmer, P.B.; Steward, W.P.; Gescher, A.J. Metabolism of the cancer chemopreventive agent curcumin in human and rat intestine. *Cancer Epidemiol. Biomarkers Prev.* **2002**, *11*, 105–111.
47. Pan, M.H.; Huang, T.M.; Lin, J.K. Biotransformation of curcumin through reduction and glucuronidation in mice. *Drug Metab. Dispos.* **1999**, *27*, 486–494.
48. Dempe, J.S.; Scheerle, R.K.; Pfeiffer, E.; Metzler, M. Metabolism and permeability of curcumin in cultured Caco-2 cells. *Mol. Nutr. Food Res.* **2013**, *57*, 1543–1549. [CrossRef]
49. Luca, S.V.; Macovei, I.; Bujor, A.; Miron, A.; Skalicka-Wozniak, K.; Aprotosoaie, A.C.; Trifan, A. Bioactivity of dietary polyphenols: The role of metabolites. *Crit. Rev. Food Sci. Nutr.* **2020**, *60*, 626–659. [CrossRef]
50. Ireson, C.; Orr, S.; Jones, D.J.; Verschoyle, R.; Lim, C.K.; Luo, J.L.; Howels, L.; Plummer, S.; Jukes, R.; Williams, M.; et al. Characterization of metabolites of the chemopreventive agent curcumin in human and rat hepatocytes and in the rat in vivo, and evaluation of their ability to inhibit phorbol ester-induced prostaglandin E2 production. *Cancer Res.* **2001**, *61*, 1058–1064.

51. Marczylo, T.H.; Steward, W.P.; Gescher, A.J. Rapid analysis of curcumin and curcumin metabolites in rat biomatrices using a novel ultraperformance liquid chromatography (UPLC) method. *J. Agric. Food Chem.* **2009**, *57*, 797–803. [CrossRef]
52. Tamvakopoulos, C.; Sofianos, Z.D.; Garbis, S.D.; Pantazis, P. Analysis of the in vitro metabolites of diferuloylmethane (curcumin) by liquid chromatography—Tandem mass spectrometry on a hybrid quadrupole linear ion trap system: Newly identified metabolites. *Eur. J. Drug Metab. Pharm.* **2007**, *32*, 51–57. [CrossRef]
53. Awasthi, H.; Tota, S.; Hanif, K.; Nath, C.; Shukla, R. Protective effect of curcumin against intracerebral streptozotocin induced impairment in memory and cerebral blood flow. *Life Sci.* **2010**, *86*, 87–94. [CrossRef] [PubMed]
54. Metzler, M.; Pfeiffer, E.; Schulz, S.I.; Dempe, J.S. Curcumin uptake and metabolism. *Biofactors* **2013**, *39*, 14–20. [CrossRef] [PubMed]
55. Edwards, R.L.; Luis, P.B.; Nakashima, F.; Kunihiro, A.G.; Presley, S.H.; Funk, J.L.; Schneider, C. Mechanistic differences in the inhibition of NF-kappaB by turmeric and its curcuminoid constituents. *J. Agric. Food Chem.* **2020**, *68*, 6154–6160. [CrossRef] [PubMed]
56. Shehzad, A.; Wahid, F.; Lee, Y.S. Curcumin in cancer chemoprevention: Molecular targets, pharmacokinetics, bioavailability, and clinical trials. *Arch. Pharm.* **2010**, *343*, 489–499. [CrossRef] [PubMed]
57. Dhillon, N.; Aggarwal, B.B.; Newman, R.A.; Wolff, R.A.; Kunnumakkara, A.B.; Abbruzzese, J.L.; Ng, C.S.; Badmaev, V.; Kurzrock, R. Phase II trial of curcumin in patients with advanced pancreatic cancer. *Clin. Cancer Res.* **2008**, *14*, 4491–4499. [CrossRef] [PubMed]
58. Lao, C.D.; Ruffin, M.T.T.; Normolle, D.; Heath, D.D.; Murray, S.I.; Bailey, J.M.; Boggs, M.E.; Crowell, J.; Rock, C.L.; Brenner, D.E. Dose escalation of a curcuminoid formulation. *BMC Complement. Altern. Med.* **2006**, *6*, 10. [CrossRef]
59. Sharma, R.A.; Euden, S.A.; Platton, S.L.; Cooke, D.N.; Shafayat, A.; Hewitt, H.R.; Marczylo, T.H.; Morgan, B.; Hemingway, D.; Plummer, S.M.; et al. Phase I clinical trial of oral curcumin: Biomarkers of systemic activity and compliance. *Clin. Cancer Res.* **2004**, *10*, 6847–6854. [CrossRef]
60. Aggarwal, B.B.; Sung, B. Pharmacological basis for the role of curcumin in chronic diseases: An age-old spice with modern targets. *Trends Pharmacol. Sci.* **2009**, *30*, 85–94. [CrossRef]
61. Kumar, A.; Ahuja, A.; Ali, J.; Baboota, S. Conundrum and therapeutic potential of curcumin in drug delivery. *Crit. Rev. Ther. Drug Carrier Syst.* **2010**, *27*, 279–312. [CrossRef] [PubMed]
62. Kato, M.; Nishikawa, S.; Ikehata, A.; Dochi, K.; Tani, T.; Takahashi, T.; Imaizumi, A.; Tsuda, T. Curcumin improves glucose tolerance via stimulation of glucagon-like peptide-1 secretion. *Mol. Nutr. Food Res.* **2017**, *61*, 1600471. [CrossRef] [PubMed]
63. Nakagawa, Y.; Mukai, S.; Yamada, S.; Matsuoka, M.; Tarumi, E.; Hashimoto, T.; Tamura, C.; Imaizumi, A.; Nishihira, J.; Nakamura, T. Short-term effects of highly-bioavailable curcumin for treating knee osteoarthritis: A randomized, double-blind, placebo-controlled prospective study. *J. Orthop. Sci.* **2014**, *19*, 933–939. [CrossRef] [PubMed]
64. Park, S.; Lee, L.R.; Seo, J.H.; Kang, S. Curcumin and tetrahydrocurcumin both prevent osteoarthritis symptoms and decrease the expressions of pro-inflammatory cytokines in estrogen-deficient rats. *Genes Nutr.* **2016**, *11*, 2. [CrossRef]
65. Chen, Y.M.; Chiu, W.C.; Chiu, Y.S.; Li, T.; Sung, H.C.; Hsiao, C.Y. Supplementation of nano-bubble curcumin extract improves gut microbiota composition and exercise performance in mice. *Food Funct.* **2020**, *11*, 3574–3584. [CrossRef]
66. Hewlings, S.J.; Kalman, D.S. Curcumin: A review of its' effects on human health. *Foods* **2017**, *6*, 92. [CrossRef]
67. Han, H.K. The effects of black pepper on the intestinal absorption and hepatic metabolism of drugs. *Expert Opin. Drug Metab. Toxicol.* **2011**, *7*, 721–729. [CrossRef]
68. Moorthi, C.; Krishnan, K.; Manavalan, R.; Kathiresan, K. Preparation and characterization of curcumin-piperine dual drug loaded nanoparticles. *Asian Pac. J. Trop. Biomed.* **2012**, *2*, 841–848. [CrossRef]
69. Shaikh, J.; Ankola, D.D.; Beniwal, V.; Singh, D.; Kumar, M.N. Nanoparticle encapsulation improves oral bioavailability of curcumin by at least 9-fold when compared to curcumin administered with piperine as absorption enhancer. *Eur. J. Pharm. Sci.* **2009**, *37*, 223–230. [CrossRef]
70. Shoba, G.; Joy, D.; Joseph, T.; Majeed, M.; Rajendran, R.; Srinivas, P.S. Influence of piperine on the pharmacokinetics of curcumin in animals and human volunteers. *Planta Med.* **1998**, *64*, 353–356. [CrossRef]

71. Ahmed Nasef, N.; Loveday, S.M.; Golding, M.; Martins, R.N.; Shah, T.M.; Clarke, M.; Coad, J.; Moughan, P.J.; Garg, M.L.; Singh, H. Food matrix and co-presence of turmeric compounds influence bioavailability of curcumin in healthy humans. *Food Funct.* **2019**, *10*, 4584–4592. [CrossRef] [PubMed]
72. Zou, L.; Liu, W.; Liu, C.; Xiao, H.; McClements, D.J. Utilizing food matrix effects to enhance nutraceutical bioavailability: Increase of curcumin bioaccessibility using excipient emulsions. *J. Agric. Food Chem.* **2015**, *63*, 2052–2062. [CrossRef] [PubMed]
73. Liu, W.; Zhai, Y.; Heng, X.; Che, F.Y.; Chen, W.; Sun, D.; Zhai, G. Oral bioavailability of curcumin: Problems and advancements. *J. Drug Target.* **2016**, *24*, 694–702. [CrossRef] [PubMed]
74. Ghosh, S.S.; He, H.; Wang, J.; Gehr, T.W.; Ghosh, S. Curcumin-mediated regulation of intestinal barrier function: The mechanism underlying its beneficial effects. *Tissue Barriers* **2018**, *6*, e1425085. [CrossRef] [PubMed]
75. Altves, S.; Yildiz, H.K.; Vural, H.C. Interaction of the microbiota with the human body in health and diseases. *Biosci. Microbiota Food Health* **2020**, *39*, 23–32. [CrossRef]
76. Belkaid, Y.; Hand, T.W. Role of the microbiota in immunity and inflammation. *Cell* **2014**, *157*, 121–141. [CrossRef]
77. Valdes, A.M.; Walter, J.; Segal, E.; Spector, T.D. Role of the gut microbiota in nutrition and health. *BMJ* **2018**, *361*, 36–44. [CrossRef]
78. Das, P.; Babaei, P.; Nielsen, J. Metagenomic analysis of microbe-mediated vitamin metabolism in the human gut microbiome. *BMC Genom.* **2019**, *20*, 208. [CrossRef]
79. LeBlanc, J.G.; Laino, J.E.; del Valle, M.J.; Vannini, V.; van Sinderen, D.; Taranto, M.P.; de Valdez, G.F.; de Giori, G.S.; Sesma, F. B-group vitamin production by lactic acid bacteria—Current knowledge and potential applications. *J. Appl. Microbiol.* **2011**, *111*, 1297–1309. [CrossRef]
80. Dunn, A.B.; Jordan, S.; Baker, B.J.; Carlson, N.S. The maternal infant microbiome: Considerations for labor and birth. *MCN Am. J. Matern. Child Nurs.* **2017**, *42*, 318–325. [CrossRef]
81. Cani, P.D. Human gut microbiome: Hopes, threats and promises. *Gut* **2018**, *67*, 1716–1725. [CrossRef] [PubMed]
82. Lozupone, C.A.; Stombaugh, J.I.; Gordon, J.I.; Jansson, J.K.; Knight, R. Diversity, stability and resilience of the human gut microbiota. *Nature* **2012**, *489*, 220–230. [CrossRef]
83. Backhed, F.; Roswall, J.; Peng, Y.; Feng, Q.; Jia, H.; Kovatcheva-Datchary, P.; Li, Y.; Xia, Y.; Xie, H.; Zhong, H.; et al. Dynamics and stabilization of the human gut microbiome during the first year of life. *Cell Host Microbe* **2015**, *17*, 852. [CrossRef]
84. Qin, J.; Li, R.; Raes, J.; Arumugam, M.; Burgdorf, K.S.; Manichanh, C.; Nielsen, T.; Pons, N.; Levenez, F.; Yamada, T.; et al. A human gut microbial gene catalogue established by metagenomic sequencing. *Nature* **2010**, *464*, 59–65. [CrossRef]
85. Backhed, F.; Ley, R.E.; Sonnenburg, J.L.; Peterson, D.A.; Gordon, J.I. Host-bacterial mutualism in the human intestine. *Science* **2005**, *307*, 1915–1920. [CrossRef] [PubMed]
86. Adak, A.; Khan, M.R. An insight into gut microbiota and its functionalities. *Cell Mol. Life Sci.* **2019**, *76*, 473–493. [CrossRef] [PubMed]
87. Moszak, M.; Szulinska, M.; Bogdanski, P. You are what you eat-the relationship between diet, microbiota, and metabolic disorders—A review. *Nutrients* **2020**, *12*, 1096. [CrossRef] [PubMed]
88. Cryan, J.F.; O'Riordan, K.J.; Cowan, C.S.M.; Sandhu, K.V.; Bastiaanssen, T.F.S.; Boehme, M.; Codagnone, M.G.; Cussotto, S.; Fulling, C.; Golubeva, A.V.; et al. The microbiota-gut-brain axis. *Physiol. Rev.* **2019**, *99*, 1877–2013. [CrossRef]
89. Kim, S.K.; Guevarra, R.B.; Kim, Y.T.; Kwon, J.; Kim, H.; Cho, J.H.; Kim, H.B.; Lee, J.H. Role of probiotics in human gut microbiome-associated diseases. *J. Microbiol. Biotechnol.* **2019**, *29*, 1335–1340. [CrossRef]
90. Li, S.; Fu, C.; Zhao, Y.; He, J. Intervention with alpha-ketoglutarate ameliorates colitis-related colorectal carcinoma via modulation of the gut microbiome. *Biomed. Res. Int.* **2019**, *2019*, 8020785.
91. Gopalakrishnan, V.; Helmink, B.A.; Spencer, C.N.; Reuben, A.; Wargo, J.A. The Influence of the gut microbiome on cancer, immunity, and cancer immunotherapy. *Cancer Cell* **2018**, *33*, 570–580. [CrossRef] [PubMed]
92. Rajagopala, S.V.; Vashee, S.; Oldfield, L.M.; Suzuki, Y.; Venter, J.C.; Telenti, A.; Nelson, K.E. The human microbiome and cancer. *Cancer Prev. Res.* **2017**, *10*, 226–234. [CrossRef] [PubMed]
93. Fujimura, K.E.; Slusher, N.A.; Cabana, M.D.; Lynch, S.V. Role of the gut microbiota in defining human health. *Expert Rev. Anti Infect. Ther.* **2010**, *8*, 435–454. [CrossRef] [PubMed]
94. Neish, A.S. Microbes in gastrointestinal health and disease. *Gastroenterology* **2009**, *136*, 65–80. [CrossRef]

95. Del Bas, J.M.; Guirro, M.; Boque, N.; Cereto, A.; Ras, R.; Crescenti, A.; Caimari, A.; Canela, N.; Arola, L. Alterations in gut microbiota associated with a cafeteria diet and the physiological consequences in the host. *Int. J. Obes.* **2018**, *42*, 746–754. [CrossRef]
96. Turnbaugh, P.J.; Hamady, M.; Yatsunenko, T.; Cantarel, B.L.; Duncan, A.; Ley, R.E.; Sogin, M.L.; Jones, W.J.; Roe, B.A.; Affourtit, J.P.; et al. A core gut microbiome in obese and lean twins. *Nature* **2009**, *457*, 480–484. [CrossRef]
97. Shen, L.; Ji, H.F. Intestinal microbiota and metabolic diseases: Pharmacological implications. *Trends Pharmacol. Sci.* **2016**, *37*, 169–171. [CrossRef]
98. Peron, G.; Sut, S.; Dal Ben, S.; Voinovich, D.; Dall'Acqua, S. Untargeted UPLC-MS metabolomics reveals multiple changes of urine composition in healthy adult volunteers after consumption of *Curcuma longa* L. extract. *Food Res. Int.* **2020**, *127*, 108730. [CrossRef]
99. Zhai, S.S.; Ruan, D.; Zhu, Y.W.; Li, M.C.; Ye, H.; Wang, W.C.; Yang, L. Protective effect of curcumin on ochratoxin A-induced liver oxidative injury in duck is mediated by modulating lipid metabolism and the intestinal microbiota. *Poult. Sci.* **2020**, *99*, 1124–1134. [CrossRef]
100. Geurts, L.; Lazarevic, V.; Derrien, M.; Everard, A.; Van Roye, M.; Knauf, C.; Valet, P.; Girard, M.; Muccioli, G.G.; Francois, P.; et al. Altered gut microbiota and endocannabinoid system tone in obese and diabetic leptin-resistant mice: Impact on apelin regulation in adipose tissue. *Front. Microbiol.* **2011**, *2*, 149. [CrossRef]
101. Shen, L.; Liu, L.; Ji, H.F. Regulative effects of curcumin spice administration on gut microbiota and its pharmacological implications. *Food Nutr. Res.* **2017**, *61*, 1361780. [CrossRef] [PubMed]
102. McFadden, R.M.; Larmonier, C.B.; Shehab, K.W.; Midura-Kiela, M.; Ramalingam, R.; Harrison, C.A.; Besselsen, D.G.; Chase, J.H.; Caporaso, J.G.; Jobin, C.; et al. The role of curcumin in modulating colonic microbiota during colitis and colon cancer prevention. *Inflamm. Bowel Dis.* **2015**, *21*, 2483–2494. [CrossRef] [PubMed]
103. Le Chatelier, E.; Nielsen, T.; Qin, J.; Prifti, E.; Hildebrand, F.; Falony, G.; Almeida, M.; Arumugam, M.; Batto, J.M.; Kennedy, S.; et al. Richness of human gut microbiome correlates with metabolic markers. *Nature* **2013**, *500*, 541–546. [CrossRef] [PubMed]
104. Liu, Z.J.; Li, Z.H.; Liu, L.; Tang, W.X.; Wang, Y.; Dong, M.R.; Xiao, C. Curcumin attenuates beta-amyloid-induced neuroinflammation via activation of peroxisome proliferator-activated receptor-gamma function in a rat model of alzheimer's disease. *Front. Pharmacol.* **2016**, *7*, 261. [CrossRef]
105. Zhang, Z.; Chen, Y.; Xiang, L.; Wang, Z.; Xiao, G.G.; Hu, J. Effect of curcumin on the diversity of gut microbiota in ovariectomized rats. *Nutrients* **2017**, *9*, 1146. [CrossRef]
106. Rashmi, R.; Santhosh Kumar, T.R.; Karunagaran, D. Human colon cancer cells differ in their sensitivity to curcumin-induced apoptosis and heat shock protects them by inhibiting the release of apoptosis-inducing factor and caspases. *FEBS Lett.* **2003**, *538*, 19–24. [CrossRef]
107. Greiner, A.K.; Papineni, R.V.; Umar, S. Chemoprevention in gastrointestinal physiology and disease. Natural products and microbiome. *Am. J. Physiol. Gastrointest. Liver Physiol.* **2014**, *307*, G1–G15. [CrossRef]
108. Peterson, C.T.; Vaughn, A.R.; Sharma, V.; Chopra, D.; Mills, P.J.; Peterson, S.N.; Sivamani, R.K. Effects of turmeric and curcumin dietary supplementation on human gut microbiota: A double-blind, randomized, placebo-controlled pilot study. *J. Evid. Based Integr. Med.* **2018**, *23*, 1–8. [CrossRef]
109. Youssef, O.; Lahti, L.; Kokkola, A.; Karla, T.; Tikkanen, M.; Ehsan, H.; Carpelan-Holmstrom, M.; Koskensalo, S.; Bohling, T.; Rautelin, H.; et al. Stool microbiota composition differs in patients with stomach, colon, and rectal neoplasms. *Dig. Dis. Sci.* **2018**, *63*, 2950–2958. [CrossRef]
110. Di Cerbo, A.; Palmieri, B.; Aponte, M.; Morales-Medina, J.C.; Iannitti, T. Mechanisms and therapeutic effectiveness of lactobacilli. *J. Clin. Pathol.* **2016**, *69*, 187–203. [CrossRef]
111. Burge, K.; Gunasekaran, A.; Eckert, J.; Chaaban, H. Curcumin and intestinal inflammatory diseases: Molecular mechanisms of protection. *Int. J. Mol. Sci.* **2019**, *20*, 1912. [CrossRef] [PubMed]
112. Guo, S.; Al-Sadi, R.; Said, H.M.; Ma, T.Y. Lipopolysaccharide causes an increase in intestinal tight junction permeability in vitro and in vivo by inducing enterocyte membrane expression and localization of TLR-4 and CD14. *Am. J. Pathol.* **2013**, *182*, 375–387. [CrossRef] [PubMed]
113. Faralli, A.; Shekarforoush, E.; Ajalloueian, F.; Mendes, A.C.; Chronakis, I.S. In vitro permeability enhancement of curcumin across Caco-2 cells monolayers using electrospun xanthan-chitosan nanofibers. *Carbohydr. Polym.* **2019**, *206*, 38–47. [CrossRef] [PubMed]

114. Wang, J.; Ghosh, S.S.; Ghosh, S. Curcumin improves intestinal barrier function: Modulation of intracellular signaling, and organization of tight junctions. *Am. J. Physiol. Cell Physiol.* **2017**, *312*, C438–C445. [CrossRef] [PubMed]
115. Hou, H.T.; Qiu, Y.M.; Zhao, H.W.; Li, D.H.; Liu, Y.T.; Wang, Y.Z.; Su, S.H. Effect of curcumin on intestinal mucosal mechanical barrier in rats with non-alcoholic fatty liver disease. *Zhonghua Gan Zang Bing Za Zhi* **2017**, *25*, 134–138.
116. Ghosh, S.S.; Bie, J.; Wang, J.; Ghosh, S. Oral supplementation with non-absorbable antibiotics or curcumin attenuates western diet-induced atherosclerosis and glucose intolerance in LDLR-/- mice—Role of intestinal permeability and macrophage activation. *PLoS ONE* **2014**, *9*, e108577. [CrossRef]
117. Moschen, A.R.; Kaser, S.; Tilg, H. Non-alcoholic steatohepatitis: A microbiota-driven disease. *Trends Endocrinol. Metab.* **2013**, *24*, 537–545. [CrossRef]
118. Chashmniam, S.; Mirhafez, S.R.; Dehabeh, M.; Hariri, M.; Azimi Nezhad, M.; Nobakht, M.G.B.F. A pilot study of the effect of phospholipid curcumin on serum metabolomic profile in patients with non-alcoholic fatty liver disease: A randomized, double-blind, placebo-controlled trial. *Eur. J. Clin. Nutr.* **2019**, *73*, 1224–1235. [CrossRef]
119. Ohno, M.; Nishida, A.; Sugitani, Y.; Nishino, K.; Inatomi, O.; Sugimoto, M.; Kawahara, M.; Andoh, A. Nanoparticle curcumin ameliorates experimental colitis via modulation of gut microbiota and induction of regulatory T cells. *PLoS ONE* **2017**, *12*, e0185999. [CrossRef]
120. Zhu, H.T.; Bian, C.; Yuan, J.C.; Chu, W.H.; Xiang, X.; Chen, F.; Wang, C.S.; Feng, H.; Lin, J.K. Curcumin attenuates acute inflammatory injury by inhibiting the TLR4/MyD88/NF-kappaB signaling pathway in experimental traumatic brain injury. *J. Neuroinflamm.* **2014**, *11*, 59. [CrossRef]
121. Fu, Y.; Gao, R.; Cao, Y.; Guo, M.; Wei, Z.; Zhou, E.; Li, Y.; Yao, M.; Yang, Z.; Zhang, N. Curcumin attenuates inflammatory responses by suppressing TLR4-mediated NF-kappaB signaling pathway in lipopolysaccharide-induced mastitis in mice. *Int. Immunopharmacol.* **2014**, *20*, 54–58. [CrossRef] [PubMed]
122. Jobin, C.; Bradham, C.A.; Russo, M.P.; Juma, B.; Narula, A.S.; Brenner, D.A.; Sartor, R.B. Curcumin blocks cytokine-mediated NF-kappa B activation and proinflammatory gene expression by inhibiting inhibitory factor I-kappa B kinase activity. *J. Immunol.* **1999**, *163*, 3474–3483. [PubMed]
123. Gan, Z.; Wei, W.; Li, Y.; Wu, J.; Zhao, Y.; Zhang, L.; Wang, T.; Zhong, X. Curcumin and resveratrol regulate intestinal bacteria and alleviate intestinal inflammation in weaned piglets. *Molecules* **2019**, *24*, 1220. [CrossRef] [PubMed]
124. Carmody, R.N.; Turnbaugh, P.J. Host-microbial interactions in the metabolism of therapeutic and diet-derived xenobiotics. *J. Clin. Invest.* **2014**, *124*, 4173–4181. [CrossRef]
125. Hassaninasab, A.; Hashimoto, Y.; Tomita-Yokotani, K.; Kobayashi, M. Discovery of the curcumin metabolic pathway involving a unique enzyme in an intestinal microorganism. *Proc. Natl. Acad. Sci. USA* **2011**, *108*, 6615–6620. [CrossRef]
126. Jazayeri, S.D. Survival of Bifidobacteria and other selected intestinal bacteria in TPY medium supplemented with Curcumin as assessed in vitro. *Int. J. Probiotics Prebiotics* **2009**, *4*, 15–22.
127. Burapan, S.; Kim, M.; Han, J. Curcuminoid demethylation as an alternative metabolism by human intestinal microbiota. *J. Agric. Food Chem.* **2017**, *65*, 3305–3310. [CrossRef]
128. Tan, S.; Calani, L.; Bresciani, L.; Dall'asta, M.; Faccini, A.; Augustin, M.A.; Gras, S.L.; Del Rio, D. The degradation of curcuminoids in a human faecal fermentation model. *Int. J. Food Sci. Nutr.* **2015**, *66*, 790–796. [CrossRef]
129. Sugiyama, Y.; Kawakishi, S.; Osawa, T. Involvement of the beta-diketone moiety in the antioxidative mechanism of tetrahydrocurcumin. *Biochem. Pharmacol.* **1996**, *52*, 519–525. [CrossRef]
130. Wu, J.C.; Tsai, M.L.; Lai, C.S.; Wang, Y.J.; Ho, C.T.; Pan, M.H. Chemopreventative effects of tetrahydrocurcumin on human diseases. *Food Funct.* **2014**, *5*, 12–17. [CrossRef]
131. Ahmed, T.; Enam, S.A.; Gilani, A.H. Curcuminoids enhance memory in an amyloid-infused rat model of Alzheimer's disease. *Neuroscience* **2010**, *169*, 1296–1306. [CrossRef] [PubMed]
132. Chen, P.T.; Chen, Z.T.; Hou, W.C.; Yu, L.C.; Chen, R.P. Polyhydroxycurcuminoids but not curcumin upregulate neprilysin and can be applied to the prevention of Alzheimer's disease. *Sci. Rep.* **2016**, *6*, 29760. [CrossRef] [PubMed]
133. Pinkaew, D.; Changtam, C.; Tocharus, C.; Govitrapong, P.; Jumnongprakhon, P.; Suksamrarn, A.; Tocharus, J. Association of neuroprotective effect of Di-O-Demethylcurcumin on Abeta25-35-Induced neurotoxicity with suppression of NF-kappaB and activation of Nrf2. *Neurotox. Res.* **2016**, *29*, 80–91. [CrossRef] [PubMed]

134. Foley, H.; Steel, A.; Cramer, H.; Wardle, J.; Adams, J. Disclosure of complementary medicine use to medical providers: A systematic review and meta-analysis. *Sci. Rep.* **2019**, *9*, 1573. [CrossRef] [PubMed]

© 2020 by the authors. Licensee MDPI, Basel, Switzerland. This article is an open access article distributed under the terms and conditions of the Creative Commons Attribution (CC BY) license (http://creativecommons.org/licenses/by/4.0/).

Review

The Use of Curcumin as a Complementary Therapy in Ulcerative Colitis: A Systematic Review of Randomized Controlled Clinical Trials

Mariana Roque Coelho [1], Marcela Diogo Romi [1], Daniele Masterson Tavares Pereira Ferreira [2], Cyrla Zaltman [3] and Marcia Soares-Mota [1,*]

1 Institute of Nutrition Josué de Castro, Federal University of Rio de Janeiro, 21941-971 Rio de Janeiro, Brazil; mariana_coelho_7@hotmail.com (M.R.C.); cela_romi@hotmail.com (M.D.R.)
2 Library of Health Sciences Center, Federal University of Rio de Janeiro, 21941-971 Rio de Janeiro, Brazil; danimasterson@yahoo.com.br
3 Division of Gastroenterology, Department of Internal Medicine, Federal University of Rio de Janeiro, 21941-913 Rio de Janeiro, Brazil; c.zaltman@gmail.com
* Correspondence: msoares@nutricao.ufrj.br; Tel.: +55-(21)-99604-9934

Received: 15 June 2020; Accepted: 16 July 2020; Published: 31 July 2020

Abstract: The objective of this study was to systematically review the literature to verify the efficacy and safety of curcumin as a complementary therapy for the maintenance or induction of remission in patients with inflammatory bowel disease (IBD). A comprehensive search was conducted by two independent authors in MEDLINE (PubMed), Scopus, Web of Science, the Cochrane Library, Lilacs, Food Science and Technology Abstracts, and ScienceDirect. The search terms "curcumin", "curcuma", "inflammatory bowel disease", "proctocolitis", "crohn disease", and "inflammation" were combined to create search protocols. This study considered randomized controlled trials (RCTs) published in any language before March 2020 that evaluated the effects of curcumin on inflammatory activity and the maintenance or remission of IBD patients. After duplicates were removed, 989 trials were identified, but only 11 met the eligibility criteria. Five of these were considered to be biased and were excluded. Therefore, six trials were considered in this review. All the studies included in the systematic review were placebo-controlled RCTs conducted on individuals with ulcerative colitis (UC). All the RCTs reported that curcumin was well tolerated and was not associated with any serious side effects. Studies show that curcumin may be a safe, effective therapy for maintaining remission in UC when administered with standard treatments. However, the same cannot be stated for Crohn's disease due to the lack of low bias risk studies. Further studies with larger sample sizes are needed before curcumin can be recommended as a complementary therapy for UC.

Keywords: inflammatory bowel disease; proctocolitis; turmeric; curcumin; complementary therapies; phytotherapy

---

1. Introduction

Inflammatory bowel disease (IBD) is a chronic condition that affects the relapsing gastrointestinal tract, with periods of exacerbation and remission [1,2]. Its main forms of presentation are ulcerative colitis (UC) and Crohn's disease (CD). Its etiopathogenesis is believed to be due to a loss of tolerance to the intestinal microbiota associated with marked immune responses and environmental factors in genetically susceptible individuals [3].

The conventional approach to IBD aims to induce and maintain clinical remission free of corticosteroids, thus minimizing the impact on quality of life [4]. Currently, corticosteroids, sulfasalazine, mesalamine (5-ASA), and immunomodulators are treatment options for patients with IBD. However, it is worth mentioning that conventional treatments cause numerous side effects due to a marked immune response suppression, which negatively impacts the quality of life of these individuals [5,6]. Studies indicate that a substantial proportion of patients do not fully respond to the conventional treatments for IBD, or that its efficacy wanes over time [7]. Corticosteroid resistance/refractoriness rates range from 8.9% to 25% in individuals with IBD [8–11].

Identifying safe and effective therapeutic agents for complementary therapies remains an unmet need for these patients. *Curcuma longa* is a plant from the Zingiberaceae family that is native to India and Southeast Asia and is well known in Asian cultures. Known commonly as turmeric, it has long been used in Ayurvedic medicine to treat inflammatory diseases. It has attracted the attention of researchers because of its compounds, called curcuminoid pigments, which are polyphenols with important medicinal properties [12,13].

Curcumin is the main pharmacologically active curcuminoid pigment in turmeric. It acts by modulating various cell-signaling pathways, producing anti-inflammatory, anti-tumor, anti-oxidant, and immunomodulatory effects [14]. The main components of commercial turmeric are curcumin I (77%), curcumin II (~17%), and curcumin III (~3%), with only 2–5% of the powdered seasoning consisting of curcumin [15]. The mechanism of its anti-inflammatory action deemed to be the most relevant is the inhibition of NF-κB, by blocking IκB kinase, which prevents cytokine-mediated phosphorylation and the degradation of IκB, an NF-κB inhibitor, thereby inhibiting the expression of pro-inflammatory cytokines (IL-1, IL-6, and TNF) [16].

Curcumin also acts by inhibiting the activity of pro-inflammatory proteins such as activated protein-1, peroxisome proliferator-activated receptor gamma, signal translators, and transcription activators, as well as the expression of b-catenin, cyclooxygenase 2, 5-lipoxygenase, and inducible nitric oxide synthase isoform, which play a key role in inflammation [17]. In addition, it acts by blocking the binding between TNF-α and its receptor, preventing the perpetuation of inflammation caused by this cytokine [18].

This systematic review aims to analyze the studies published so far, to review the positive or negative effects of the use of curcumin, and to determine whether it is safe and effective as a complementary therapy in the management of IBD, offering fewer side effects than conventional therapies.

2. Methods

2.1. Protocol and Registration

To conduct the study, we used the PRISMA checklist, composed of 27 items [19]. The study protocol was registered in the PROSPERO database under the registration number CRD42019104827.

2.2. Information Sources and Search Strategies

A literature review was performed by two independent authors (M.R.C. and M.D.R.) on the following databases: MEDLINE (PubMed), Scopus, Web of Science, Cochrane Library, Lilacs, Food Science and Technology Abstracts, and Science Direct. Studies published before March 2020 were included. All the databases were monitored periodically until the study's completion. Divergences between the researchers retrieving the data were resolved by consensus.

The controlled vocabulary and keywords used in the search strategy were defined based on the PICOS questions [20]:

1. Population: individuals with IBD (UC or CD) of either sex and from any age group;
2. Intervention: curcumin supplementation in the form of spice, capsule, or enema;
3. Comparison: placebo or conventional drug therapy;
4. Outcomes: disease activity, clinical, or endoscopic inflammatory activity;
5. Study design: randomized clinical trials (RCTs).

The search strategy was designed following the guidance of an expert librarian (D.M) and according to the specificity of each database, whenever possible, using the controlled vocabulary of the subject descriptors (Mesh/Medline and DeCs/VHL). The following subject headings and free-text terms were used in the search: "curcumin", "curcuma", "inflammatory bowel disease", "proctocolitis", "crohn's disease", and "inflammation" (Table 1).

Table 1. Electronic Databases and Respective Search Strategies.

PubMed	
#1 (Inflammatory Bowel Disease [Mesh] or Inflammatory Bowel Disease [Tiab] or Crohn Disease [Mesh] or Crohn Disease [Tiab] or Proctocolitis [Mesh] or Proctocolitis [Tiab])	#2 (Curcuma [Mesh] or Curcuma [Tiab] or Curcumin [Mesh] or Curcumin * [Tiab])
#1 AND #2	
Scopus	
#1 (TITLE-ABS-KEY (("Inflammatory Bowel Disease" or "Crohn Disease" or proctocolitis)))	#2 (TITLE-ABS-KEY ((curcuma or curcumin *)))
#1 AND #2	
Web of Science	
#1 ("Inflammatory Bowel Disease" or "Crohn Disease" or Proctocolitis)	#2 (Curcuma or Curcumin *)
#1 AND #2	
Lilacs	
#1 tw: (tw: ((mh: "inflamatory bowel diseases" or "doenças inflamatórias intestinais" or mh: "crohn disease" or "doença de crohn" or mh: proctocolitis or "retocolite ulcerativa")))	#2 (tw: (tw: ((mh: curcumin or curcumina or curcuma))))
#1 AND #2	
Food Science and Technology Abstracts	
#1 ("Inflammatory Bowel Disease" or "Crohn Disease" or Proctocolitis)	#2 (Curcuma or Curcumin *)
#1 AND #2	
ScienceDirect	
#1 ("Inflammatory Bowel Disease" or "Crohn Disease" or Proctocolitis)	# (curcumin or curcumina or curcuma)
#1 AND #2	
Cochrane Library	
#1 MeSH descriptor: [Inflammatory Bowel Diseases] explode all trees #2 "Inflammatory Bowel Disease" #3 #1 or #2 #4 MeSH descriptor: [Crohn Disease] explode all trees #5 "Crohn Disease" #6 #4 or #5 #7 MeSH descriptor: [Proctocolitis] explode all trees	#8 Proctocolitis #9 #7 or #8 #10 #3 or #6 or #9 #11 MeSH descriptor: [Curcuma] explode all trees #12 (Curcuma or Curcumin or Curcumin *) #13 #11 or #12 #14 #10 and #13

#- represents the combination of searches conducted previously; *- matches one or more occurrences of any character or group of characters, including no character.

2.3. Eligibility Criteria

The criteria for the inclusion of the RCTs in this study were that they used curcumin for the maintenance or remission of IBD in patients of both sexes and of any age who were in remission or who had mild or moderate activity at the time of recruitment, and that they evaluated the effects of curcumin on the inflammatory activity. Studies published in any language were accepted, and no minimum follow-up period was established. Review articles, animal studies, editorial letters, in-vitro studies, observational, and descriptive studies, such as case reports and case series, were excluded. In addition, studies that did not describe the curcumin dose or did not meet the minimum bias risk assessment score were also excluded.

2.4. Study Selection and Data Collection Process

Initially, the articles were selected by title and abstract. Articles that appeared in more than one database were considered only once, using the EndNote bibliography management software to exclude duplicate articles. Full articles were read when there was not enough information in the title and abstract to make a clear decision about whether to include or exclude the study.

2.5. Risk of Bias Assessment

Two independent reviewers performed the quality assessment of the trials using the Cochrane Collaboration tool for assessing the risk of bias in RCTs [21]. There were seven assessment criteria: random sequence generation; allocation concealment; blinding of participants and personnel; blinding of the outcome assessors; incomplete outcome data; selective outcome reporting, and other possible sources of bias. The potential risk of bias for each criterion was rated at low, uncertain, or high, as described in the *Cochrane Handbook for Systematic Reviews of Interventions*, version 5.1.0 (http://handbook.cochrane.org). Studies with a high risk of bias in three or more items were excluded from the systematic review.

The Oxford quality scoring system, the Jadad scale [22], was also used to assess the study quality. This scale provides a score for each individual study ranging from 0 to 5 points, with 5 being the highest quality score. Studies were given one point if they were described as randomized, one if they were described as double-blind, and one if a description of the withdrawals and dropouts from the study was provided. An additional point was awarded if the randomization method was described and considered appropriate, and another point if the blinding method was described and also considered appropriate. If any of the randomization or blinding methods were considered inappropriate, a point was deducted from the sum for each item.

Each criterion was judged by one of three answers: "yes" to indicate a low risk of bias, "no" to indicate a high risk of bias, and "not described" to indicate a lack of information or uncertainty about potential bias. Studies with scores < 3 were considered to have a high risk of bias and were excluded from the study, while scores ≥ 3 indicated studies with a low risk of bias, which were retained in the analysis [22].

3. Results

3.1. Search Results

After database screening and duplicate removal, 989 studies were identified (Figure 1). The title analysis resulted in the exclusion of 951 of these, while a further 26 were excluded after reading the abstracts because they failed to meet the eligibility criteria. The twelve remaining articles were read in full and only one of these was ruled out as it was a case report. Eleven studies were included in the bias risk analysis, after which only six remained in this systematic review.

3.2. Assessment of the Risk of Bias and Excluded RCTs

After assessing the risk of bias, two studies [23,24] were excluded because they presented three items with a high risk of bias, as recommended in the Cochrane Handbook [21], and did not reach the minimum score necessary on the Jadad scale [22]. The three other studies [25–27], excluded at this point, also failed to reach the minimum score on the Jadad scale [22] and had four or more items classified as uncertain according to the Cochrane Handbook [21] because they did not have enough information for the analysis, which put them at a high risk of bias.

The assessment of the risk of bias in the selected studies is presented in Figure 2. Most of the articles clearly described the randomization method used [26,28–33], as well as the blinding method [26–33]. Some studies [25–27] reported incomplete outcome data and others [27,29,32,33] presented the outcomes selectively, which indicates a description bias. With regard to the allocation concealment, two studies [25,27] did not describe this and two others [23,24] stated that there was none.

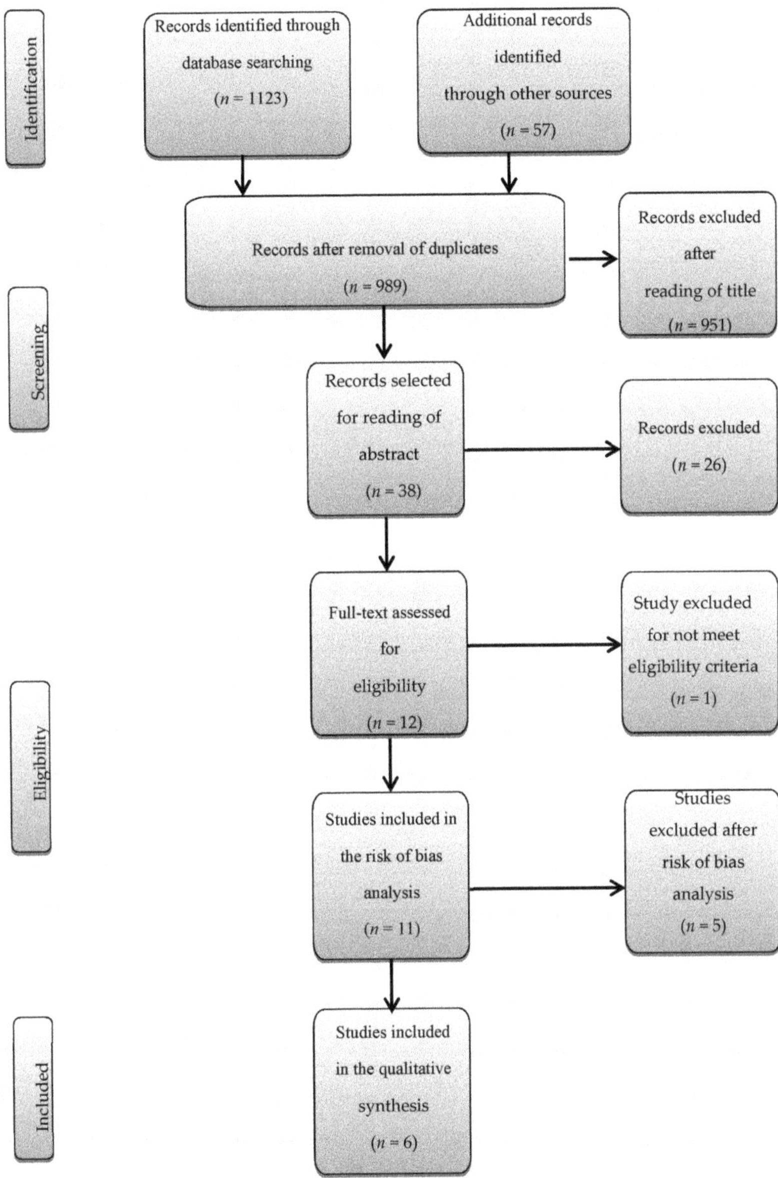

Figure 1. Flowchart of articles selected for the systematic review.

All the studies included in the systematic review presented the main elements recognized to minimize the risk of bias, according to the Cochrane Handbook [21]: randomization, blinding, and reporting of dropouts/withdrawals.

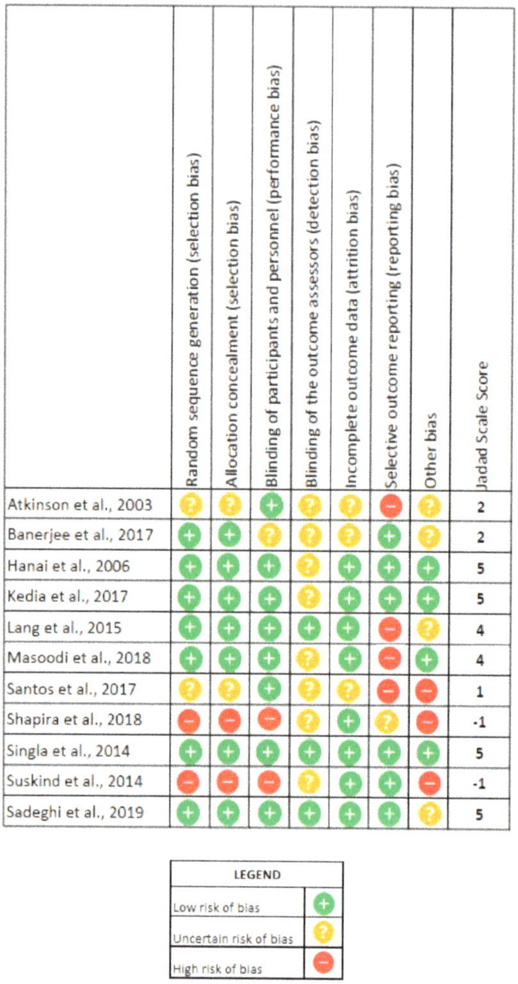

	Random sequence generation (selection bias)	Allocation concealment (selection bias)	Blinding of participants and personnel (performance bias)	Blinding of the outcome assessors (detection bias)	Incomplete outcome data (attrition bias)	Selective outcome reporting (reporting bias)	Other bias	Jadad Scale Score
Atkinson et al., 2003	?	?	+	?	?	−	?	2
Banerjee et al., 2017	+	+	?	?	?	+	?	2
Hanai et al., 2006	+	+	+	?	+	+	+	5
Kedia et al., 2017	+	+	+	?	+	+	+	5
Lang et al., 2015	+	+	+	+	+	−	?	4
Masoodi et al., 2018	+	+	+	?	+	−	+	4
Santos et al., 2017	?	?	+	?	?	−	−	1
Shapira et al., 2018	−	−	−	?	+	?	−	-1
Singla et al., 2014	+	+	+	+	+	+	+	5
Suskind et al., 2014	−	−	−	?	+	+	−	-1
Sadeghi et al., 2019	+	+	+	+	+	+	?	5

LEGEND	
Low risk of bias	+
Uncertain risk of bias	?
High risk of bias	−

Figure 2. Bias risk assessment according to the Cochrane Handbook tool and the Oxford Quality Scoring System, the Jadad Scale.

3.3. Characteristics of Selected Articles

All six of the studies included in the systematic review were placebo-controlled RCTs performed on individuals with UC and conducted between 2006 and 2019. No studies involving CD met all the eligibility criteria. Table 2 details the characteristics and results of each study. They were all conducted in outpatient settings, including a total of 372 subjects, ranging in age from 23 to 61 years. There were no significant differences regarding the number of male and female patients who participated in these trials.

Table 2. Summary of studies included in the systematic review.

Author, Year and Country	Study Design	Characterization of UC Population	Intervention	Variables of Interest Analyzed	Results
Hanai et al. 2006 [28] Japão	Multicenter, randomized, double-blind, placebo-controlled	In remission n = 89 (49♂/30♀) 25–61 years	Curcumin (capsule) 2 g + 1.5–3 g 5ASA or 1–3 g sulfasalazine/day (n = 45) or placebo + 5ASA/sulfasalazine (n = 44) 6 months	Clinical Activity Index and Endoscopic Index assessed at baseline and every 2 months up to 12 months	• 7 participants did not complete the protocol. • 2/43 (4.65%) curcumin group and 8/39 (20.51%) placebo group relapsed at 6 months. • Curcumin significantly improved Clinical Activity Index and Endoscopic Index. • Curcumin was well tolerated and not associated with any SAE.
Singla et al. 2014 [29] India	Pilot study, double-blind, randomized, placebo-controlled	Mild/moderate proctitis and proctosigmoiditis n = 45 (22♂/23♀) 23–49 years	140 mg NCB-02 (standardized extract curcumin) enema + oral 1.6 g 5ASA/day (n = 28) or placebo enema + oral 1.6 g 5ASA/day (n = 22) 8 weeks	UCDAI and endoscopic activity by mucosal appearance score at baseline and after 8 weeks	• 9 NCB-02 group and 6 placebo group did not complete the protocol. • At the end of 8 weeks, Clinical response: 13/14 (92.9%) NCB-02 vs. 50% placebo group. • Clinical remission: 71.4% NCB-02 vs. 31.3% placebo group. • Improvement of Endoscopic activity: 85.7% NCB-02 vs. 50% placebo group. • No SAE.
Lang et al. 2015 [30] Israel, Hong Kong and Cyprus.	Multicenter, randomized, double-blind, placebo-controlled	Mild/moderate proctitis/left colitis/pancolitis n = 50 (17♂/33♀) 27–55 years	95% pure curcumin (capsule)—3 g + 4 g 5ASA/day (n = 26) or placebo + 4 g 5ASA/day (n = 24) 4 weeks	SCCAI and Mayo endoscopic score assessed at baseline and after 4 weeks	• 2 participants did not complete the protocol. • Clinical remission (SCCAI): 54% curcumin vs. 0% placebo group. • Clinical response: 65.3% curcumin vs. 12.5% placebo group. • Endoscopic remission (Mayo): 38 curcumin vs. 0% placebo group.
Masoodi et al. 2018 [31] Iran	Single-center, Double-blind, randomized, placebo-controlled	Mild/moderate left colitis/pancolitis n = 56 (28♂/28♀) 25–54 years	Nanomicellar curcumin (capsule) 80 mg 3x/day = 240 mg + 3 g 5ASA/day (n = 28) or placebo + 3 g 5ASA/day (n = 28) 4 weeks	SCCAI assessed at baseline, and at 2 and 4 weeks	• 2 participants did not complete the protocol. • Significant reduction of fecal urgency: 60% curcumin vs. 28.6% placebo group. • Improvement of general well-being: 64% curcumin vs. 39.3% placebo group). • Significant reduction in SCCAI score (1.71 points curcumin group vs. 2.68 points placebo group).
Kedia et al. 2017 [32] India	Single-center, Double-blind, randomized, placebo-controlled	Mild/moderate proctitis/left colitis/pancolitis n = 62 (41♂/21♀) 24–48 years	Curcumin (capsule) 450 mg/day + 2.4 g 5-ASA/day (n = 29) or placebo + 2.4 g 5-ASA/day (n = 33) 8 weeks	UCDAI and endoscopic Baron score evaluation assessed at baseline, and at 4 and 8 weeks	• 21 participants did not complete the study. • No significant difference between the clinical remission and endoscopic remission rates of curcumin and placebo groups (UCDAI).
Sadeghi et al. 2019 [33] Iran	Double-blind, randomized, placebo-controlled	Mild/moderate proctitis/left colitis/pancolitis) n = 70 (21♂/49♀) 27–53 years	Curcumin (capsule-turmeric extract) 1.500 mg/day + routine drugs (n = 35) or placebo + routine drugs (n = 35) 8 weeks	SCCAI, IBDQ-9, ESR, hs-CRP, anthropometric indices and dietary intakes were assessed at baseline and after 8 weeks	• 4 patients curcumin group and 3 from placebo group withdrew from the study. • Clinical remission 83.9% curcumin vs. 43.8% placebo group. • Significant reduction hs-CRP concentrations in Curcumin compared to placebo group. • Significant decrease of ESR in curcumin group. • Significant increase of mean of IBDQ-9 score in curcumin group compared to placebo group.

LEGEND: 5-ASA—mesalamine; UC—ulcerative colitis; SCCAI—simple clinical colitis activity index; UCDAI—ulcerative colitis disease activity index, IBDQ-9—Inflammatory Bowel Disease Questionnaire; ESR—erythrocyte sedimentation rate; hs-CRP—high-sensitivity C-reactive protein; SAE—serious adverse effects.

The studies compared the use of curcumin as a complementary therapy given in combination with mesalamine (5-ASA), a conventional drug regularly prescribed for patients with UC, with placebos also in conjunction with 5-ASA. The oral capsule curcumin dosage ranged from 450 mg to 3 g/day. One study reported using NCB-02, a standardized extract with 72% curcumin, 18.08% demethoxicurcumin, and 9.42% bisdemetoxicurcumin as an enema at a dosage of 140 mg/day [31]. One study used capsules of a nanomicellar curcumin formulation (SinaCurcumin®) at doses of 240 mg/day [33]. The duration of the interventions ranged from 4 weeks to 12 months.

Different methods were used to evaluate the clinical activity of the disease, but in general the parameters evaluated by the different scores included: number of bowel movements, fecal urgency, bloody stools, self-reported general well-being, abdominal pain, and extra-intestinal manifestations. The endoscopic scores used evaluated the following parameters: vascular pattern, presence of erythema, friability of the mucosa, erosions, spontaneous bleeding and presence of ulcerations.

3.4. Outcomes after Intervention

In most of the studies, positive outcomes were reported after the interventions. Hanai et al. (2006) reported a lower number of relapses in the intervention group than in the control group, while other studies showed a higher proportion of remission in the intervention group [29,30]. Significant clinical responses measured through the disease activity indices and an improved endoscopic activity were also reported in four of the five studies, except for the study conducted by Kedia et al. [32], which showed no significant difference between the clinical and endoscopic remission rates of the intervention and placebo groups.

In Masoodi's study [31], in addition to a significant reduction in the simple clinical colitis activity index (SCCAI), a higher proportion of the intervention group patients reported improved general well-being and decreased fecal urgency than did the patients from the placebo group after four weeks. Curcumin was well tolerated in all the RCTs and was not associated with any serious side effects. However, Hanai et al. [28] did report some mild adverse events (AE), such as: feeling of abdominal distension, nausea, and a transient increase in the number of bowel movements.

The only one to use a quality of life questionnaire [33] showed a significant increase in the mean score of IBDQ-9 in the curcumin group compared to the placebo group. In the same study, curcumin supplementation induced a significant reduction in the high-sensitivity C-reactive protein (hs-CRP) concentrations after eight weeks vs. no significant reduction seen in the placebo group.

4. Discussion

This systematic review included six RCT studies [28–33] that compared the use of curcumin as a complementary therapy given in combination with 5-ASA in patients with UC. In four of the five studies included, there was a significant improvement in the clinical response with curcumin, allied with no serious AEs. Five studies [28–31,33] reported that curcumin was able to reduce the symptoms of the disease, achieve clinical remission, and/or prevent relapse when used as a complementary therapy to mesalamine.

The study by Kedia et al. [32], using lower doses of oral capsule curcumin than other studies [28,30,33], which used 2000 mg/day, 3000 mg/day and 1500 mg/day, respectively, was the only study in which no significant difference was found between clinical remission and the endoscopic remission rates in the intervention and placebo group (UCDAI). It is understood that very low doses of curcumin may not achieve the desired effect unless administered locally in the form of an enema, as in the study by Singla et al. [29], or in more bioavailable nanoformulations such as SinaCurcumin®, used in the study by Masoodi et al. [31].

The curcuminoid gelatin capsule is dissolved in the acidic environment of the stomach and the nanomicelles are released, which are stable for up to 6 h and are absorbed into the intestine [34]. Importantly, curcumin absorption may be lower when it is administered orally than when it is administered in an enema when it comes to IBD, due to the direct delivery to the site-of-action.

Therefore, RCTs should also explore the enema administration of curcumin in order to define the most effective route of administration, since only one RCT [29] has evaluated this route so far.

Most of the studies did not describe the purity of the curcumin used. Singla et al. [29] only described the percentage of curcumin in the extract used, while Lang et al. [30] reported using Cur-Cure® (a preparation containing 95% pure curcumin); the other four studies did not provide any such information. Any comparison of this study with others that do not describe the composition of the capsule is hampered by the fact that it is not known exactly what the intervention groups in the other studies were given. An important caveat is that capsule curcumin found on pharmaceutical and nutritional store shelves often contains numerous chemical additives, as does the turmeric powder sold commercially as a spice.

Only Hanai et al. reported some mild AEs, such as abdominal distension, nausea and an increased number of bowel movements in a Japanese population. These non-specific symptoms may be related to factors not controlled by the researchers, such as dietary factors (lactose, fodmaps, gluten) and the presence of associated functional diseases, such as IBS (irritable bowel syndrome). Therefore, it cannot be confirmed that these symptoms experienced by only 7 of 89 patients are related to curcumin use.

The Food and Drug Administration states that curcumin is "generally recognized as safe" and has no known toxic effects. According to the Food and Agriculture Organization of the United NationsFAO/WHO Joint Food Additives Expert Committee and the European Food Safety Authority, the acceptable daily intake of curcumin is 0–3 mg/kg/day [35,36]. Lao et al. [37] administered 500–12,000 mg of curcumin (95% standardized extract of curcumin) in healthy subjects to examine its maximum tolerance and safety dose, and found that an ingestion of up to 12 g/day of curcumin brought about no ill effects. Based on this recommendation, a healthy 70 kg individual could consume 4–10 g turmeric powder, or two tablespoons, per day, which is well above the usual consumption in western countries.

The bioavailability of orally ingested unformulated curcumin is low. High doses of curcumin (2–4 g) are usually required to improve bioavailability due to its hydrophobic nature, but recently, in addition to nanoformulations, studies have been conducted into a self-micro emulsifying drug delivery system (SMEDDS) for curcumin: hydrophilic drug droplets that can diffuse easily into the bloodstream, resulting in higher intraintestinal concentrations than when conventional curcumin is used [26]. Although SMEDDS has been used successfully in a study, the current evidence on its usefulness is still scant. In a single-blind crossover study on healthy adults, the bioavailability of the curcumin micellar formulation was found to be 185 times higher than that of the same dose of unformulated curcumin [38]. Despite promising results concerning curcumin micellization, the study was conducted on healthy subjects; there are no studies on individuals with IBD. This nanometer-sized drug delivery system could become an effective strategy for treating IBD [39].

In a recent systematic review, Jamwal [40] compared the pharmacokinetic effect of different curcumin formulations in healthy subjects. The formulations that exhibited a better bioavailability than unformulated curcumin were: NovaSol®, CurcuWin®, and LongVida®, which were found to achieve a 185, 136, and 100 times higher bioavailability, respectively. The studies cited report promising preliminary results from in vitro and in vivo experimental trials by developing specific curcumin delivery systems, protecting against rapid degradation, and targeting the inflamed colon. However, it is not yet known which would be the best option from those among the most bioavailable formulations for treating IBD.

Another strategy for improving bioavailability, the use of piperine (a component of black pepper, *Piper nigrum*, and long pepper, *Piper longum*) as an adjunct to curcumin, has been described for a long time. Shoba et al. [41] demonstrated that 20 mg of piperine administered concomitantly with 2 g of curcumin increases the bioavailability of curcumin 20-fold in humans. One experimental study by Li et al. [42], using CUR-PIP-SMEDDS (an emulsified curcumin and piperine formulation), reported that the administration of this formulation through an enema in rats had an effect similar to 5-ASA in maintaining remission in dextran sulfate sodium DSS-induced colitis. As a substance that inhibits the metabolism of several active compounds, piperine may cause toxicity when associated with some drugs [43]. Moreover, the gastrointestinal transit time has been reported to be significantly reduced by the consumption of 20 mg of piperine by rats [44], which in individuals with IBD may not be beneficial.

Despite the small number of studies, the average intervention time that appears to be sufficient to present the outcomes is four to eight weeks. However, this time may vary according to the pharmaceutical formulation used. Shorter intervention periods seem to be associated with more bioavailable curcumin formulations.

It is noteworthy that phytochemicals are not evenly distributed among plant parts: the curcumin content of mother rhizomes is higher than the curcumin content of primary and secondary rhizomes, both fresh and dried [45]. The seasonality, altitude, temperature, solar incidence, water availability, and nutrient content in the soil may also alter the presence or concentration of phytochemicals [46]. Post-harvest processing for the production of dry powder results in a reduced curcumin content in the mother's rhizomes and fingers, as it is easily decomposed when exposed to light and is sensitive to high temperatures (>60 °C) [47].

In addition, the pre-harvest management, specifically cultivation practices, also play a significant role in the quantity and quality of turmeric phytochemicals. In the study by Choudhury et al. [47], the application of nitrogen, phosphorus, potassium (NPK)inorganic fertilizers resulted in a 31–43% increase in the curcumin content of the mother rhizome compared to the non-application of NPK fertilizers. Similarly, the application of organic fertilizers (pig and poultry manure) also increased the curcumin content of mother rhizomes by 18–36%. All the factors mentioned may have influenced the curcumin content used in the studies and, consequently, the results obtained.

Efforts are currently being made by different research groups to improve curcumin's bioavailability, and further efforts will be needed to answer questions related to curcumin therapy in IBD. New well-designed, long-term RCTs with a large enough sample size to demonstrate clinically significant effects and to determine the efficacy of new pharmaceutical formulations are required.

Considering the six trials included, five of them demonstrated good results related to the clinical and/or endoscopic remission/response. The findings suggest that curcumin may be a safe, effective therapy for maintaining or inducing UC remission when administered with standard treatments; the same cannot be said for CD, due to the absence of RCTs with a low risk of bias when investigating patients with this condition. However, based on these findings it is not yet possible to establish the best protocol for the use of curcumin, considering that the determination of the therapeutic dose and the duration of treatment depend on the identification of the ideal pharmaceutical formulation, which has not yet been determined for this population.

We recognize that this study inevitably has some limitations and so its results should be interpreted with caution. Firstly, it was based on only six RCTs, all with a relatively small sample size. Additionally, although the studies included were well designed, randomized, placebo-controlled studies that rated highly on the Jadad scale [22] and in the Cochrane Handbook bias analysis [21], they used different doses and formulations of curcumin and different administration routes. This itself may add bias to the results of our study.

Author Contributions: M.R.C. and M.S.-M. study concept and development. M.R.C., M.D.R. and D.M.T.P.F. acquisition of data; M.R.C. and M.D.R. analysis and interpretation of data. M.R.C. and M.S.-M. drafting of the manuscript; C.Z. and M.S.-M. critical revision of the manuscript and study supervision. All authors have read and agreed to the published version of the manuscript.

Funding: This research received no specific grant from any funding agency in the public, commercial, or not for-profit sectors.

Conflicts of Interest: The authors declare no conflict of interest.

References

1. Magro, F.; Gionchetti, P.; Eliakim, R.; Ardizzone, S.; Armuzzi, A.; Barreiro-de Acosta, M.; Burisch, J.; Gecse, K.B.; Hart, A.L.; Hindryckx, P.; et al. Third European evidence-based consensus on diagnosis and management of ulcerative colitis. Part 1: Definitions, diagnosis, extra-intestinal manifestations, pregnancy, cancer surveillance, surgery, and ileo-anal pouch disorders. *J. Crohn's Colitis* **2017**, *11*, 649–670. [CrossRef] [PubMed]
2. Stapley, S.A.; Rubin, G.P.; Alsina, D.; Shephard, E.A.; Rutter, M.D.; Hamilton, W.T. Clinical features of bowel disease in patients aged <50 years in primary care: A large case-control study. *Br. J. Gen. Pract.* **2017**, *67*, e336–e344. [CrossRef] [PubMed]
3. Ilhara, S.; Hirata, Y.; Koike, K. TGF-β in inflammatory bowel disease: A key regulator of immune cells, epithelium and the intestinal microbiota. *J. Gastroenterol.* **2017**, *52*, 777–787. [CrossRef]
4. D'Haens, G.R.; Sartor, R.B.; Silverberg, M.S.; Petersson, J.; Rutgeerts, P. Future directions in inflammatory bowel disease management. *J. Crohn's Colitis* **2014**, *8*, 726–734. [CrossRef] [PubMed]
5. Kawalec, P.; Malinowski, K.P. Indirect health costs in ulcerative colitis and Crohn's disease: A systematic review and meta-analysis. *Expert Rev. Pharm. Outcomes Res.* **2015**, *15*, 253–266. [CrossRef] [PubMed]
6. Harbord, M.; Eliakim, R.; Bettenworth, D.; Karmiris, K.; Katsanos, K.; Kopylov, U.; Kucharzik, T.; Molnár, T.; Raine, T.; Sebastian, S.; et al. Third European evidence-based consensus on diagnosis and management of ulcerative colitis. Part 2: Current management. *J. Crohn's Colitis* **2017**, *11*, 769–784. [CrossRef] [PubMed]
7. Ben-Horin, S.; Mao, R.; Chen, M. Optimizing biologic treatment in IBD: Objective measures, but when, how and how often? *BMC Gastroenterol.* **2015**, *15*, 178. [CrossRef] [PubMed]
8. Munkholm, P.; Langholz, E.; Davidsen, M.; Binder, V. Frequency of glucocorticoid resistance and dependency in Crohn's disease. *Gut* **1994**, *35*, 360–362. [CrossRef]
9. Faubion, W.A., Jr.; Loftus, E.V., Jr.; Harmsen, W.S.; Zinsmeister, A.R.; Sandborn, W.J. The natural history of corticosteroid therapy for inflammatory bowel disease: A population-based study. *Gastroenterology* **2001**, *121*, 255–260. [CrossRef]
10. Ho, G.T.; Chiam, P.; Drummond, H.; Loane, J.; Arnott, I.D.; Satsangi, J. The efficacy of corticosteroid therapy in inflammatory bowel disease: Analysis of a 5-year UK inception cohort. *Aliment. Pharmacol. Ther.* **2006**, *24*, 319–330. [CrossRef] [PubMed]
11. Li, J.; Wang, F.; Zhang, H.J.; Sheng, J.Q.; Yan, W.F.; Ma, M.X.; Fan, R.Y.; Gu, F.; Li, C.F.; Chen, D.F.; et al. Corticosteroid therapy in ulcerative colitis: Clinical response and predictors. *World J. Gastroenterol.* **2015**, *21*, 3005–3015. [CrossRef] [PubMed]
12. Zhou, H.; Beevers, S.C.; Huang, S. Targets of curcumin. *Curr. Cancer Drug Targets* **2012**, *12*, 332–347. [CrossRef] [PubMed]
13. Kocaadam, B.; Sanlier, N. Curcumin, an active component of turmeric (*Curcuma longa*), and its effects on health. *Crit. Rev. Food Sci. Nutr.* **2017**, *57*, 2889–2895. [CrossRef] [PubMed]
14. Gupta, S.C.; Patchva, S.; Koh, W.; Aggarwal, B.B. Discovery of curcumin, a component of the golden spice, and its miraculous biological activities. *Clin. Exp. Pharmacol. Physiol.* **2012**, *39*, 283–299. [CrossRef] [PubMed]
15. Gupta, S.C.; Prasad, S.; Kim, J.H.; Patchva, S.; Webb, L.J.; Priyadarsini, I.K.; Aggarwal, B.B. Multitargeting by curcumin as revealed by molecular interaction studies. *Nat. Prod. Rep.* **2011**, *2011*, 1937–1955. [CrossRef]
16. Baliga, M.S.; Joseph, N.; Venkataranganna, M.V.; Saxena, A.; Ponemone, V.; Fayad, R. Curcumin, an active component of turmeric in the prevention and treatment of ulcerative colitis: Preclinical and clinical observations. *Food Funct.* **2012**, *3*, 1109–1117. [CrossRef]
17. Taylor, R.A.; Leonard, M.C. Curcumin for inflammatory bowel disease: A review of human studies. *Altern. Med. Rev.* **2011**, *16*, 152–156.
18. Khalaf, H.; Jass, J.; Olsson, E.P. Differential cytokine regulation by NF-κB and AP-1 in Jurkat T-cells. *BMC Immunol.* **2010**, *11*, 26. [CrossRef]
19. Hutton, B.; Salanti, G.; Caldwell, D.M.; Chaimani, A.; Schmid, C.H.; Cameron, C.; Ioannidis, J.P.; Straus, S.; Thorlund, K.; Jansen, J.P.; et al. The PRISMA extension statement for reporting of systematic reviews incorporating network meta-analyses of health care interventions: Checklist and explanations. *Ann. Intern. Med.* **2015**, *162*, 777–784. [CrossRef]
20. O'Connor, D.; Green, S.; Higgins, P.T.J. Chapter 5: Defining the review question and developing criteria for including studies. In *Cochrane Handbook for Systematic Review A of Interventions Version 5.0*; Higgins, J.P.T., Greeen, S., Eds.; The Cochrane Collaboration: London, UK, 2008.

21. Higgins, J.P.T.; Green, S. *Cochrane Handbook for Systematic Reviews of Interventions Version 5.1.0. Updated*; The Cochrane Collaboration: London, UK, 2011.
22. Jadad, A.R.; Moore, R.A.; Carroll, D.; Jenkinson, C.; Reynolds, D.J.; Gavaghan, D.J.; McQuay, H.J. Assessing the quality of reports of randomized clinical trials: Is blinding necessary? *Control. Clin. Trials* **1996**, *17*, 1–12. [CrossRef]
23. Shapira, S.; Leshno, A.; Katz, D.; Maharshak, N.; Hevroni, G.; Jean-David, M.; Kraus, S.; Galazan, L.; Aroch, I.; Kazanov, D.; et al. Of mice and men: A novel dietary supplement for the treatment of ulcerative colitis. *Ther. Adv. Gastroenterol.* **2018**, *11*, 1–10. [CrossRef] [PubMed]
24. Suskind, D.L.; Wahbeh, G.; Burpee, T.; Cohen, M.; Christie, D.; Weber, W. Tolerability of curcumin in pediatric inflammatory bowel disease: A forced dose titration study. *J. Pediatric Gastroenterol. Nutr.* **2013**, *56*, 277. [CrossRef] [PubMed]
25. Atkinson, R.J.; Hunter, J.O. Double blind, placebo controlled randomized trial of Curcuma extract in the treatment of steroid dependent inflammatory bowel disease. *Gastroenterology* **2003**, *124*, A205. [CrossRef]
26. Banerjee, R.; Medaboina, K.; Boramma, G.G.; Amsrala, S.; Reddy, D.N. Novel bio-enhanced curcumin with mesalamine forinduction of remission in mild to moderate ulcerative colitis. *Gastroenterology* **2017**, *152*, S587. [CrossRef]
27. Santos, L.R. Perfil Inflamatório de Pacientes com Colite Ulcerativa em uso de Cúrcuma Longa. Master's Thesis, UFG-Faculdade de Nutrição, Goiânia, Brazil, 2017; p. 34.
28. Hanai, H.; Iida, T.; Takeuchi, K.; Watanabe, F.; Maruyama, Y.; Andoh, A.; Tsujikawa, T.; Fujiyama, Y.; Mitsuyama, K.; Sata, M.; et al. Curcumin maintenance therapy for ulcerative colitis: Randomized, multicenter, double-blind, placebo-controlled trial. *Clin. Gastroenterol. Hepatol.* **2006**, *4*, 1502–1506. [CrossRef]
29. Singla, V.; Pratap Mouli, V.; Garg, S.K.; Rai, T.; Choudhury, B.N.; Verma, P.; Deb, R.; Tiwari, V.; Rohatgi, S.; Dhingra, R.; et al. Induction with NCB-02 (curcumin) enema for mild-to-moderate distal ulcerative colitis—A randomized, placebo-controlled, pilot study. *J. Crohn's Colitis* **2014**, *8*, 208–214. [CrossRef]
30. Lang, A.; Salomon, N.; Wu, J.C.; Kopylov, U.; Lahat, A.; Har-Noy, O.; Ching, J.Y.; Cheong, P.K.; Avidan, B.; Gamus, D.; et al. Curcumin in combination with mesalamine induces remission in patients with mild-to-moderate ulcerative colitis in a randomized controlled trial. *Clin. Gastroenterol. Hepatol.* **2015**, *13*, 1444–1449. [CrossRef]
31. Masoodi, M.; Mahdiabadi, M.A.; Mokhtare, M.; Agah, S.; Kashani, A.H.; Rezadoost, A.M.; Sabzikarian, M.; Talebi, A.; Sahebkar, A. The efficacy of curcuminoids in improvement of ulcerative colitis symptoms and patients'self-reported well-being: A randomized double-blind controlled trial. *J. Cell. Biochem.* **2018**, *119*, 9552–9559. [CrossRef] [PubMed]
32. Kedia, S.; Bhatia, V.; Thareja, S.; Garg, S.; Mouli, V.P.; Bopanna, S.; Tiwari, V.; Makharia, G.; Ahuja, V. Low dose oral curcumin is not effective in induction of remission in mild to moderate ulcerative colitis: Results from a randomized double blind placebo controlled trial. *World J. Gastrointest. Pharmacol. Ther.* **2017**, *8*, 147–154. [CrossRef]
33. Sadeghi, N.; Mansoori, A.; Shayesteh, A.; Hashemi, S.J. The effect of curcumin supplementation on clinical outcomes and inflammatory markers in patients with ulcerative colitis. *Phytother. Res.* **2019**, *34*, 1123–1133. [CrossRef]
34. Yallapu, M.M.; Jaggi, M.; Chauhan, S.C. Curcumin nanoformulations: A future nanomedicine for cancer. *Drug Discov. Today* **2012**, *17*, 71–80. [CrossRef] [PubMed]
35. European Food Safety Authority (EFSA). Refined exposure assessment for curcumin (E 100). *EFSA J.* **2014**, *12*, 3876. [CrossRef]
36. JECFA. Curcumin. (Prepared by Ivan Stankovic). Chemical and Technical Assessment Compendium Addendum 11/Fnp 52 Add.11/29. *Monographs* **2014**, *1*, 417.
37. Lao, C.D.; Ruffin, M.T.; Normolle, D.; Heath, D.D.; Murray, S.I.; Bailey, J.M.; Boggs, M.E.; Crowell, J.; Rock, C.L.; Brenner, D.E. Dose escalation of a curcuminoid formulation. *BMC Complement. Altern. Med.* **2006**, *6*, 10. [CrossRef] [PubMed]
38. Schiborr, C.; Kocher, A.; Behnam, D.; Jandasek, J.; Toelstede, S.; Frank, J. The oral bioavailability of curcumin from micronized powder and liquid micelles is significantly increased in healthy humans and differs between sexes. *Mol. Nutr. Food Res.* **2014**, *58*, 516–527. [CrossRef] [PubMed]

39. Ohno, M.; Nishida, A.; Sugitani, Y.; Nishino, K.; Inatomi, O.; Sugimoto, M.; Kawahara, M.; Andoh, A. Nanoparticle curcumin ameliorates experimental colitis via modulation of gut microbiota and induction of regulatory T cells. *PLoS ONE* **2017**, *12*, e0185999. [CrossRef] [PubMed]
40. Jamwal, R. Bioavailable curcumin formulations: A review of pharmacokinetic studies in healthy volunteers. *J. Integr. Med.* **2018**, *16*, 367–374. [CrossRef] [PubMed]
41. Shoba, G.; Joy, D.; Joseph, T.; Majeed, M.; Rajendran, R.; Srinivas, P.S. Influence of piperine on the pharmacokinetics of curcumin in animals and human volunteers. *Planta Med.* **1998**, *64*, 353–356. [CrossRef]
42. Li, Q.; Zhai, W.; Jiang, Q.; Huang, R.; Liu, L.; Dai, J.; Gong, W.; Du, S.; Wu, Q. Curcumin-piperine mixtures in self-microemulsifying drug delivery system for ulcerative colitis therapy. *Int. J. Pharm.* **2015**, *490*, 22–31. [CrossRef]
43. Kumar, G.; Mittal, S.; Sak, K.; Tuli, H.S. Molecular mechanisms underlying chemopreventive potencial of curcumin: Current challenges and future perspectives. *Life Sci. J.* **2016**, *148*, 313–328. [CrossRef]
44. Platel, K.; Srinivasan, K. Studies on the influence of dietary spices on food transit time in experimental rats. *Nutr. Res.* **2001**, *21*, 1309–1314. [CrossRef]
45. Tanko, H.; Carrier, D.J.; Duan, L.; Clausen, E. Pre-and post-harvest processing of medicinal plants. *Plant Genet. Resour.* **2005**, *3*, 304–313. [CrossRef]
46. Figueiredo, A.C.; Barroso, J.G.; Pedro, L.G.; Scheffer, J.J. Factors affecting secondary metabolite production in plants: Volatile components and essential oils. *Flavour Fragr. J.* **2008**, *23*, 213–226. [CrossRef]
47. Choudhury, B.U.; Nath, A.; Hazarika, S.; Ansari, M.A.; Buragohain, J.; Mishra, D. Effects of pre-harvest soil management practices and post-harvest processing on phytochemical qualities of turmeric (Curcuma longa). *Indian J. Agric. Sci.* **2017**, *87*, 1002–1007.

© 2020 by the authors. Licensee MDPI, Basel, Switzerland. This article is an open access article distributed under the terms and conditions of the Creative Commons Attribution (CC BY) license (http://creativecommons.org/licenses/by/4.0/).

MDPI
St. Alban-Anlage 66
4052 Basel
Switzerland
Tel. +41 61 683 77 34
Fax +41 61 302 89 18
www.mdpi.com

Nutrients Editorial Office
E-mail: nutrients@mdpi.com
www.mdpi.com/journal/nutrients

www.ingramcontent.com/pod-product-compliance
Lightning Source LLC
LaVergne TN
LVHW071014170726
843515LV00006B/1131

* 9 7 8 3 0 3 6 5 6 0 8 7 8 *